T0290900

"This book by Schmidt & Staffell is sure to be a game-changer for professionals entering this space. It provides the groundwork to understanding technologies, applications, cost developments, revenue potentials and conducting own assessments and comparisons. This is essential for me as an investor to navigate this complex, fast-paced energy storage industry."

Gerard Reid, Co-founder and Partner,
Alexa Capital

"An indispensable resource for anyone seeking to understand the rapidly changing landscape of energy storage, and the role it plays in complementing intermittent renewable power. While intermittency is often cited as the Achilles heel of non-dispatchable power, storage goes a long way in mitigating it. The exhibits are extremely helpful in visualizing what is often a topic mired in energy jargon, and I strongly recommend it to anyone focused on the speed and dynamics of the energy transition."

Michael Cembalast, Chairman of Market and Investment Strategy,
JP Morgan

"This book is a great resource for anyone trying to get their head around the energy system of the future. I particularly love some of Schmidt and Staffell's charts – the collection of experience curves and the strategic gameboard by likely storage technology – they make complex trade-offs much easier to understand."

Michael Liebreich, CEO and Principal,
Liebreich Associates

"Battery storage is an essential part of the path to Net Zero Emissions, helping to ensure electricity security as renewables form the foundation of tomorrow's power systems."

Brent Wanner, Head of Power Systems, International Energy Agency

"Now that Energy Storage is more competitive and the cost and revenue issue is getting more and more attention, this book sheds light in the jungle of stacking revenue streams and defining the right applications and the right technologies to make an Energy Storage project work. This book is a "must read" for anyone who wants to understand the Energy Storage business and a milestone to the Energy Storage sector."

Patrick Clerens, Secretary General,
European Association for Storage of Energy (EASE)

"Energy storage is a key technology to improve energy access to millions in Africa by supporting weak power grids and enabling remote microgrids. By skilfully navigating the complexity of storage technologies and applications with their outstanding book, Oliver Schmidt and Iain Staffell have done a huge service towards making that vision a reality. The clarity and authority of their research will build much-needed confidence among governments, entrepreneurs, NGOs and investors."

Chinnan Maclean Dikwal, Vice Chair Board of Directors,
African Energy Council

"This is a must-read for industry and policy professionals. It clearly outlines the diversity of energy storage technologies available and provides real world applications and critical information for decision makers. Most importantly, it highlights the essential role of long duration energy storage (LDES) to provide flexibility and reliability to the grid. Readers will walk away with the ability to understand and assess the wide range of LDES technologies that are available today."

Julia Souder, Executive Director,
Long Duration Energy Storage Council

"This is a full, detailed and comprehensive resource on how to navigate the complex world of energy storage economics from two people who know what they're talking about. Full of clear insight, data and charts."

Andrew Turner, Head of Modelling,
Bloomberg New Energy Finance

"This book is unique in spanning a broad field covering energy storage in the context of the energy transition, strategic high-level assessments, future cost estimate methodologies and project specific considerations. This coverage enables the reader to "connect the dots" between policies, strategy and implementation. Together with the practical examples that can be reproduced via the specifically created online platform the book connects theory and practice. Overall, I highly recommend it to policy makers, business leaders, project developers and engineers wanting to make informed decisions based on the delicate sensitivities in the economics of energy storage systems."

Benjamin Sternkopf, Founder and Managing Director,
IFE Sternkopf

"A must-read for anyone who wants to understand the critical role that energy storage is starting to play in the global energy ecosystem. The clear and concise description of technical concepts backed up by solid research and data allows the reader to develop a holistic baseline for creating their own framework to assess energy storage projects. The authors provide an unbiased and insightful assessment of the potential of each technology, as well as the challenges that must be overcome to realize their full potential. I highly recommend the book on energy storage to anyone looking to deepen their understanding of this critical field. The author's technical expertise and engaging writing style make it an essential beginner's resource for researchers, investors, policymakers, and anyone interested in the intersection of energy and sustainability."

Anoop Poddar, Managing Director, Eversource

"A comprehensive text covering virtually everything useful there is to know on energy storage, it's benefits – and importantly the sensitivity to the evolution of the energy storage landscape over the next 20 years to technological improvements, discount rates and other factors. Not only do Schmidt and Staffell set this all out coherently, they also provide the reader the open-access tools to make their own analyses and to draw their own conclusions."

Marek Kubik, Managing Director, Fluence

"Monetizing Energy Storage is what I would consider essential reading for anyone that is new to the energy storage industry, or simply for those of us that have been in the industry for a while and need a refresher. This comprehensive text captures the dynamics of a complex market in a clearly laid out and easy to digest way. The book along with the associated website and worked examples, provides readers with all the tools they need to become an expert on energy storage and understand the merits of different energy storage technologies. Oliver and Iain have done a huge service to the storage industry in writing this book, I wish it had been available when I first started covering the sector."

James Frith, Principal,
Volta Energy Technologies

"The grid storage Bible: Schmidt and Staffell provide a well-grounded, comprehensive, insightful analysis of electricity storage across the entire value chain, full of real-world examples and complemented by a user-friendly theoretical framework with which to explore the growing role energy storage will play in systems and in markets."

Jeffrey Douglass, Markets and Research Manager,
Invinity Energy Systems plc

"Monetizing Energy Storage is THE new must-read within the booming field of storage technologies. For us as project developers, it helps us to keep an eye on the big picture, while also providing an impressive amount of well-researched detail insights in technological and market aspects. The book is red-hot at the moment, but its clarity and structure will continue to enrich the storage industry for many years to come."

Benedikt Deuchert, Head of Regulatory Affairs, Kyon Energy

"Whether you are an energy storage novice or expert, "Monetizing Energy Storage" is an indispensable toolkit. It has been brilliantly conceived and written to offer critical and timely insights into possibly the most interesting enabler of sustainable energy system transformation. Schmidt & Staffell bring years of experience and research to bear in such a deep, yet accessible and practical guide. I love that they break down complex, technical concepts into easily digestible ideas."

Mervin Ekpen Azeta, Global Flexible Work & Culture Project Manager,
SLB

"In order to mitigate climate change, a rapid and deep transition to net-zero emission energy systems is needed. Renewable energy will play a critical role in this transition, and as a result, energy storage will be among the most critical factors that determine how easy or difficult our pathways to net zero will be. In this context, this book is extremely relevant, providing a valuable resource for practitioners, policymakers, and academics interested in the deployment potential and economic value of energy storage. The book discusses all relevant aspects and its interactive online tool enables easy reproduction of the presented analyses with custom data. It is a must-read for those interested in understanding how energy storage can play a role in accelerating the energy transition."

Professor Detlef van Vuuren,
PBL Netherlands, Universiteit Utrecht

"Energy storage is the key to unlocking access to many low carbon energy technologies. For practitioners and academics alike, this book is the go-to resource for understanding the principles and practicalities of energy storage in its many forms. The clear coverage of economics and finance, at both the technology and system level, is exemplary in terms of academic rigour set in a real-world context."

Professor Jim Skea,
Chair of the Intergovernmental Panel on Climate Change

"For more than a decade, I have been searching for a book that combines the science, technology, economics, and financial aspects of energy storage - and now I have found it. Oliver and Iain's book is perfect for readers interested in understanding how energy storage can capture value as the world accelerates into the clean energy transition. It is easy to digest yet highly technical; it uses many examples, and has an accompanying website to play around with the numbers. This is a great resource for energy storage technology scientists transitioning into applied systems work, for industry practitioners who want to understand how the technology affects use cases and value, and for policymakers who want to encourage investments in storage. I will definitely use this book for my graduate course on energy storage."

Professor Joey Ocon,
University of the Philippines Diliman

"This book is ground-breaking. Schmidt and Staffell bring structure to the complex set of metrics used to describe the finances of energy storage. They use latest data to bring these metrics to life and derive key insights for the industry. More importantly, they empower readers to perform their own analyses. This is an essential read for all professionals and academics who want to engage with the industry."

Professor Dan Kammen, Lau Distinguished Professor of Sustainability,
University of California, Berkeley

"Some books were born "made Handbook". Whatever you do when crossing their territory, you need them with you continuously for that journey: they are handy! Try to navigate the area of electricity storage, and you will come back over and again to learn from this wonderful map and guide book. You will be introduced to all the relevant basics on alternative technologies of electricity storage. Plus all the robust methodologies for assessing cost and valuing the multiple revenues that these storage systems can earn, especially as they are "revenue stackers" by destiny and choice. Coming from two certified experts in their fields, this book is unique, remarkable and outstanding. It is both the ultimate Swiss Army knife of energy storage applied economics and the golden compass for storage investment. Incredibly well written, conceived, documented and illustrated; a great volume contributing to the Energy Transition Encyclopedia of the 21 Century. In my 30 years of electricity economic research I have not seen such a wonderful book: this will easily get the "Applied Book Decade 2020" Award. So well-conceived, so comprehensive, so useful: really great!"

Professor Jean-Michel Glachant
Florence School of Regulation,
President of the International Association
of Energy Economics (IAEE)

Monetizing Energy Storage

Monetizing Energy Storage

A Toolkit to Assess Future Cost and Value

Oliver Schmidt
and
Iain Staffell

OXFORD
UNIVERSITY PRESS

Great Clarendon Street, Oxford, OX2 6DP,
United Kingdom

Oxford University Press is a department of the University of Oxford.
It furthers the University's objective of excellence in research, scholarship,
and education by publishing worldwide. Oxford is a registered trade mark of
Oxford University Press in the UK and in certain other countries

Published in the United States of America by Oxford University Press
198 Madison Avenue, New York, NY 10016, United States of America

British Library Cataloguing in Publication Data
Data available

Library of Congress Control Number: 2023932438

ISBN 978–0–19–288817–4

DOI: 10.1093/oso/9780192888174.001.0001

Printed and bound by

CPI Group (UK) Ltd, Croydon, CR0 4YY

In loving memory of Robert Staffell

1941–2020

To my parents

who gave me all the energy

to write this book

Preface

Motivation for this book

The energy sector is transforming rapidly as efforts to reduce carbon emissions intensify and renewables become the cheapest source of electricity. Energy storage can provide the required flexibility to balance variable and inflexible low-carbon power generation with demand. Recent years have also seen electric vehicles break through into the mainstream and stationary electricity storage deployment grow at more than 30% each year. In fact, this industry is projected to grow to hundreds of times its current size in the coming decades. We are therefore just at the beginning of a significant overhaul in how energy is produced, stored, and consumed worldwide.

In light of this transformation, businesses, policy-makers, and academics need to assess the future cost and value of energy storage. However, this is complicated by the rapidly falling investment cost, the wide range of technologies with different performance characteristics, the wide range of use cases with different performance requirements, and the vastly different market structures around the world. Together, these lead to significant uncertainty regarding the expected commercial viability of energy storage and its potential roles in the future that prevent policy and investment decisions.

Our hope is that this book will increase the transparency around the future commercial viability and potential roles of energy storage and enable readers to assess these confidently.

The online version of this book is free. We did not write it to make money. We wrote it because the transition of the energy system needs to happen rapidly, and energy storage will play a crucial enabling role. We believe that helping as many people as possible to understand the economic case for energy storage is our best route to driving its deployment and helping to accelerate the energy transition. A free digital copy of the book will be available to download from: https://global.oup.com/academic/product/monetizing-energy-storage-9780192888174

What this book covers

This book is targeted explicitly at practitioners from industry like strategists, investors, consultants; at policy-makers like policy analysts, regulators; and at academics like graduate students and researchers. It aims to help these stakeholders understand the cost reduction and deployment potential of energy storage as well as assess its economic value by:

- Introducing key cost and performance parameters, and energy storage technologies and applications
- Developing and explaining quantitative methods for assessing the future investment and lifetime cost, along with the economic and system value of energy storage

- Presenting cutting-edge research insights to exemplify these methods with current data
- Introducing an interactive online tool that enables the presented analyses to be easily reproduced with custom data

There are many topics this book does not cover. It is not an engineering textbook with detailed technical analyses of energy storage technologies. Nor is it a market research report with in-depth discussions on the future size and requirements of specific applications within different world markets. While this book covers many topics relating to the economic performance of energy storage (costs, revenues, financial appraisal, revenue stacking), it is not an economics textbook. It does not cover the wider topics of market structures, regulation, taxation, or competition. Many of these topics are market-specific and readers can incorporate their own insights using the practical methods and interactive tools we introduce.

This book focuses on *electricity* storage technologies: technologies which consume *electrical* energy and then provide *electrical* energy again at a later time. The rapid expansion of low-cost, low-carbon electricity is driving the electrification of the wider energy system (especially heat and transport), which is positioning electricity storage technologies centrally within the whole energy sector transformation. Complete decarbonization of the energy system will also require other forms of energy storage technologies, for example technologies that consume and provide thermal energy. The concepts introduced in this book can also be applied to these other energy storage technologies, even though they are introduced here with data and examples referring to *electricity* storage.

How to use this book

Transforming the global energy system is no easy task, so you may not have a lot of time available for reading books. That is why we have collected together 'Key insights' at the beginning of each chapter. If you have only an hour to spare, you will get the most out of this book by reading through the key insights from all the chapters. Not only will this give you a broad overview of the future cost and value of energy storage but you will also understand the overall structure of the book so that when you want to come back for more detail, you will know where to go.

This book is split into three parts. The first part (Chapters 1–3) introduces the role of energy storage in energy system transformation, its key parameters and components, the main technologies and applications. The second part (Chapters 4–7) provides methods to assess the future investment cost, lifetime cost, market value, and the system value of energy storage, and key insights derived from these. The third part, Chapter 8, provides a detailed documentation of the methods behind the presented analyses.

The first part is relevant for everyone without prior knowledge of energy storage. The second part is aimed at practitioners from industry, policy-makers, and academics that already work with or are planning to work with the energy storage industry. It provides a broad overview of current and future economics of energy storage applications and enables you to perform initial analyses to scope out business cases, the impact of distinct policies, or the roles that storage may play. The third part is relevant for academics and analysts working across the realm of techno-economics, including technology costs, sector trends, and market size. It transparently describes the methods used throughout the book to enable you to build your own models, explains the relative advantages and disadvantages of these methods, and explores the quality of economic forecasts.

A range of features are also designed to help you navigate the content of this book and identify the most relevant information. These are:

- A summary of the key insights at the beginning of each chapter
- Key messages and warnings highlighted throughout the chapters to spotlight the most relevant insights and complexities for you to be aware of
- Frequently asked questions in the chapters to provide real-world examples and to resolve industry myths
- Worked examples at the end of Chapters 4–7 which guide you step-by-step through how to perform your own analysis of energy storage costs, value, and financial viability, using the interactive tools that we provide alongside the book

A special feature of this book is that it comes with the companion website <www.EnergyStorage.ninja>. This allows you to reproduce the analyses conducted in this book with different input parameters or assumptions easily. You can apply the knowledge you gain from the book directly to assess new technological developments, new markets, or new applications for storage. Technology costs are notorious for changing quickly, for wrong-footing analysts, and going quickly out of date, so this website also provides latest cost, performance, and deployment data for energy storage technologies as they come in.

Nobody is perfect. We have conducted the analyses with the highest level of diligence, however there may be errors and we welcome any suggestions for corrections or improvements. Likewise, we are happy to engage in any discussion regarding the derived insights.

If you would like to cite the book, please refer to it as: Schmidt O and Staffell I. *Monetizing Energy Storage: A Toolkit to Assess Future Cost and Value* (Oxford: Oxford University Press, 2023).

Acknowledgements

We want to thank four anonymous reviewers, the delegates of Oxford University Press, and reviewers across the energy storage industry and academia who were so kind to provide detailed feedback on earlier drafts of this book: Jenny Baker, Inga Beyers, Andrea Biancardi, Aditya Chhatre, Julien Demoustier, Jeffrey Douglass, Vincent Filter, Richard Green, Stefan Häselbarth, Sebastian Ljungwaldh, Javier López Prol, Francesco Marasco, Nathan Murray, Fabio Oldenburg, Joel Omale, Münür Sacit Herdem, Benjamin Silcox, Benjamin Sternkopf, and Benedikt Ziegert.

We also thank Benedikt Deuchert, Jeffrey Douglass, Adrien Lebrun, Marek Kubik, and Benjamin Sternkopf for fruitful discussions on various topics covered in this book during its creation.

We thank Bundesverband Solarwirtschaft e.V. (BSW Solar) for providing data on historical cost reductions for selected energy storage products in Germany.

This book has been generated in part via support from the Integrated Development of Low-carbon Energy Systems (IDLES) programme. We gratefully acknowledge funding from the Engineering and Physical Sciences Research Council for this 5-year interdisciplinary programme (EP/R045518/1). For more information visit: <http://www.imperial.ac.uk/energy-futures-lab/idles>.

Contents

Dr Oliver Schmidt

Oliver Schmidt is an entrepreneur and management consultant in the clean energy sector. He has previous experience as project manager at the strategy and financial transaction advisory firm Apricum, where he supported top management with strategic advice in the energy storage, solar PV, and hydrogen industries. Oliver also worked as an energy analyst at the International Energy Agency and as a management consultant at E.ON. He has a PhD on the future cost and value of energy storage from Imperial College London and developed the online platform <www.EnergyStorage.ninja>. His background is in engineering and renewable energy, which he studied at Imperial and the Swiss Federal Institute of Technology (ETH). He currently lives in Berlin.

Dr Iain Staffell

Iain Staffell lives in London with his wife and three children. He is an Associate Professor at Imperial College London, where he teaches energy economics and policy and leads a sustainable energy research group. He holds degrees in Physics, Chemical Engineering, and Economics from the University of Birmingham. His research has won the Baker Medal and President's Award for Excellence in Research, and has featured in over 120 national and international media articles. Iain is passionate about making energy research transparent and openly available to all. He is a developer of the <www.Renewables.ninja> platform for modelling renewable energy supply and demand, and the <www.EnergyStorage.ninja> platform which accompanies this book.

List of Abbreviations

°C	Degrees Celsius
A-CAES	Adiabatic Compressed Air Energy Storage
APAC	Asia Pacific
BMS	Battery management system
bn	Billion
CAES	Compressed Air Energy Storage
Cap	Capacity
CCGT	Combined cycle gas turbine
CO_2	Carbon dioxide
DD	Minimum discharge duration
Deg	Degradation
DoD	Depth-of-discharge
E/P	Energy-to-power ratio
EU	European Union
EUR	Euro
EV	Electric Vehicle
GB	Great Britain
GBP	Great Britain Pound
$GtCO_2$	Gigatonnes of carbon dioxide
GW	Gigawatt
GWh	Gigawatt hour
HVDC	High Voltage Direct Current
IPCC	Intergovernmental Panel on Climate Change
kg	Kilogram
km	Kilometre
kW	Kilowatt
kWh	Kilowatt hour
LCO	Lithium cobalt oxide
LFP	Lithium iron phosphate
LMO	Lithium manganese oxide
LTO	Lithium titanate
m	Metre
m^3	Cubic metre
min	Minute
mn	Million
ms	Millisecond
Mt	Megatonnes
MW	Megawatt
MWh	Megawatt hour
NCA	Nickel Cobalt Aluminium
NIMH	Nickel Metal Hydride
NMC	Nickel Manganese Cobalt

OCGT	Open cycle gas turbine
Pa	Per annum
PEM	Proton exchange membrane
PHES	Pumped Hydro Energy Storage
PV	Photovoltaic
R&D	Research and development
RE	Renewable energy
s	Second
SoC	State-of-charge
SoH	State-of-health
SSP	Shared Socio-economic Pathways
TW	Terawatt
TWh	Terawatt hour
UK	United Kingdom
US	United States
USD	US Dollar
V	Voltage
VRE	Variable renewable energy
W	Watt
Wh	Watt hour
ρ	Density

PART I: Introducing Energy Storage

1 Introduction
Looking at the big picture

KEY INSIGHTS	WHAT IT MEANS
The global energy sector must transform rapidly, reducing net annual carbon emissions from 20 billion tonnes of CO_2 in 2020 to zero by 2050 to keep global warming to below 2 °C.	Fossil-fuel power plants must be replaced with low-carbon renewables, nuclear, or fossil with carbon capture and storage.
Global wind and solar power generation capacity has grown from under 20 GW in 2000 to ~700 GW each in 2020, increasing the overall share of renewable energy in the electricity sector from 21% to 29%.	Wind and solar farms are already being deployed at scale, driving the low-carbon transformation of the energy sector.
Electric vehicles (EV) are entering the mainstream with global stock nearly doubling each year. EV sales make up more than 15% in core markets like Europe and China and need to reach 100% globally by 2035 to limit global warming to below 2 °C.	Electric vehicles are also being deployed at scale, meaning battery storage is directly driving the low-carbon transformation of the transport sector.
On the supply side, wind and solar power generation is variable and nuclear generation is inflexible. On the demand side, electricity demand will become less predictable and more variable due to the electrification of transport and heat.	Technologies that provide flexibility to match energy supply and demand are essential in low-carbon power systems, and demand for them will grow substantially.
The four key options to provide flexibility to the electricity system are flexible power generation (~3,400 GW as of 2020), electricity network interconnection (~180 GW), demand-side response (~40 GW), and electricity storage (~180 GW).	Besides electricity storage, there are alternatives that can provide flexibility to the electricity system, potentially at lower cost.
Flexible power generation is retiring due to the phase-out of fossil fuels. Increases in network interconnection and demand-side response are uncertain. Electricity storage capacity is projected to more than triple to > 600 GW by 2030.	The role of electricity storage is increasing rapidly, and it is set to become the second most widely deployed flexibility technology after flexible power generation.
In terms of energy capacity, countries require terawatt-hours (TWhs) of electricity storage to replace at least part of the storage capacity that fossil fuel reserves currently provide and enable future time-shifting of energy use.	Substantial amounts of low-cost, low-carbon electricity storage energy capacity are required to enable cost-efficient, net-zero carbon energy systems.

Monetizing Energy Storage. Oliver Schmidt & Iain Staffell, Oxford University Press. © Oliver Schmidt & Iain Staffell (2023).
DOI: 10.1093/oso/9780192888174.003.0001

1.1 Energy system transformation

Pathways limiting global warming to 1.5 °C with no or limited overshoot would require rapid and far-reaching transitions in energy . . . and industrial systems.
—The IPCC.[1]

Human activities, primarily the burning of fossil fuels, have unequivocally caused the concentration of greenhouse gases in the atmosphere to increase since pre-industrial times.[2] Annual human-made emissions reached 40 gigatonnes of CO_2-equivalent ($GtCO_2$ or billion tonnes of CO_2), its highest level in history.[3] As a result, atmospheric CO_2 concentration increased by 50% since 1850, and global mean surface temperature was 1.1 °C above pre-industrial levels by 2020.[2]

In order to limit climate change, the international community agreed in December 2015 to hold the increase in global mean surface temperature to well below 2 °C above pre-industrial levels and to pursue efforts to limit temperature increase even further to 1.5 °C.[4] This difference matters because an extra half a degree of warming makes the loss of almost all coral reefs likely, along with increasing intensity and frequency of extreme weather events like droughts, heatwaves, and heavy precipitation.[5,6]

Total cumulative CO_2 emissions since 1850 must remain below 3,500 $GtCO_2$ to limit temperature increase to well below 2 °C (see Figure 1.1).[2] Given that by 2020, total human-made carbon emissions amounted to ~2,400 $GtCO_2$, the remaining carbon budget for a 67% chance of staying below 2 °C is 1,150 $GtCO_2$ and only 400 $GtCO_2$ for a 67% probability of staying below 1.5 °C.[2] At current levels of annual emissions, this 1.5 °C budget would be fully used up by 2030. Therefore, to limit temperature increase to below 2 °C, annual carbon emissions should reach net zero by 2050 and become negative thereafter.

This requires a deep and rapid transformation of the energy sector. Annual emissions from energy supply need to reduce from 20 billion tonnes of CO_2 to zero by 2050 in pathways likely to limit global warming to 2 °C. For the electricity sector, this means that by 2030, 50–70%, and by 2050 > 90% of electricity must come from low-carbon sources like renewables or nuclear, compared to 39% in 2020 (see Figure 1.2).[7]

KEY INSIGHT

The global energy sector must transform rapidly, reducing net annual carbon emissions from 20 billion tonnes of CO_2 in 2020 to zero by 2050 to keep global warming to below 2 °C.

This transformation is already taking place, and while climate change was an early motivation for deploying renewable energy technologies, it is by no means the sole driver of the energy transition today. A range of policy instruments have combined with investment cost reductions for wind and solar photovoltaic (PV) technologies and favourable financing conditions to deliver a significant increase in installed renewable energy capacity. Wind turbines and solar panels have achieved cost reductions of around 60–90% since 2010,[9] and in many parts of the world they are now the cheapest source of new-built electricity generation, undercutting conventional coal and gas generation (see Figure 1.3).

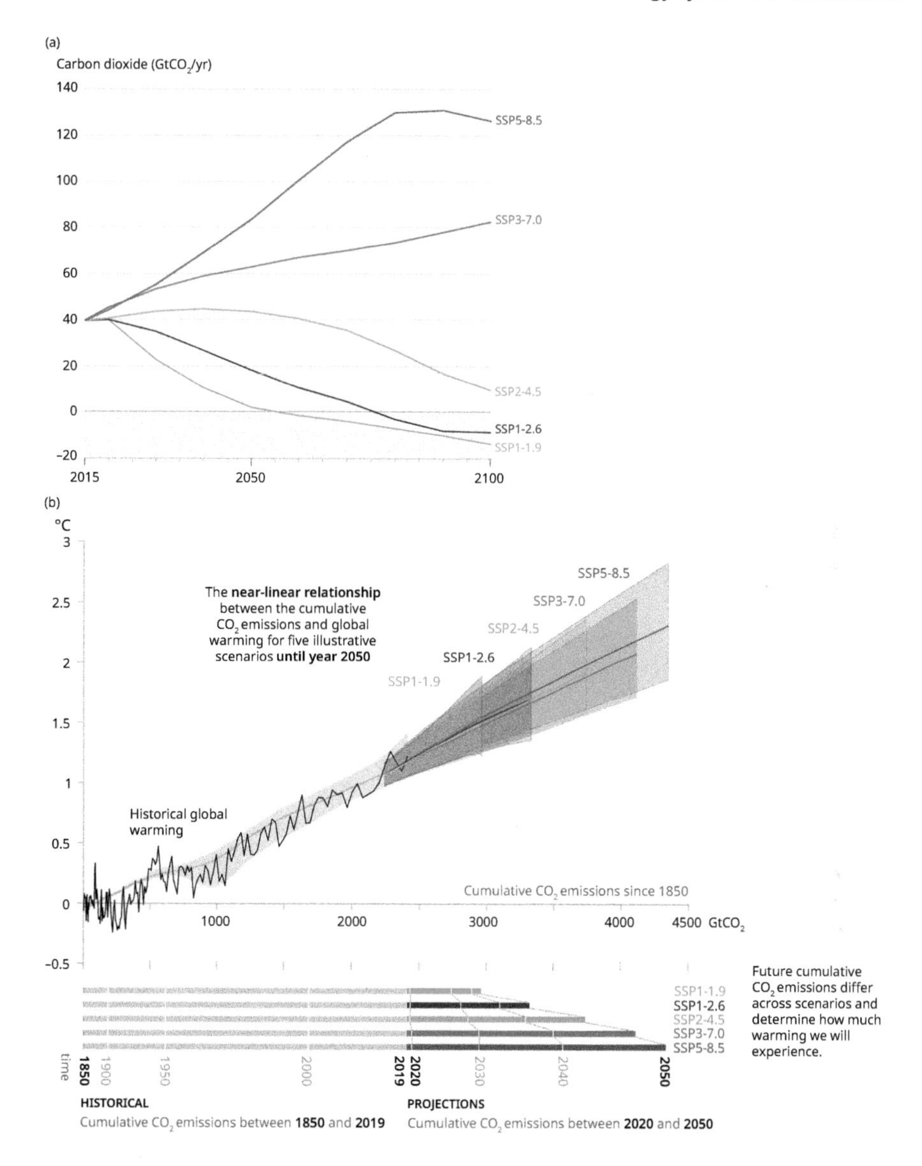

Figure 1.1 Global carbon emissions and expected temperature increase due to climate change. a) Emissions of carbon dioxide (CO_2) in five Shared Socio-economic Pathways (SSP).[2] b) Global mean surface temperature increase since 1850–1900 as a function of cumulative global CO_2 emissions. Coloured ranges and timelines refer to the temperature ranges and timelines of respective cumulative CO_2 emissions for the respective SSP scenarios. Figure reproduced with permission from the Intergovernmental Panel on Climate Change.[2]

Wind and solar PV reached installed capacities of 740 and 720 GW by 2020 respectively, compared to < 20 GW in total in 2000. As a result, the share of renewable electricity (including hydropower, geothermal, and biomass) increased from 21% to 29% of total global electricity production,[8] and annual capacity additions of wind and solar PV are continuously increasing.

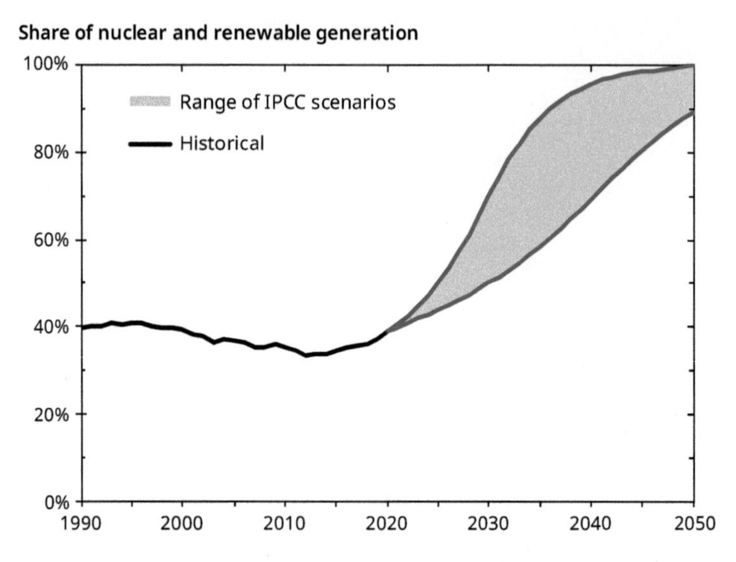

Figure 1.2 Global requirement for the share of low-carbon electricity from nuclear and renewables in pathways likely to limit global warming to 2 °C. Data source: IPCC.[7]

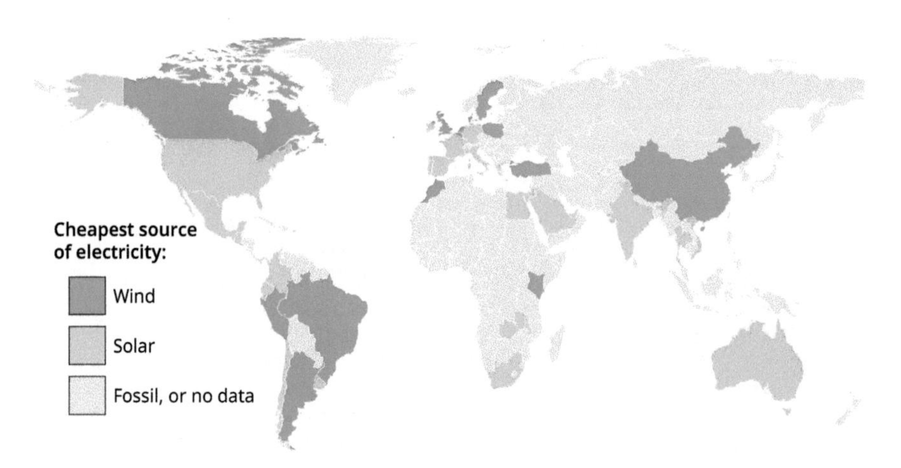

Figure 1.3 World map showing the cheapest source of new-built electricity generation by country. Data source: BNEF.[10,11]

KEY INSIGHT

Global wind and solar power generation capacity has grown from under 20 GW in 2000 to ~700 GW each in 2020, increasing the overall share of renewable energy in the electricity sector from 21% to 29%.

In the transport sector, the global stock of electric vehicles (EVs) has grown by almost a factor of 1,000 between 2010 and 2021, driven by the near-90% reduction in the cost of lithium-ion batteries.[12,13] Figure 1.4 shows that the number of EVs on the world's roads has almost doubled each year for the last decade, to reach ~16 million globally by 2021. As with wind and solar PV farms, EVs are becoming competitive with conventional vehicles and are rapidly entering the mainstream. In selected markets like China and Europe, EVs now represent more than 15% of new car sales, and worldwide more than two-fifths of two- and three-wheeler sales are now electric.[12] For context, limiting global temperature rise to 1.5 °C requires every car sold by 2035 to be electric.[1,12]

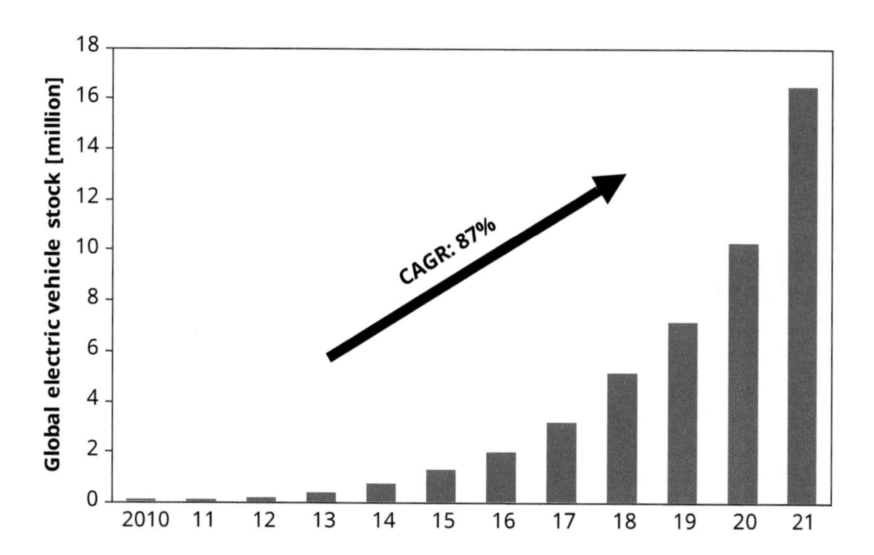

Figure 1.4 Global stock of electric light-duty passenger vehicles, including battery and plug-in hybrid electric vehicles.[12] CAGR—Compound annual growth rate.

KEY INSIGHT

Electric vehicles (EV) are entering the mainstream with global stock nearly doubling each year. EV sales make up more than 15% in core markets like Europe and China and need to reach 100% globally by 2035 to limit global warming to below 2 °C.

1.2 The need for power system flexibility

The physical characteristics of electricity imply that supply and demand must be in balance at all times.[14] Electricity demand varies (e.g., hourly, daily, weekly) and the potential electrification of heating and transport with heat pumps and electric vehicles will further increase the variation between minimum and maximum demand. Currently, flexible fossil fuel-powered generators ensure that electricity supply precisely follows demand. But these must be replaced with low-carbon generators to reduce carbon emissions. The replacement capacity is likely to come from wind, solar PV, or nuclear power plants, based on resource availability, production cost, and public preferences. However, these generate electricity in a variable pattern (wind and solar PV are weather-dependent) or are inflexible (nuclear power is operated at near-constant output for economic and safety reasons in most countries).[15]

Figure 1.5 highlights the projected mismatch when scaling existing solar PV, wind, and nuclear capacity to meet the required demand in Great Britain's power system for the first week of July 2020. The lack of flexibility in the system means there would be continuous overproduction for the first three days and near-continuous shortfall in the following four days. With no possibility of time-shifting energy supply or demand this power system would collapse, resulting in nationwide blackouts.

The temporal mismatch between the variation of electricity demand and low-carbon generation highlights the need for power system flexibility, which is 'the extent to which a power system can modify electricity production or consumption in response to variability, expected or otherwise'.[16] Power system flexibility is further categorized along relevant timescales in Table 1.1.

Very short-term flexibility at the sub-second to minute timescale is driven by technical power system characteristics. It is required to stabilize the power system voltage and frequency in case of an unexpected event leading to an immediate change in electricity supply or demand. It is also used more generally to keep frequency within normal limits. Short- to very long-term flexibility is required to ensure the availability of sufficient generation capacity to meet electricity demand at hourly, daily, or up to seasonal timescales.

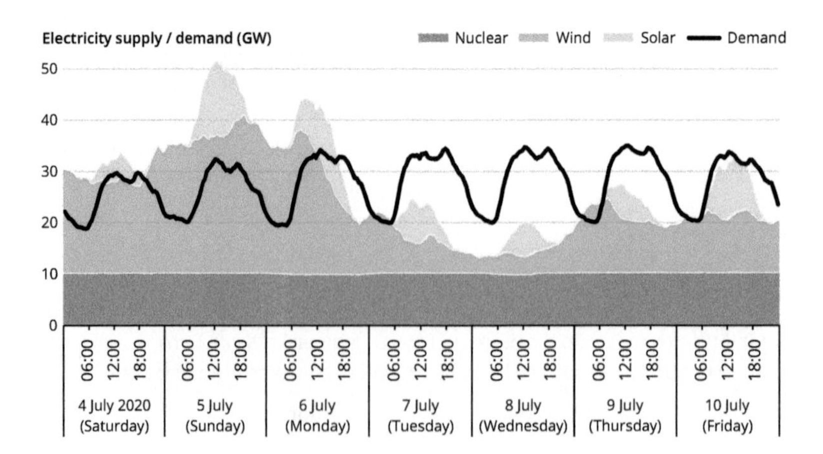

Figure 1.5 Time series of electricity demand with generation from nuclear, wind, and solar in Great Britain for the first week of July in 2020. Generation is scaled so that weekly production of all three generator types would meet weekly demand. Data Source: Electric Insights.[17]

Table 1.1 Categorization of power system flexibility types and resources (adapted from IEA, 2018).[18]

Flexibility timescale	Very short term (sub-seconds to minutes)	Short term (minutes to hours)	Medium term (hours to days)	Long term (days to months)	Very long term (months to years)
Challenges	Ensuring system stability by keeping voltage and frequency in required range	Meeting frequent, rapid, and non-predictable changes in electricity supply and/or demand	Determining operation schedule of generation capacity to balance electricity supply and demand	Addressing longer periods of surplus or deficit in electricity generation	Balancing seasonal and interannual availability of generation resources with electricity demand
System relevance	Dynamic system stability and frequency control	Real-time balancing	Hour-ahead and day-ahead planning	Generation adequacy guarantee	Power system planning

KEY INSIGHT

On the supply side, wind and solar power generation is variable and nuclear generation is inflexible. On the demand side, electricity demand will become less predictable and more variable due to the electrification of transport and heat.

1.3 Options for power system flexibility

Figure 1.6 highlights the four key technical options to provide power system flexibility in low-carbon power systems to match largely variable or inflexible low-carbon electricity supply with variable electricity demand.

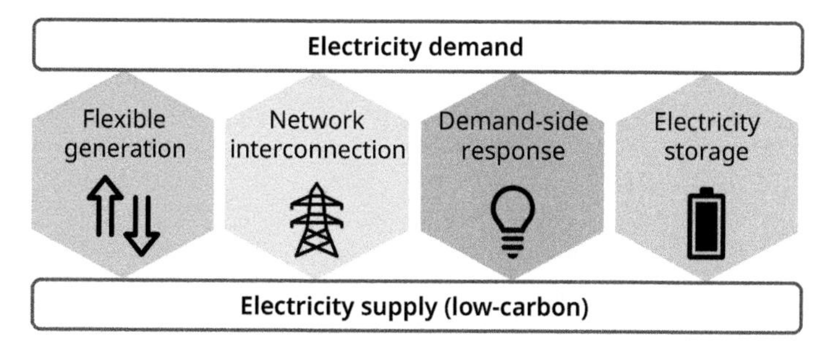

Figure 1.6 The four technical flexibility options that can help to match variable or inflexible low-carbon electricity supply with demand.

KEY INSIGHT

The four key options to provide flexibility to the electricity system are flexible power generation (~3,400 GW as of 2020), electricity network interconnection (~180 GW), demand-side response (~40 GW), and electricity storage (~180 GW).

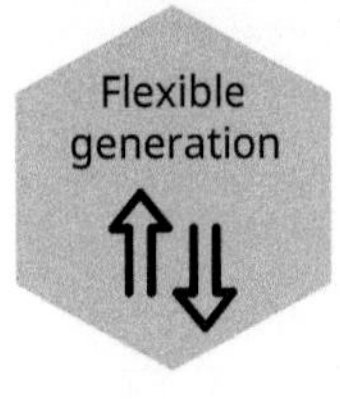

Flexibility is the ability of power plants to deliberately adjust their output. This is the main source of flexibility today with roughly 3,400 GW of the 7,300 GW installed generation capacity worldwide. Gas-fired power (29%) and hydropower plants (28%) provide most flexible generation capacity.[19] Very short-term flexibility is provided passively by the mechanical inertia of the generators, where their physical spinning mass contributes to mitigating any change in power system frequency.[18] Short-term flexibility comes from the automated or manual change in power output. Longer term flexibility is ensured by the start-up of plants, improvements to operation criteria enabled through appropriate monitoring equipment, and the general availability of flexible generation capacity (e.g., new-build, retrofit, reserve). Variable renewable generators can contribute to power system flexibility requirements through:[18]

- Synthetic inertia: Programming inverters to emulate behaviour of generators
- Curtailment: Deliberate output reduction in times of oversupply
- Reserve operation: Deliberate operation below output potential to enable output increase in times of undersupply
- Strategic location: Reduction of flexibility requirements by locating renewable generators in areas with different weather patterns
- Output forecasting: Improved anticipation of renewable energy output

However, the majority of flexible generation capacity is based on fossil fuels, which must be phased out to mitigate severe climate change. Therefore, other flexibility options become more important.

Network
interconnection

The electricity network comprises all assets that connect electricity generation to demand locations. While it is foremost a means of overcoming the geographic mismatch between the two, the aggregation of diverse demand and variable generation smooths overall demand and generation patterns. It also expands the pool of available flexibility options, thereby reducing the need for active power system flexibility.[19,20] Therefore, network expansion with increased interconnection between regions with different weather patterns is seen as a cost-effective option to decarbonize power systems.[21] However, global interconnection between countries (a proxy for electricity network interconnection between regions with different weather patterns) is only 180 GW and the deployment of further capacity is slow due to significant up-front investment and interregional coordination requirements, as well as possible resistance from local residents.[19,20]

Demand-side response (DSR) affects the pattern and magnitude of end-use electricity consumption by reducing, increasing, or rescheduling demand.[20] It has the potential to be the most cost-effective option to provide flexibility due to limited requirements for new hardware infrastructure.[22] However, there is only ~40 GW of demand-side response capacity installed globally, and further deployment is slow. The installed capacity is mostly restricted to large industrial or commercial consumers and night-time tariffs for residential consumers.[19,20] In addition to a lack of required information and communication technology, this option also faces multiple non-technical barriers, namely the implementation of incentive structures like price signals and reluctant behavioural change from domestic consumers.[20,22]

Electricity storage encompasses all technologies that can consume electricity (e.g., in times of oversupply) and return it later (e.g., in times of undersupply). Electricity storage technologies provide flexibility by time-shifting both energy production and consumption. Most options can be deployed anywhere in the electricity network, for example at the generator or consumer site. Different technologies are suitable for different flexibility requirements, with supercapacitors and flywheels most suitable for very short-term, and pumped hydro or hydrogen storage best suited for longer-term flexibility. Various types of electrochemical batteries can be used for intermediate timescales. Following flexible generation and network interconnection, electricity storage was the third most widely deployed flexibility option in 2020 with 158 GW of pumped hydro and around 20 GW of other technologies deployed.[23,24] Falling investment costs and the wide range of use cases drive increasing annual deployment rates to cumulative projections of > 600 GW by 2030 (additional ~65 GW for pumped hydro and > 400 GW for other stationary storage technologies).[25,26] However, electricity storage is rated by energy professionals to be one of the most essential, yet most poorly understood technologies for the transformation of the energy industry.[27]

KEY INSIGHT

Flexible power generation is retiring due to the phase-out of fossil fuels. Increases in electricity network interconnection and demand-side response are uncertain. Electricity storage capacity is projected to more than triple to > 600 GW by 2030.

1.4 The scale of the required transformation

The scale of the transformation becomes evident when shifting perspective from power capacities (GW and TW) to energy capacities (GWh and TWh). Two key advantages of fossil fuels are that they are relatively easy to store and have very high energy densities, meaning that countries can store weeks or even months of supply to account for supply and demand imbalances in the energy system. Taking the UK for example, fossil fuel reserves at the system level amount to ~250 TWh of calorific energy (i.e. the amount of heat released when combusted) (see Table 1.2).[28,29]

Table 1.2 Energy content of UK fossil fuel reserves on system level in calorific terms in 2020 for coal, crude oil, petroleum product, and natural gas stocks.[28,29]

Fuel	$TWh_{calorific}$
Coal	30
Crude oil	75
Petroleum products	130
Natural gas	15
Total	**250**

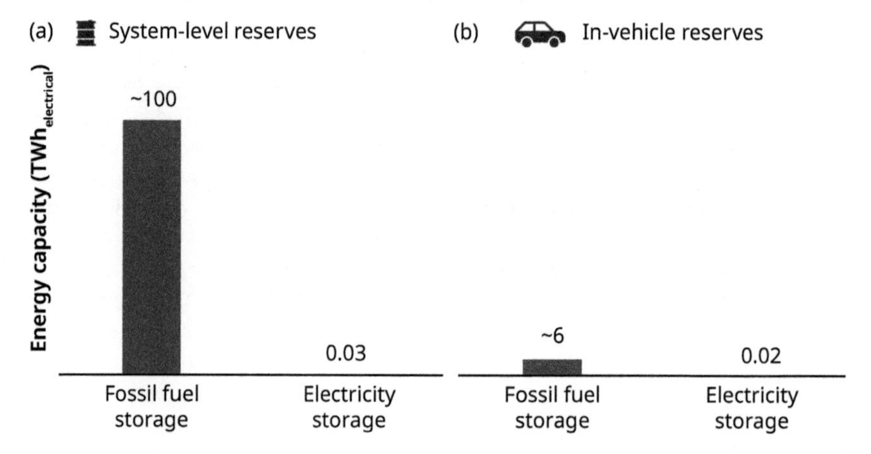

Figure 1.7 Energy storage capacity in the UK as of 2020. a) System-level reserves of fossil fuel storage includes coal, crude oil, petroleum product, and natural gas stocks. Equivalent electrical energy capacity is derived by applying a conversion efficiency of 40% to their calorific energy content. Electricity storage energy capacity includes operational pumped hydro and battery storage systems. b) In-vehicle reserves of fossil fuels refers to capacity of petrol tanks in conventional cars. Equivalent electrical energy capacity is derived by applying a conversion efficiency of 30% to the calorific energy content of petrol. Electricity storage energy capacity refers to battery capacity of all electric vehicles, including hybrids.

The energy stored in these fossil fuels could produce approximately 100 TWh of electrical energy, assuming 40% conversion efficiency. For comparison, the electrical energy storage capacity of the UK in 2020 was approximately 30 GWh in pumped hydro and 2 GWh in battery storage systems.[28] The UK therefore stores around 3,000 times more energy in fossil fuels than it does in the form of electrical energy storage (Figure 1.7).

Similarly, the fuel tanks of the 32 million cars in the UK can store around 17 TWh of calorific energy in the form of petrol and diesel. This equates to approximately 6 TWh of electrical energy, assuming 30% conversion efficiency. In contrast, the ~400,000 electric vehicles in 2020 could just 20 GWh.[30]

In a future with net-zero carbon emissions these fossil fuel reserves must be replaced by low-carbon alternatives. The UK is required by law to achieve net-zero carbon emissions by 2050.[31]

This simple, high-level analysis suggests upper bounds for the amount of electricity storage the UK might need for a net-zero energy system with complete electrification of other sectors: 100 TWh at the system level and 6 TWh within personal transport, which are two to three orders of magnitude greater than current levels. There are several flexibility options that can balance mismatches in energy supply and demand on the system-level, as discussed in section 1.3. These will reduce the overall storage capacity required. Also, sustainable fuels such as biomass or synthetic fuels derived from electricity can be stored directly, and not all energy sectors will be electrified. Alternative forms of energy storage, such as hot water tanks for heat, will further reduce the amount of electricity storage capacity needed.[32]

Still, this cursory analysis provides an initial, high-level understanding of the orders-of-magnitude growth of electricity storage that will be required in future low-carbon energy systems with little or no reliance on fossil fuels for matching supply with demand. Chapter 7 revisits this topic with a more detailed analysis.

KEY INSIGHT

In terms of energy capacity, countries require terawatt-hours (TWhs) of electricity storage to replace at least part of the storage capacity that fossil fuel reserves currently provide and enable future time-shifting of energy use.

1.5 References

1. Masson-Delmotte V, Zhai P, Pörtner H-O, Roberts D, *et al.* IPCC Special Report: Global Warming of 1.5 C (Geneva: Intergovernmental Panel on Climate Change, 2018).
2. Masson-Delmotte V, Zhai P, Chen Y, Goldfarb L, *et al.* (eds). Working Group I Contribution to the Sixth Assessment Report of the Intergovernmental Panel on Climate Change (Geneva: Intergovernmental Panel on Climate Change, 2021).
3. Friedlingstein P, Jones MW, O'Sullivan M, Andrew RM, *et al.* 'Global Carbon Budget 2021' (2022) 14 *Earth System Science Data* 1917–2005.
4. UN Conference of the Parties to the UNFCC. *Paris Agreement* (Paris, United Nations, 2015).
5. Core Writing Team, Pachauri, RK, and Meyer, LA (eds). Climate Change 2014: Synthesis Report. Contribution of Working Groups I, II and III to the Fifth Assessment Report of the Intergovernmental Panel on Climate Change (Geneva: Intergovernmental Panel on Climate Change, 2014).
6. Figueres C, Le Quéré C, Mahindra A, Bäte O, *et al.* 'Emissions Are Still Rising: Ramp Up the Cuts' (2018) 564 *Nature* 27–30.
7. Skea J, Shukla PR, Reisinger A, Slade R, *et al.* Climate Change 2022: Working Group III Contribution to the Sixth Assessment Report of the Intergovernmental Panel on Climate Change (Geneva: Intergovernmental Panel on Climate Change, 2022).

8. U.S. Energy Information Administration (EIA). International. Available at: <https://www.eia.gov/international/overview/world> accessed 5 November 2022.
9. IRENA. *Renewable power generation costs in 2021* (Masdar City: IRENA, 2022).
10. BloombergNEF. 1H 2020 LCOE Update (2020). Available at: <https://www.bnef.com/insights/22829>. accessed 5 November 2022.
11. BloombergNEF. Cost of new renewables temporarily rises as inflation starts to bite (2022). Available at: <https://about.bnef.com/blog/cost-of-new-renewables-temporarily-rises-as-inflation-starts-to-bite/> accessed 5 November 2022.
12. International Energy Agency. *Global EV Outlook 2022* (Paris: IEA, 2022).
13. BloombergNEF. Battery pack prices fall to an average of $132/kWh, but rising commodity prices start to bite (2021). Available at: <https://about.bnef.com/blog/battery-pack-prices-fall-to-an-average-of-132-kwh-but-rising-commodity-prices-start-to-bite/> accessed 5 November 2022.
14. International Energy Agency. *The Power of Transformation* (Paris: IEA, 2014).
15. Lokhov A. *Technical and Economic Aspects of Load Following with Nuclear Power Plants* (Paris: OECD & NEA, 2011).
16. International Energy Agency. *Harnessing Variable Renewables: A Guide to the Balancing Challenge* (Paris: IEA, 2011).
17. Staffell I, Green R, Gross R, Green T, and Clark L. *Electric Insights Quarterly—July to September 2018* (Selby: Drax Group, 2018).
18. International Energy Agency. *Status of Power System Transformation 2018—Advanced Power Plant Flexibility* (Paris, IEA, 2018).
19. International Energy Agency. *World Energy Outlook* (Paris: IEA, 2018).
20. Lund PD, Lindgren J, Mikkola J, and Salpakari J. 'Review of Energy System Flexibility Measures to Enable High Levels of Variable Renewable Electricity' (2015) 45 *Renewable and Sustainable Energy Reviews* 785–807.
21. MacDonald AE, Clack CTM, Alexander A, Dunbar A, *et al.* 'Future Cost-Competitive Electricity Systems and Their Impact on US CO2 Emissions' (2016) 6 *Nature Climate Change* 526–31.
22. Sanders D, Hart A, Brunert J, Strbac G, *et al. An Analysis of Electricity System Flexibility for Great Britain* (London: Carbon Trust, 2016).
23. International Hydropower Association (2022). Available at: <https://www.hydropower.org/> accessed 5 November 2022.
24. Wood Mackenzie. *Global Energy Storage Outlook H1 2021* (Edinburgh, Wood Mackenzie, 2021).
25. International Energy Agency. *Hydropower Special Market Report* (Paris: IEA, 2021).
26. BloombergNEF. Global energy storage market to grow 15-fold by 2030. (2022). Available at: <https://about.bnef.com/blog/global-energy-storage-market-to-grow-15-fold-by-2030/> accessed 5 November 2022.
27. World Energy Council. *World Energy Issues Monitor* (London: World Energy Council, 2019).
28. Harris K, Michaels C, Rose S, Ying D, *et al. Digest of UK Energy Statistics* (London: UK Government, 2021).
29. Staffell I. *The Energy and Fuel Data Sheet* (2011). Available at: <https://www.academia.edu/1073990/The_Energy_and_Fuel_Data_Sheet> accessed 5 November 2022.

30. Next Green Car. Electric vehicle market statistics (2022). Available at: <https://nextgreencar.com/electric-cars/statistics/> accessed 5 November 2022.

31. *Climate Change Act 2008* (London: UK Government, 2008).

32. Llewellyn-Smith C. 'The Need for Energy Storage in a Net Zero World' in *Talk at Virtual Conference on Medium-Duration Energy Storage in the Net-Zero UK* (Oxford University: Energy Research Accelerator & Supergen Energy Storage Network+, 2020), 1–39.

2 Energy storage toolkit
Separating the wheat from the chaff

KEY INSIGHT	WHAT IT MEANS
Energy storage technologies are unique assets in the energy system because they consume, store, *and* deliver energy.	This versatility leads to a wide range of performance parameters that must be considered when assessing energy storage technologies.
The cost of energy storage technologies is assessed with the metrics of investment cost and lifetime cost. Investment cost usually makes up a major share of lifetime cost.	These metrics differ as lifetime cost considers all costs incurred and the amount of energy or power provided over the technology's lifetime.
Quotations for investment cost may refer to different system scopes (e.g. single cell vs fully installed system).	Stakeholders should ensure quotations refer to the appropriate system scope to avoid comparing apples to oranges.
Specific investment cost gives the cost of adding an additional unit of power capacity *or* energy capacity. Total investment cost includes the cost for both power capacity *and* energy capacity.	Quotations should always mention which cost components are included since both metrics can be given in USD/kW or USD/kWh, i.e. when total investment cost is given relative to total power capacity or energy capacity.
For long discharge durations, technologies with low specific energy cost have lowest total cost (even with high specific power cost). For short discharge durations, low specific power cost (even with high specific energy cost) will give lowest total cost.	The ratio of specific power and specific energy investment cost provides initial insights on the suitability of energy storage technologies to certain applications.
Application-specific lifetime cost accounts for all cost and performance parameters relevant throughout the lifetime of the technology.	Lifetime cost should be used to assess storage projects with a specific technology and application.
The key metrics for lifetime cost are levelized cost of storage (LCOS) for applications that value the provision of energy, and annuitized capacity cost (ACC) for applications that value the provision of power.	The first step in determining lifetime cost is to choose the relevant metric. This requires identifying the desired output for the given application (i.e. what it is paid for)

Monetizing Energy Storage. Oliver Schmidt & Iain Staffell, Oxford University Press. © Oliver Schmidt & Iain Staffell (2023).
DOI: 10.1093/oso/9780192888174.003.0002

2.1 Performance parameters

Performance parameters characterize energy storage technologies from a technical perspective and quantify their performance (Table 2.1). The ability to consume, store, and deliver energy makes storage technologies more complex than other energy technologies and so a wider range of performance parameters need to be considered.

KEY INSIGHT

Energy storage technologies are unique assets in the energy system because they consume, store, *and* deliver energy.

Table 2.1 Technical parameters of energy storage technologies with example values for lithium-ion energy storage systems.

Parameter	Symbol	Description	Unit	Example
Design parameters				
Nominal power capacity	$Cap_{p, nom}$	Rated amount of power that can be charged and discharged.	kW	1,000
Power density—gravimetric	$\rho_{p, gra}$	Nominal power capacity divided by system mass.	kW/kg	0.3
Power density—volumetric	$\rho_{p, vol}$	Nominal power capacity divided by system volume.	kW/m³	6,000
Nominal energy capacity	$Cap_{e, nom}$	Rated amount of energy that can be discharged.	kWh	4,000
Energy density—gravimetric	$\rho_{e, gra}$	Nominal energy capacity divided by system mass	kWh/kg	0.2
Energy density—volumetric	$\rho_{e, vol}$	Nominal energy capacity divided by system volume	kWh/m³	450
Depth-of-discharge	DoD	Energy capacity that can be charged/discharged without severely degrading nominal energy capacity, measured relative to full capacity	$\%_{cap}$	100%

(Continued)

Table 2.1 (*Continued*)

Parameter	Symbol	Description	Unit	Example
Usable energy capacity	$Cap_{e, use}$	Energy capacity that can be discharged accounting for depth of discharge	kWh	4,000
Energy-to-power ratio	E/P	Usable energy capacity divided by nominal power capacity	hours	4
Discharge duration	DD	Time to discharge usable energy capacity at nominal power. *Same as E/P ratio*	hours	4
Max. C-rate	C	Maximum rate to discharge storage system relative to its usable energy capacity. *Inverse of E/P ratio or minimum DD*	1/hours	0.25
Response time	T_{res}	Time between idle state and maximum power	seconds	< 1
Operational parameters				
State of charge	SoC	Fraction of energy stored at any moment in time, measured relative to full capacity	$\%_{cap}$	80%
Round-trip efficiency	η_{RT}	Proportion of energy discharged over energy required to charge for a full charge–discharge cycle	%	80%
Self-discharge	η_{self}	Unavoidable loss of state of charge when a storage system is idle (highly dependent on usage profile - can be measured per cycle or averaged across all cycles per year)	$\%_{cap}$	1%
Degradation	Deg_t Deg_c	Rate of loss in usable energy capacity incurred by cycles and/or time lapse due to e.g. changes in state of charge or operating temperature	$\%_{cap}$ per year; $\%_{cap}$ per cycle	1% 0.01%
Cycle life	$Life_{cyc}$	Number of full charge–discharge cycles before end of usable life	#	4,000

(*Continued*)

Table 2.1 (*Continued*)

Parameter	Symbol	Description	Unit	Example
Calendar life	$Life_{cal}$	Number of years before end of usable life with no operation	years	20
State-of-health	SoH	Actual energy capacity relative to nominal energy capacity	%	90%
End-of-life threshold	EoL	Actual energy capacity relative to nominal energy capacity at which the storage system is taken out of service	%	80%

2.1.1 Design parameters

The first group of parameters in Table 2.1 can be labelled *design parameters* because they describe the energy storage system design in terms of energy and power metrics, and the interplay between the two.

Nominal energy capacity and **power capacity** describe the size of a storage system in terms of the energy that can be discharged, and the rate at which it can be discharged (or charged) respectively. They can both be normalized by system mass or volume. The resulting **energy density** and **power density** are common metrics to compare technologies. For example, in transport applications like electric vehicles, a high gravimetric energy density is desired, because the weight of the storage system has a direct impact on the energy consumption per kilometre (km) driven. High volumetric energy density is desired because the volume of the storage system has a direct impact on vehicle design and passenger comfort.

Depth-of-discharge (DoD) is the energy discharged during a cycle relative to the nominal energy capacity. Some technologies degrade faster at high depth-of-discharge and operators may decide to not charge/discharge the full nominal energy capacity. If DoD is pre-defined to be < 100%, the energy capacity that is charged and discharged in each cycle is called **usable energy capacity**. If DoD is 100%, then usable energy capacity equals nominal energy capacity.

The energy-to-power (E/P) ratio is obtained by dividing usable energy capacity by nominal power capacity. This is equal to the **discharge duration**: how long a storage system can discharge energy in one discharge cycle at nominal power output. Minimum discharge duration would be a more accurate term since storage systems can also be discharged at less than nominal power. However, the word *minimum* is omitted here for simplicity. The inverse of discharge duration is the **maximum C-rate**, that is, the maximum rate at which a storage system can be discharged relative to its usable energy capacity. **Response time** describes the time between requesting and delivering nominal power output. It is a key parameter for storage systems providing short-term power system flexibility to immediately correct for diversions in crucial electrical system parameters.

Care must be taken when considering the energy capacity and power capacity of a storage system. Energy capacity can refer to the total physical capacity, or to the available capacity (also called usable or accessible capacity) which will be lower in systems which restrict depth of discharge to preserve lifetime. A '10 kWh system' with 80% DoD could refer to 10 kWh total capacity and thus 8 kWh available, or to 10 kWh available and thus 12.5 kWh total. The latter system would be more expensive with all else held equal but would offer greater utility as it should be modelled as 10 kWh at 100% DoD, rather than 10 kWh at 80% DoD.

The power capacity of a storage system can be measured in terms of the power limit for charging or for discharging, and in terms of nominal or peak power throughput. This is not an issue for many systems, as they are sized with symmetrical input/output capacity and designed for steady-state operation. This adds complexity for electric vehicles though, as several definitions for their power capacity are possible, with no widely-held convention.

Taking the 2022 Nissan Leaf as an example,[1] the battery can be charged at a maximum of 46 kW but can supply the motors with a peak power of 110 kW. Given the battery's energy capacity is 40 kWh (39 kWh usable), this gives an energy/power ratio of either 0.85:1 or 0.35:1. The car's battery would be fully discharged within just 21 minutes if supplying the motor's peak power constantly, but this does not reflect real-world usage. Real-world driving is intermittent and highly dependent on road conditions and driver behaviour. An estimate of the nominal power capacity comes from energy consumption statistics: driving constantly at highway speeds in cold weather (at –10°C) consumes an estimated 325 Wh per mile (20.2 kWh per 100 km), which equates to 23 kWh per hour (at 70 mph / 110 km/h). This is the most energy-intensive steady-state driving mode, and implies a 1.7:1 energy to power ratio (the battery would be fully discharged after 1.7 hours). Whether the energy-to-power ratio is modelled as 1.7, 0.85 or 0.35:1 has major implications for the economic assessment of the storage system. A simplified ratio of 1:1 is chosen for further analysis of electric vehicles in this book.

2.1.2 Operational parameters

The second group can be labelled *operational parameters* because they describe the state of or define the efficiency of the storage system operation, or quantify the degradation process and resulting lifetime metrics.

State of charge describes how 'full' the battery is relative to its usable energy capacity at any point in time. This is shown by our phones when we check how much charge is left. **Round-trip efficiency** describes the proportion of energy discharged relative to the energy required to charge. This metric may differ significantly across storage technologies. **Self-discharge** quantifies the spontaneous loss in usable energy of a storage system. It happens through unwanted electrochemical reactions (e.g. batteries) or mechanical resistance (e.g. flywheels) and reduces the state of charge without any productive use of the discharged energy. In contrast to degradation, it does not affect energy capacity and is not permanent.

Degradation quantifies the rate of loss in usable energy capacity, which some storage technologies suffer from. It can be further specified into cycle degradation, that is, loss incurred through each charge–discharge cycle, or temporal degradation, that is, loss incurred by time lapse. **Cycle life** (see Attention box) and **calendar life** are determined by cycle and temporal degradation respectively and quantify lifetime of a storage technology in charge–discharge cycles or years. Both degradation effects can be additive. The **end-of-life threshold** is reached when energy capacity has degraded to a pre-defined value and the storage system is taken out of service, for example 80%. Such a point-in-time value of nominal energy capacity is also called **state of health**.

ATTENTION Cycle life refers to *full equivalent charge–discharge cycles*. The energy discharged in these cycles is set by the usable energy capacity. In practice, storage systems operate flexibly and may not discharge the full amount in each cycle. However, cycle degradation can be approximated by the amount of energy that passes through a storage system. Thus, full equivalent charge–discharge cycles represent a common proxy for this energy throughput and form a common basis to determine/compare cycle life.

2.2 Cost parameters

Cost parameters are essential for determining the financial attractiveness of energy storage systems. The key metrics are investment cost and lifetime cost. Investment cost is an input to lifetime cost and usually makes up the major share of it. For investment cost, it is important to clearly define the scope and not mix up specific and total cost.

KEY INSIGHT

The cost of energy storage technologies is assessed with the metrics of investment cost and lifetime cost. Investment cost usually makes up a major share of lifetime cost.

2.2.1 Scope of costs

The energy storage product value chain consists of multiple upstream and downstream steps, each of which adds to the final product cost. Figure 2.1 and Table 2.2 describe this value chain for stationary energy storage systems. The upstream section lists the key components required for a complete stationary storage system. The downstream section lists the steps required to fully install and operate the system.

KEY INSIGHT

Quotations for investment cost may refer to different system scopes (e.g. single cell vs fully installed system).

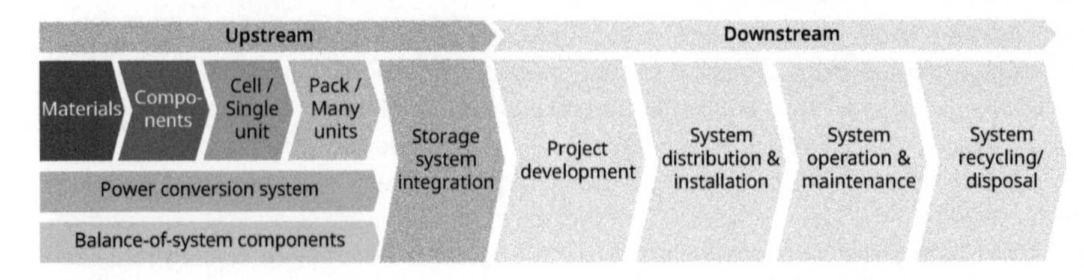

Figure 2.1 Product value chain for stationary energy storage systems. Shades of grey indicate upstream and downstream sections. Colours highlight the physical components that form part of a stationary storage system.

Table 2.2 Description of energy storage product value chain steps with examples for stationary lithium-ion battery system.

Value chain step	Description	Examples
Materials	Mining of raw materials and processing to manufacturing-grade materials	Spodumene ore Lithium hydroxide
Components	Manufacturing of subcomponents and assembly to components	Electrode powder Coated electrodes
Storage unit(s)	Manufacturing/assembly of single and multiple units of energy storage technology; *(can be characterized by energy capacity and/or nominal power capacity)*	Battery cell Battery pack
Power conversion system	Manufacturing of products which ensure that the electricity delivered by the storage technology fulfils the requirements of the target application	Inverter/converter Data management
Balance-of-system components	Manufacturing of products that are required for the operation of the storage technology	Container Thermal control
Storage system integration	Integration of storage unit(s), power conversion system and balance-of-system components into a full storage system and tailoring to the target application	Turnkey stationary storage system
Project development	Identification and financial, technical, and regulatory assessment of opportunities for energy storage technology deployment	Land acquisition Regulatory permits Financial studies
System distribution & installation	Technology distribution (e.g. through wholesalers for small systems), detailed engineering planning and installation at customer site	Engineering studies Procurement Installation
System operation & maintenance	Day-to-day operation, regular maintenance, and asset management for the storage system	Dispatch schedule Financial statements
System recycling & disposal	Collection and disassembly of storage systems Recycling or disposal of used materials	Recycled metals for electrode powder

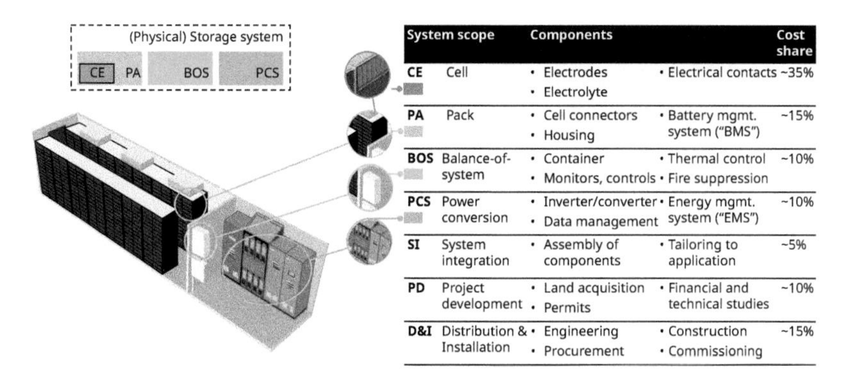

System scope		Components		Cost share
CE	Cell	• Electrodes • Electrolyte	• Electrical contacts	~35%
PA	Pack	• Cell connectors • Housing	• Battery mgmt. system ("BMS")	~15%
BOS	Balance-of-system	• Container • Monitors, controls	• Thermal control • Fire suppression	~10%
PCS	Power conversion	• Inverter/converter • Data management	• Energy mgmt. system ("EMS")	~10%
SI	System integration	• Assembly of components	• Tailoring to application	~5%
PD	Project development	• Land acquisition • Permits	• Financial and technical studies	~10%
D&I	Distribution & Installation	• Engineering • Procurement	• Construction • Commissioning	~15%

Figure 2.2 Exemplary lithium-ion stationary storage system split into system scope, components, and indicative cost contributions. Energy storage material costs are included in the cell and pack components (see Figure 2.1).

The upstream value chain shows that the physical energy storage system consists of multiple system scopes. Figure 2.2 visualizes the key scopes and lists the respective components and cost shares for stationary lithium-ion systems as an example. This highlights how quotations for investment cost may differ in terms of components included. For lithium-ion systems the cost of the cell could be quoted (~35% of total cost) or the cost of the fully installed stationary system (100% of total cost). Stakeholders should be aware of the different components to avoid comparing apples to oranges. For example, lithium-ion pack prices recently crossed the 150 USD/kWh threshold.[3] However, that does not mean that solar-plus-storage systems can already provide cost-effective baseload power as suggested at this price point.[2] This is because the pack level only makes up ~50% of the cost of fully installed stationary lithium-ion systems, which still cost over 300 USD/kWh in 2020.[4] It is also important to note that distribution and installation cost and project development cost are both highly location dependent and may vary significantly between projects.

2.2.2 Specific vs total cost

Another misconception associated with energy storage cost parameters is the difference between specific and total investment costs. Specific investment cost describes either:

- Specific power cost: the cost of components that enable the charging and discharging of energy (e.g. inverters, turbines), or
- Specific energy cost: the cost of components that enable the storage of energy (e.g. battery cells, water reservoirs).

These costs can also be called marginal power or energy costs because they describe the cost required to add an additional unit of power capacity or energy capacity.

Total investment cost refers to the cost of the entire storage system, including all components. This is obtained by multiplying the specific power cost by the nominal power capacity and adding the specific energy cost multiplied by the energy capacity.

To complicate matters further, the total investment cost can be specified relative to the total power capacity or energy capacity of an energy storage system (dividing by those capacities). This is where confusion can occur because then the units for total and specific cost are the same (USD/kW or USD/kWh).

Figure 2.3 shows the difference between specific and total investment costs with a simple example. Stakeholders may advertise the cost of the displayed energy storage system at 1 USD/kWh, but this only refers to the specific cost of adding an additional unit of energy capacity. In reality, the average cost of the total system is 10 USD/kWh.

KEY INSIGHT

Specific investment cost gives the cost of adding an additional unit of power capacity *or* energy capacity. Total investment cost includes the cost for both power capacity *and* energy capacity.

While total investment cost is used to describe the all-in cost of an energy storage system, specific investment cost may be used to highlight the cost-effectiveness of adding additional power capacity or energy capacity. In fact, the ratio between specific power and specific energy investment cost indicates whether storage technologies are more cost effective in short-duration or long-duration applications. This is because total cost will scale differently with increasing discharge duration due to this ratio.

Figure 2.4 compares four energy storage technologies with different ratios of specific power and energy cost. While the specific power cost for pumped hydro is highest, its specific

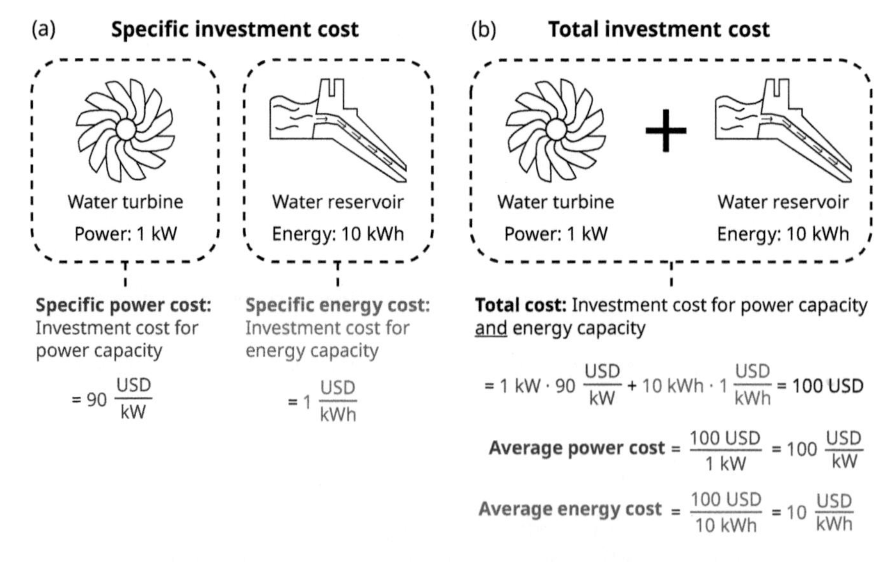

Figure 2.3 Schematic with sample values for (a) specific investment cost and (b) total investment cost for the same energy storage system consisting of components to deliver/consume power (water turbine) and to store energy (water reservoir).

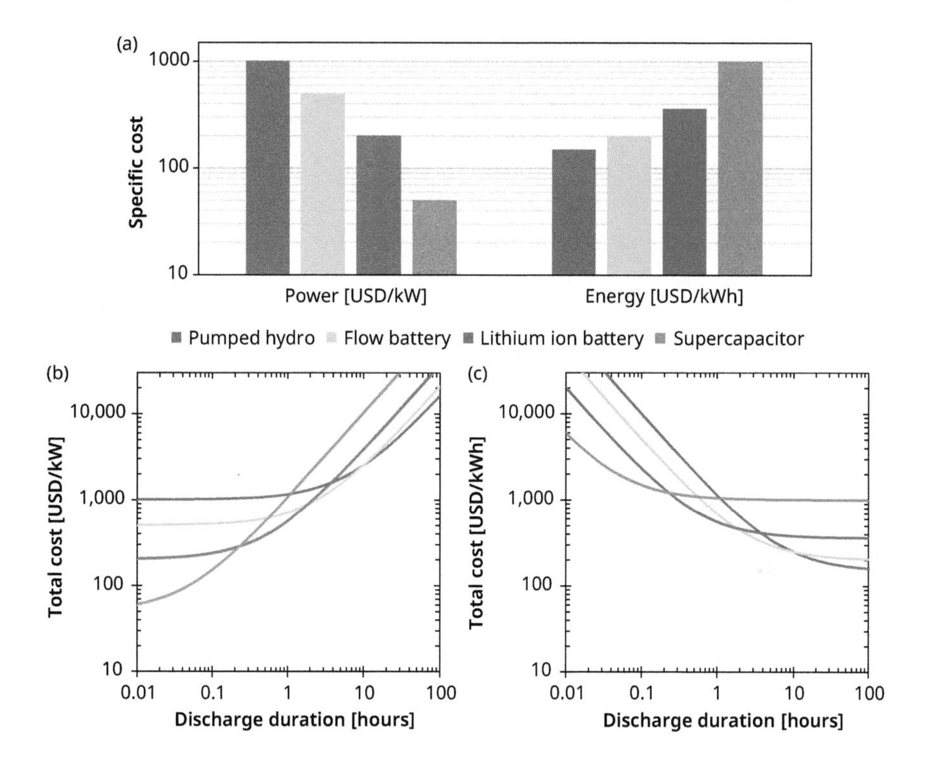

Figure 2.4 Comparison of (a) specific cost to (b) total cost in terms of power capacity and (c) energy capacity for four exemplary energy storage technologies. Cost assumptions are for illustrative purposes only.

energy cost is lowest. For supercapacitors it is the opposite. As a result of the illustrative input values, supercapacitors are cheapest at discharge durations below ~0.25 hours (15 minutes). Pumped hydro plants are cheapest above 10 hours. The more balanced ratio of specific power and energy investment cost of lithium-ion and flow batteries make those technologies most cost efficient at 0.25–2 and 2–10 hours respectively.

KEY INSIGHT

For long discharge durations, technologies with low specific energy cost have lowest total cost (even with high specific power cost). For short discharge durations, low specific power cost (even with high specific energy cost) will give lowest total cost.

2.2.3 Lifetime cost

Investment cost alone cannot determine the commercial viability of energy storage technologies or their competitiveness against each other. A comprehensive assessment must consider *lifetime cost*. This accounts for all costs incurred during the lifetime of a technology as well as the technology's performance parameters. A comparative example is car ownership. While purchase costs (or investment costs) make up a large proportion of the overall economics, repair costs and refuelling costs (or operational costs) are decisive as well. Thus, comparison should not be based on purchase price alone but the cost per km driven over the lifetime of the car.

Lifetime cost assessment is particularly relevant for energy storage due to the diversity of technologies, all with different cost and performance characteristics, and the wide range of applications, all with different requirements. Using the car example again, different models will be cost optimal for city, long-distance, or off-road usage.

KEY INSIGHT

Application-specific lifetime cost accounts for all cost and performance parameters relevant throughout the lifetime of the technology.

Levelized cost of storage (LCOS) quantifies the overall cost per unit of discharged electricity for a storage technology serving a specific application. It takes all costs incurred during the lifetime and divides them by the cumulative delivered electricity.[5–7] This metric could be used when assessing which technology is more competitive in wholesale energy market arbitrage, for example.

Annuitized capacity cost (ACC) divides all costs incurred during the lifetime of a storage system by its power capacity and lifetime (USD/kW-year). This is useful for applications that value the provision of power instead of energy like frequency regulation.

$$LCOS\left[\frac{USD}{MWh}\right] = \frac{Lifetime\ cost\ incurred}{Lifetime\ energy\ discharged} = \frac{Investment + O\&M + Charging + End\ of\ life}{Energy\ capacity \cdot Cycles\ p.a. \cdot Lifetime}$$

$$ACC\left[\frac{USD}{kWyear}\right] = \frac{Lifetime\ cost\ incurred}{Lifetime\ power} = \frac{Investment + O\&M + Charging + End\ of\ life}{Power\ capacity \cdot Lifetime}$$

KEY INSIGHT

The key metrics for lifetime cost are levelized cost of storage (LCOS) for applications that value the provision of energy, and annuitized capacity cost (ACC) for applications that value the provision of power.

Table 2.3 describes the key cost parameters of energy storage technologies, all of which are considered in lifetime cost assessments.

Lifetime energy discharged accounts for the key performance parameters like nominal energy storage capacity, depth of discharge, self-discharge, degradation, and lifetime in years. In addition it accounts for the number of annual charge–discharge cycles, which is given by the respective application.

Table 2.3 Cost parameters of energy storage technologies with example values for lithium-ion battery systems.

Parameter	Symbol	Description	Unit	Example
Investment cost (specific)	$C_{e,inv}$ $C_{p,inv}$	Cost to add energy capacity and power capacity respectively. This should cover all cost components required to set up the storage system for a specific application.	USD/kWh USD/kW	300 250
Construction time	T_{con}	Time to construct a storage system from breaking ground to completion.	years	1
Replacement cost	$C_{e,rep}$ $C_{p,rep}$	Cost to replace major technology components not accounted for in O&M (e.g. inverter). *Also called augmentation cost.*	USD/kWh USD/kW	n/a
Replacement interval	Cyc_{rep}	Time interval for replacement of major technology components.	cycles	n/a
O&M cost	$C_{e,om}$ $C_{p,om}$	Cost to operate (e.g. on-site staff), insure (e.g. fire, damage, theft), and periodically service technology components (e.g. servicing mechanical parts, scheduled inspections, minor replacements). These costs can be incurred per unit of electricity discharged (USD/MWh) or scale with the power capacity (USD/kW-year).	USD/MWh$_{el}$ USD/kW-year	0.4 5
Charging cost	P_{el}	Cost to charge a storage system with energy. It accounts for the amount of electricity charged and its price. It is a type of operating cost but treated separately due to its quantitative significance, like fuel costs are for vehicles.	USD/MWh$_{el}$	50
End-of-life cost	$C_{e,eol}$ $C_{p,eol}$	Cost to dispose technology at its end of life. Can be negative if technology still has value (e.g. raw materials like copper, individual components like inverters, or repurposing the system for a 'second life' application).	USD/kWh USD/kW	0 20

(Continued)

Table 2.3 (*Continued*)

Parameter	Symbol	Description	Unit	Example
Discount rate	r	Rate to discount future cost or revenues. It can be based on the weighted cost of capital (debt and equity) and thereby reflect the return requirement or 'hurdle rate' of companies. Nominal discount rates also account for inflation.	%	8%

Lifetime power is the product of power capacity and lifetime of the system. Depending on whether lifetime is given in years, weeks, or hours, the result for capacity cost is USD/kW-year, USD/kW-week, or USD/kW-hour. When given in years, it is defined as annuitized capacity cost (ACC).

ATTENTION Investment cost should include all energy-related (e.g. battery pack) and power-related (e.g. inverter) components and EPC costs (e.g. installation). A common system boundary for stationary storage is all components 'up to the transformer'. This ensures comparability of energy storage system cost quotes.

FREQUENTLY ASKED QUESTION

Why do you account for replacement or augmentation cost separately and not include them in Operation and Maintenance (O&M) cost?
Replacement cost can be quite significant and occur at a specific point in time (e.g. 40% of original investment after 10 years). In that case, it may be more accurate to model these costs explicitly rather than within time-averaged O&M cost.

The key issues to consider when including the relevant cost and performance parameters in lifetime cost assessments are outlined in Figure 2.5 (for LCOS calculation). These considerations highlight how the respective application affects storage lifetime cost. Required power capacity and energy capacity as well as response time influence investment cost. Annual cycle requirements determine degradation speed and lifetime. Applicable power prices determine the charging cost.

ATTENTION When considering a project with a specific remuneration period, for example a Power Purchase Agreement, often the duration of that period is considered as lifetime. That is because any revenues following that period are uncertain.

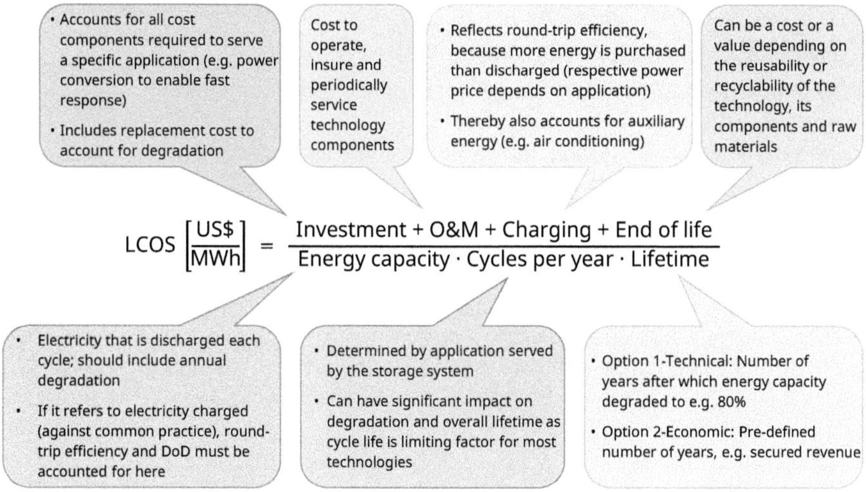

- Accounts for all cost components required to serve a specific application (e.g. power conversion to enable fast response)
- Includes replacement cost to account for degradation

Cost to operate, insure and periodically service technology components

- Reflects round-trip efficiency, because more energy is purchased than discharged (respective power price depends on application)
- Thereby also accounts for auxiliary energy (e.g. air conditioning)

Can be a cost or a value depending on the reusability or recyclability of the technology, its components and raw materials

$$\text{LCOS}\left[\frac{\text{US\$}}{\text{MWh}}\right] = \frac{\text{Investment} + \text{O\&M} + \text{Charging} + \text{End of life}}{\text{Energy capacity} \cdot \text{Cycles per year} \cdot \text{Lifetime}}$$

- Electricity that is discharged each cycle; should include annual degradation
- If it refers to electricity charged (against common practice), round-trip efficiency and DoD must be accounted for here

- Determined by application served by the storage system
- Can have significant impact on degradation and overall lifetime as cycle life is limiting factor for most technologies

- Option 1-Technical: Number of years after which energy capacity degraded to e.g. 80%
- Option 2-Economic: Pre-defined number of years, e.g. secured revenue

Figure 2.5 Key considerations when including parameters in lifetime cost assessments using the example of levelized cost of storage (LCOS).

The concept of levelized cost of storage is analogous to levelized cost of energy (LCOE) for energy generation technologies. For LCOE, the denominator would be the energy generated during the lifetime of the technology. The numerator would include fuel and carbon cost instead of charging cost. Similarly, annuitized capacity cost can be determined for energy storage and energy generation technologies.

However, the lifetime cost methodology is more complex for energy storage than for energy generation technologies. While generation technologies 'only' deliver energy, storage technologies consume and store energy, in addition to delivering it. This adds complexity to consider parameters like round-trip efficiency, self-discharge, or usable energy storage capacity.

As with LCOE, the incurred lifetime cost as well as the lifetime energy discharged (or power provided for ACC) are discounted over time. While intuitive for the cost components, it may seem counterintuitive to discount energy discharged or power provided. However, it is required so that LCOS and ACC correctly identify the minimum revenue requirement for a profitable project. The lifetime cost represents a net present value (NPV) of zero, so LCOS and ACC represent the minimum revenue requirement for a project to be economically viable. As such, the energy discharged/power provided are representative for any future revenues and must be discounted.

ATTENTION Because lifetime cost represents the revenue requirement for a project to have a net present value of zero, both the incurred costs and the discharged energy/power capacity must be discounted over the lifetime. This is because energy/power provide revenues in the future, which are worth less in today's money.

FREQUENTLY ASKED QUESTION

Why do you need to discount energy discharged or power provided?

It might at first seem counterintuitive to discount a physical metric like energy or power. This is not suggesting that a MWh of stored energy will degrade to 0.9 MWh after a year, as it is not the physical energy which matters, but rather the financial value of that energy when it is sold. The revenue from selling a MWh of stored energy next year may only be worth the revenue from selling 0.9 MWh today. These revenues from energy sales should be discounted just as the costs of operating the system, so that the resulting levelized cost represents the revenue requirement for achieving a project NPV of zero.

Please find below the derivation of the LCOS formula, which highlights this point:

NPV	=	0	*NPV: Net present value*
$NPV\ of\ cost$	=	$NPV\ of\ remuneration$	
$\sum_n^N \frac{cost(n)}{(1+r)^n}$	=	$\sum_n^N \frac{remuneration(n)}{(1+r)^n}$	*n: year, r: discount rate, N: lifetime in years*
$\sum_n^N \frac{cost(n)}{(1+r)^n}$	=	$\sum_n^N \frac{E_{out}(n) \cdot LCOS}{(1+r)^n}$	*E_{out}: Electricity discharged in year n, LCOS: price for electricity discharged*
$\sum_n^N \frac{cost(n)}{(1+r)^n}$	=	$LCOS \cdot \sum_n^N \frac{E_{out}(n)}{(1+r)^n}$	*LCOS is constant over time and can be taken out of the sum*
$LCOS$	=	$\frac{\sum_n^N \frac{cost(n)}{(1+r)^n}}{\sum_n^N \frac{E_{out}(n)}{(1+r)^n}}$	*Cost and E_{out} need to be discounted*

Inflation should also be accounted for by determining a nominal discount rate. In case only a real discount rate is given (i.e. not including inflation), it can be converted into a nominal discount rate with the following formula:[8]

$$r_{nominal} = \left[(1+r_{real}) \cdot (1+inflation\ rate)\right] - 1$$

2.3 References

1. EV Database. Nissan Leaf price and specifications. Available at: <https://ev-database.uk/car/1656/Nissan-Leaf> accessed 5 November 2022.
2. BloombergNEF. Battery Pack Prices Fall to an Average of $132/kWh, but Rising Commodity Prices Start to Bite (2021). Available at: <https://about.bnef.com/blog/battery-pack-prices-fall-to-an-average-of-132-kwh-but-rising-commodity-prices-start-to-bite/> accessed 5 November 2022.
3. Ziegler MS, Mueller JM, Pereira GD, Song J, *et al.* 'Storage Requirements and Costs of Shaping Renewable Energy Toward Grid Decarbonization' (2019) 3 *Joule* 2134–53.
4. Frith J. *Lithium-Ion Batteries: The Incumbent Technology Platform for Coal Regions in Transition* (London: BloombergNEF, 2019).
5. Schmidt O, Melchior S, Hawkes A, and Staffell I. 'Projecting the Future Levelized Cost of Electricity Storage Technologies' (2019) 3 *Joule* 81–100.
6. Pawel I. 'The Cost of Storage—How to Calculate the Levelized Cost of Stored Energy (LCOE) and Applications to Renewable Energy Generation' (2014) 46 *Energy Procedia* 68–77.
7. Jülch V. 'Comparison of Electricity Storage Options Using Levelized Cost of Storage (LCOS) Method' (2016) 183 *Applied Energy* 1594–606.
8. Aldersey-Williams J and Rubert T. 'Levelised Cost of Energy—A Theoretical Justification and Critical Assessment' (2019) 124 *Energy Policy* 169–79.

3 Technologies and applications

Entering the maze

KEY INSIGHT	WHAT IT MEANS
Energy storage technologies are categorized by the form in which energy is stored: *chemical*, *thermal*, *mechanical*, *electrical*, or the type of conversion reaction that takes place: *electrochemical*.	These categories shape the technical characteristics of energy storage technologies. Those from the same category often have common features and suitability for use cases.
Chemical storage is suited to large energy capacities and long discharge durations. *Thermal*, *mechanical*, and *electrochemical* storage are suitable for medium energy capacities and short to long durations. *Electrical* storage is best suited to small energy capacities and short durations.	The category of an electricity storage technology can already indicate its suitability to specific use cases. A key driver is specific energy investment cost, which depends on the materials required for storing energy.
Almost all stationary electricity storage is either pumped hydro or lithium ion. These made up 89% and 10% of global capacity in 2020. Both are also growing rapidly, by at least 5 GW per year.	Both technologies dominate stationary electricity storage applications in the 2020s. The market share of lithium ion will further increase in future.
Energy storage is a competitive industry with no structural drivers for monopolies to form. The number of active players in each step of the value chain depends on the technology's market size.	The energy storage industry is structurally similar to the wider energy industry. Many new players will enter the market as it grows.
There are 23 unique electricity storage applications that create economic value through increasing power quality, power reliability, asset utilization, or arbitrage at different locations in the power system. When focusing on discharge duration and annual cycle frequency requirements alone, these can be reduced to 13 archetype applications.	A wide range of electricity storage applications can be monetized. Lifetime cost must be determined specifically for each, since this depends on the suitability of a technology's cost and performance to the application's requirements.
Electricity storage deployment can be clustered into four phases. These are characterized by the share of low-carbon power generation in a given market, and influence the required storage discharge duration. Archetype applications are frequency regulation in phase 1, peaking capacity in 2, renewables integration in 3, and seasonal storage in 4.	This conceptual categorization can help to assess which applications become relevant in different markets over time. It also allows market size to be quantified from decarbonization progress. Phase 3 is the 'holy grail' for storage deployment potential.

Monetizing Energy Storage. Oliver Schmidt & Iain Staffell, Oxford University Press.
DOI: 10.1093/oso/9780192888174.003.0003

3.1 Energy storage categories

Energy storage is commonly classified into five categories: chemical, thermal, mechanical, electrical, and electrochemical (see Figure 3.1). The first four categories refer to the form in which energy is stored. *Electrochemical* is a separate category that is used to classify the wide range of battery technologies and refers to the type of reaction on which they are based. Within each of those categories there is a range of concepts that utilize the respective energy forms or reactions to store energy. Almost all energy storage technologies are based on one of these concepts.

KEY INSIGHT

Energy storage technologies are categorized by the form in which energy is stored: *chemical, thermal, mechanical, electrical*; or the type of conversion reaction that takes place: *electrochemical.*

Technologies within the *chemical* category store energy in the form of chemical bonds. These chemical bonds are multi-purpose energy carriers that can be converted to electricity or used in other energy sectors, such as transport, heating, or as feedstock in industry.[1] They can occur naturally as fossil fuels or be produced synthetically with electricity. The concepts refer to the chemical that is produced. The production of hydrogen gas from electricity and

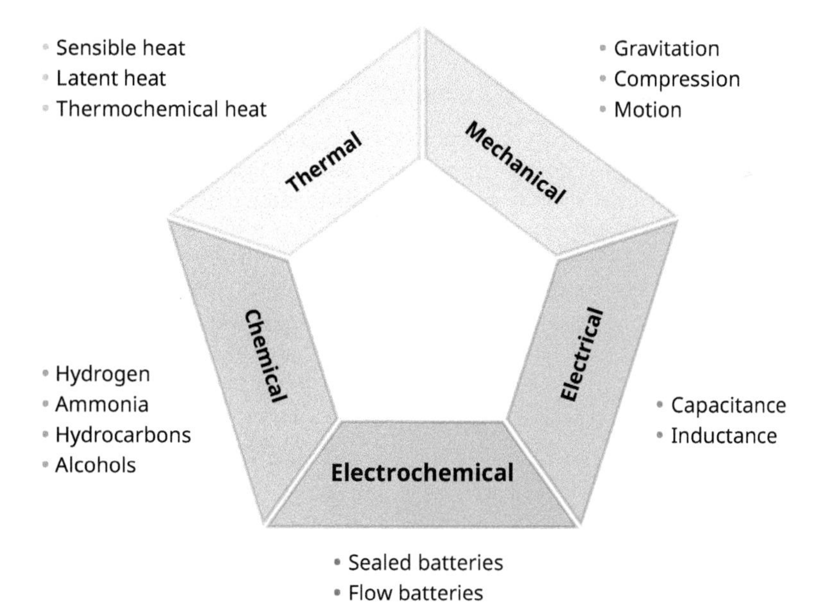

Figure 3.1 The five categories of energy storage with respective energy storage concepts. Coloured shapes show categories. Bullet points name concepts that utilize each energy form or reaction. Chemical, Thermal, Mechanical, and Electrical refer to the form of energy that is stored. Electrochemical refers to the type of reaction that takes place in battery technologies.

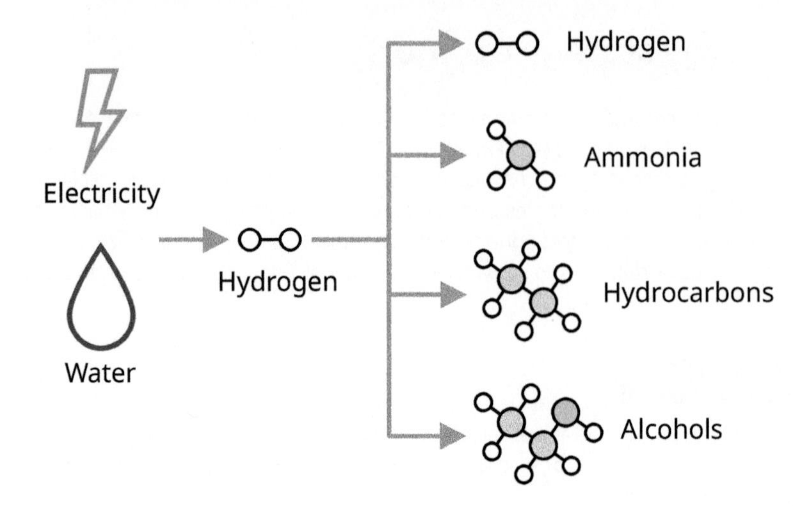

Figure 3.2 Schematic of electricity storage through conversion of electricity into chemical energy. The amount of energy stored is given by the heating value of the respective chemical compounds.

water through electrolysis is the key enabler of all electricity storage concepts that convert the electrical input energy into chemical energy, as can be seen in Figure 3.2. Hydrogen can be processed to ammonia by the addition of nitrogen, hydrocarbons by adding carbon dioxide (e.g. methane, kerosene) or alcohols, also with carbon dioxide (e.g. methanol, ethanol). These chemical compounds have higher volumetric energy density than hydrogen, making them easier to store and transport. The amount of chemical energy stored is given by the heating value of the respective chemical compounds.

Technologies within the *thermal* category store energy as heat. Electricity may be used to generate the heat that is stored. The heat can be removed later and used directly or re-converted to electricity. There are three concepts to store thermal energy: sensible heat, latent heat, and thermochemical heat.[2] Sensible heat is the thermal energy associated with heating or cooling of a material without changing its physical state. Latent heat is the energy associated with the phase change of a material between the solid, liquid, and gaseous states. Thermochemical heat is associated with a reversible chemical reaction or sorption process that releases or consumes large amounts of thermal energy.[2,3] Figure 3.3 depicts the concepts and shows the formulas to quantify the amount of thermal energy stored.

Mechanical energy is a collective term for gravitational, elastic, and motion energy. The respective concepts that utilize these forms of energy storage are gravitation, compression, and linear or rotational motion. Figure 3.4 depicts them graphically and displays the formulas to quantify the mechanical energy stored. Pumps, turbines, and electric motors/generators are the machines that convert electrical into mechanical energy and vice versa for the respective electricity storage technologies.

Electric energy can also be stored directly using the concepts of capacitance or inductance. Capacitance separates positive and negative charges on two conductive plates. The energy is stored in the electric field between them. Inductance uses the magnetic field generated

(a) **Sensible heat**

Energy associated with heating or cooling a material

Example: Hot water tank

$$E_{th} = mC_p\Delta T$$

m: mass of material
C_p: specific heat capacity
ΔT: temperature difference

(b) **Latent heat**

Energy associated with phase change of a material

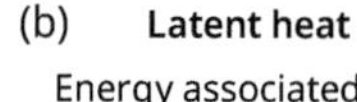

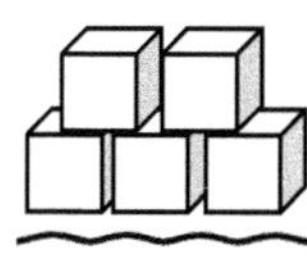

Example: Melting ice cubes

$$E_{th} = mL$$

m: mass of material
L: specific latent heat capacity

(c) **Thermochemical heat**

Energy associated with a reversible chemical reaction or sorption process

Example: Zeolite-water reaction

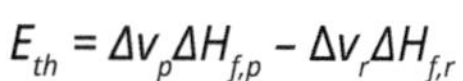

$$E_{th} = \Delta v_p \Delta H_{f,p} - \Delta v_r \Delta H_{f,r}$$

v_p: amount of products
v_r: amount of reactants
ΔH_f: standard enthalpy of formation

Figure 3.3 Schematics of three concepts of thermal energy storage: (a) sensible heat, (b) latent heat, and (c) thermochemical heat, including formulas to calculate the amount of thermal energy stored.

(a) **Gravitation**

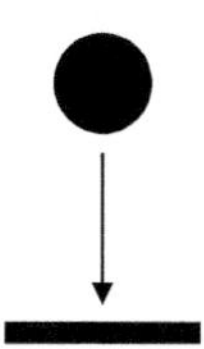

Example: Pumped hydro

$$E_{mech.} = mgh$$

m: mass of material
g: gravitational field strength
h: height

(b) **Compression**

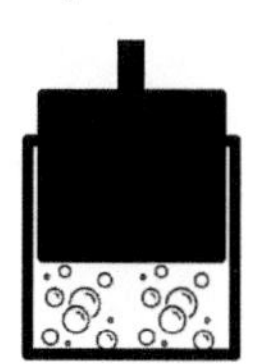

Example: Compressed air

$$E_{mech.} = p_2V_2 \ln\left(\frac{p_1}{p_2}\right) + (p_2 - p_1)V_2$$

p: pressure outside (1) and inside (2)
V: volume outside (1) and inside (2)

(c) **Linear motion**

Example: Rolling wheel

$$E_{mech.} = \frac{1}{2}mv^2$$

m: mass of material
v: velocity

(d) **Rotational motion**

Example: Flywheel

$$E_{mech.} = \frac{1}{2}mr^2\omega^2$$

m: mass of wheel
r: radius of wheel
ω: angular velocity

Figure 3.4 Schematics of the different mechanical energy storage concepts: (a) gravitation, (b) compression, (c) linear motion, and (d) rotational motion, including formulas to calculate the amount of mechanical energy stored.

by an electric current flowing through a superconducting coil to keep it flowing until needed.[2] Figure 3.5 shows sketches of both concepts and the respective formulas to calculate the amount of electrical energy stored.

The final category classifies battery technologies based on the *electrochemical* reactions that take place within them. Energy is stored as the electrochemical potential between two materials that could react to form a new one. The net chemical energy of forming the new material (Gibbs free energy) is balanced by the electrostatic energy between the two separated materials.[2] The dominant concepts are sealed and flow batteries. In sealed batteries, electrodes constitute the active material separated by an ion-conducting electrolyte. All components are in a confined battery cell. In flow batteries, the active material is not the electrode itself but two liquid electrolytes that can be circulated and stored outside of the system.[4] Figure 3.6 shows both concepts with the examples of lithium-iodine (sealed) and vanadium flow batteries.

In panel a), lithium (Li) and iodine (I_2) electrodes are separated by an electrolyte. Both materials naturally want to react to form lithium iodine (LiI). The net chemical energy of that reaction is directly converted into electrical energy during discharge in the form of electrons that travel through the closed connector from the lithium anode (i.e. releasing electrons during discharge) to the iodine cathode (i.e. consuming electrons during discharge). This enables iodine ions to move to the lithium anode through the electrolyte where lithium iodine is formed.

In panel b), a similar electrochemical process takes place with ions of vanadium pentoxide. The difference is that the vanadium pentoxide electrolyte is the active material. It is liquid and can be pumped to the electrodes for the electrochemical reactions to take place, thus the system is not confined. Electrolyte tanks and electrode cell can be scaled fully independently.

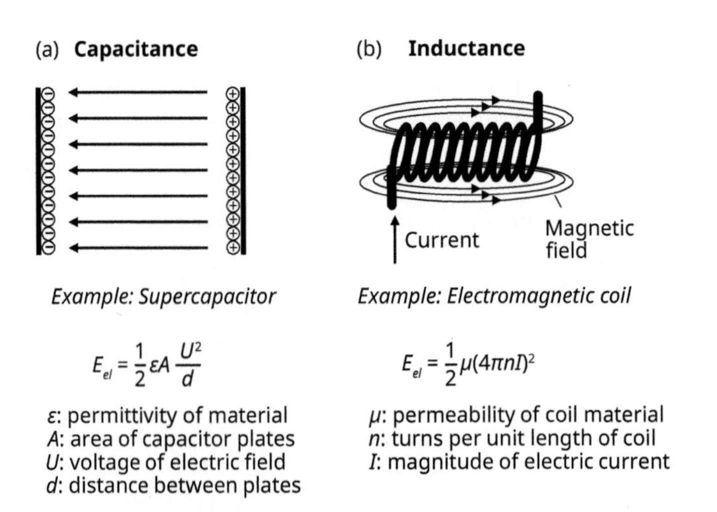

Figure 3.5 Schematics of electrical energy storage through (a) capacitance and (b) inductance, including formulas to calculate the amount of electrical energy stored.

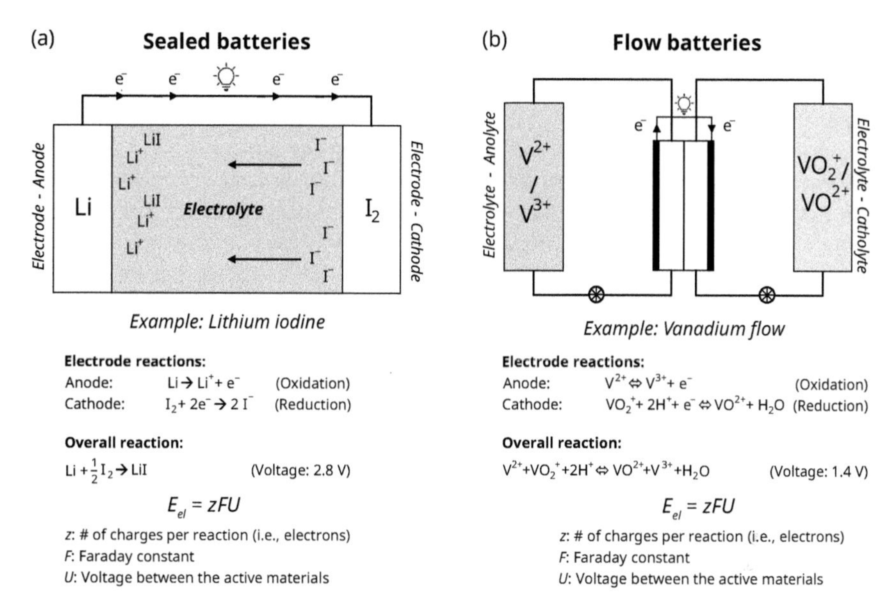

Figure 3.6 Schematics of (a) a primary lithium-iodine battery cell and (b) a secondary vanadium redox-flow battery, including formulas to calculate the amount of electrical energy stored.

Electrochemical technologies are further classified as primary or secondary batteries. In primary batteries, the electrochemical reaction cannot be reversed, so they are not rechargeable. In secondary batteries, applying an external potential that is higher than the battery's potential reverses the electron flow. This restores the battery's initial electrochemical potential, enabling it to be fully recharged.

3.2 Energy storage technologies

This section introduces the most widely deployed stationary electricity storage technologies. These technologies have a fixed location to support power system flexibility by consuming electricity and discharging it at a later point in time. All these technologies rely on the energy storage categories and concepts outlined in the previous section. The technologies covered are:

- Pumped hydro
- Adiabatic compressed air
- Flywheel
- Lead-acid battery
- Lithium-ion battery
- Sodium-sulphur battery
- Vanadium redox-flow battery
- Supercapacitor
- Hydrogen

3.2.1 Pumped hydro

A pumped hydro energy storage system (PHES) relies on gravitational energy using the difference in height between two water reservoirs to store energy (Figure 3.7). During periods when electricity demand is low, electricity is used to pump water from the lower reservoir to the higher one. During periods of high demand this water is released through the pumps now acting as turbines to generate electricity. Systems can operate using reversible pump turbines or separate turbines and pumps.[5–7] The size of the reservoirs determines the charge/discharge duration.

Pumped hydro is the most widely deployed stationary storage technology with 158 GW installed in 2020[8]. This is ~90% of the installed stationary storage capacity globally. The first PHES plant was built in Switzerland in 1907. Significant deployment started in the 1960s in line with nuclear power plant deployment to charge with low-cost nuclear electricity overnight and discharge during peak demand mid-day. With deployments of ~5 GW each year, pumped hydro is still one of the fastest growing stationary storage technologies.[9]

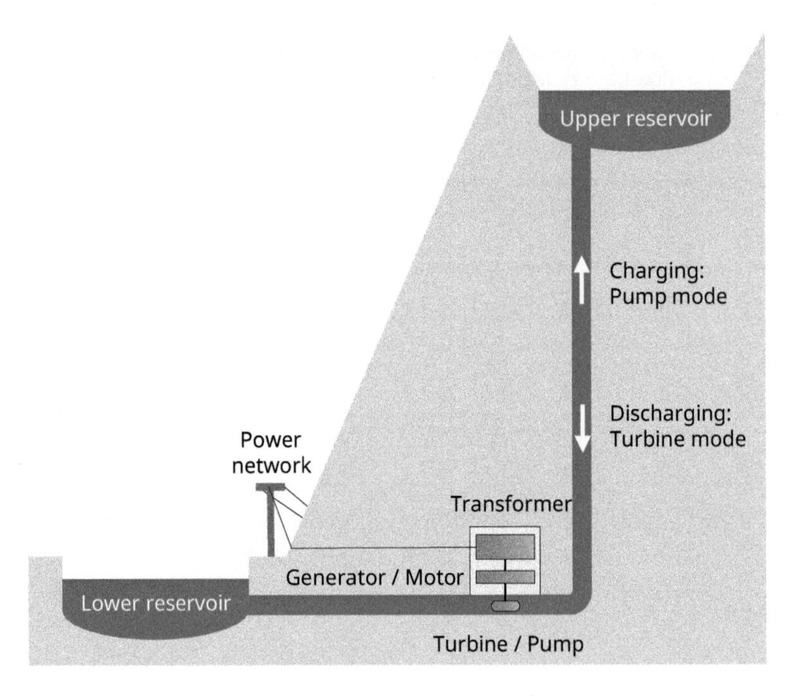

Figure 3.7 Schematic of a pumped hydro energy storage plant.

Category	Mechanical energy storage—Gravitation
Status	Commercial stage—most widely deployed stationary storage technology with 158 GW installed by 2020
Application	PHES is used for various stationary storage applications. Two prominent examples are • Peak capacity: Deliver electricity during peak demand periods • Energy arbitrage: Consume electricity during low-price periods and provide during high-price periods
Variations	Variations of this technology are based on the type of turbines used (e.g. Francis, Kaplan, Pelton), and whether they operate at fixed or variable rotational speeds. Pumped hydro storage can be realized through dedicated plants with two reservoirs using fresh or sea water, or a pump-back functionality in traditional hydropower plants.
Formula	$E_{PHES} = V\rho g H \eta$ E: Nominal energy capacity* [J] V: Reservoir volume [m^3] ρ: Fluid density [kg/m^3] g: Gravity acceleration [m/s^2] H: Head height in metres [m] η: Component efficiency [%]
Advantages	+ Technical maturity (~160 GW deployed) + Low specific energy capacity cost (< 50 USD/kWh) + Independent sizing of energy capacity and power capacity + High round-trip efficiency (> 80%) + Long lifetime (> 30,000 cycles)
Disadvantages	– Need for suitable geographical conditions – Long lead-time to build (multiple years) – Low energy density, thus large footprint (< 2 kWh/m^3) – Potential negative environmental and social impacts through creation of water reservoirs – Economical only at large scale (> multiple hundred MW) and long discharge duration (> 4 hours)
Sample players (turbine supplier)	• VOITH (Germany) • General Electric (US) • Andritz (Austria) • Hinac (China)
Innovation focus	• Reservoir selection: Underground structures or open sea • Fluids: Higher density fluids

* To convert from joules (J) to kilowatt hours (kWh), divide by 3,600,000.

3.2.2 Adiabatic compressed air

Adiabatic compressed air energy storage plants (A-CAES) compress and store air either using geological underground voids or purpose-made tanks (Figure 3.8). They rely on the internal kinetic energy within a compressed gas relative to ambient conditions. When electricity is available, air is compressed and stored underground or in a tank. When electricity is needed, the air is expanded to ambient pressure, driving a turbine. Since gases heat up during compression and cool down during expansion, adiabatic A-CAES plants store the generated heat from the compressor and provide it later to the turbine in the expansion stage.[10] This additional thermal storage is needed for round-trip efficiencies > 70%. In diabatic CAES plants, the generated heat is lost and natural gas is burned to generate heat at the expansion stage, resulting in round-trip efficiencies < 50%.

The only two large-scale CAES plants are diabatic, one in Germany from 1978 and one in the US from 1991, with a total of 400 MW power capacity. Therefore, the cost and performance data used in the analyses in this book refer to diabatic CAES plants. A 2 MW commercial adiabatic CAES plant has been operational in Canada since 2019 with larger ones at development stage.[11] However, many recently announced large-scale CAES projects have been cancelled.[12]

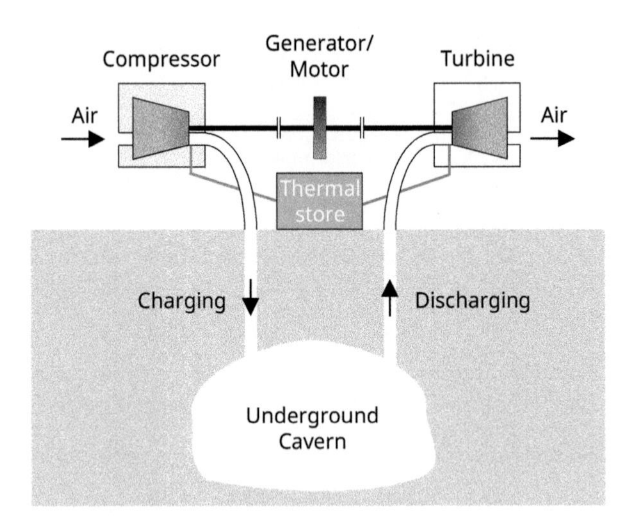

Figure 3.8 Schematic of an adiabatic CAES plant.

Category	Mechanical energy storage—Compression
Status	Commercial stage—400 MW deployed
Application	Low specific energy investment cost make CAES suited for long-duration energy storage applications, for example: • Renewable integration: Storing large amounts of excess renewable electricity supply to be used at a later time • Seasonal storage: Balancing longer-term supply disruption or seasonal variability in supply and demand
Variations	Technology variations are diabatic and isothermal CAES plants. Diabatic plants require fuel to heat the gas during expansion, which significantly decreases round-trip efficiency.[5] Isothermal CAES plants compress and expand the gas at constant temperature, for example through the parallel injection of a liquid.[13]
Formula	$E_{CAES} = p_2 V_2 \ln\left(\frac{p_1}{p_2}\right) + (p_2 - p_1) V_2 \eta$ E: Nominal energy capacity* [J] p_2: Pressure of compressed gas [N/m²] V_2: Volume of compressed gas [m³] p_1: Ambient pressure [N/m²] V_1: Volume at ambient state [m³] η: Component efficiency [%]
Advantages	+ Low specific energy capacity cost (< 50 USD/kWh) + Independent sizing of energy capacity and power capacity + Long lifetime (> 15,000 cycles) + Modular and location independent when using storage tanks
Disadvantages	– Cost-efficient underground CAES plants are geographically limited by the availability of caverns – Diabatic plants have low round-trip efficiency (< 50%) and require fuel for discharge – Low energy density (~4 kWh/m³) – Only economic at large scale (> multiple hundred MW) and long discharge duration (> 4 hours)
Sample players (project developer)	• Hydrostor (Canada) • RWE (Germany) • APEX CAES (US)
Innovation focus	• Adiabatic: Efficient heat store and process integration • Isothermal: Process implementation • Storage reservoir: Carbon fibre cylinders for overground storage

* To convert from joules (J) to kilowatt hours (kWh), divide by 3,600,000.

3.2.3 Flywheel

Flywheel energy storage makes use of the mechanical inertia contained within a rotating mass. To charge, electricity is used in an electric motor to spin the flywheel (Figure 3.9). The process is reversed when electricity is needed, with the motor that accelerated the flywheel acting as a generator extracting energy from the rotating flywheel to discharge. To reduce friction losses, it is common to place flywheels inside a vacuum with magnetic levitation.[5-7]

Flywheels are widely deployed in transport applications (e.g. regenerative braking and acceleration for trains). For stationary applications they are still relatively niche with < 100 MW deployed. Nonetheless, the technology is mature with multiple companies offering stationary systems for short-term power flexibility services. Deployment is still hindered by the relatively high cost compared to other energy storage technologies.

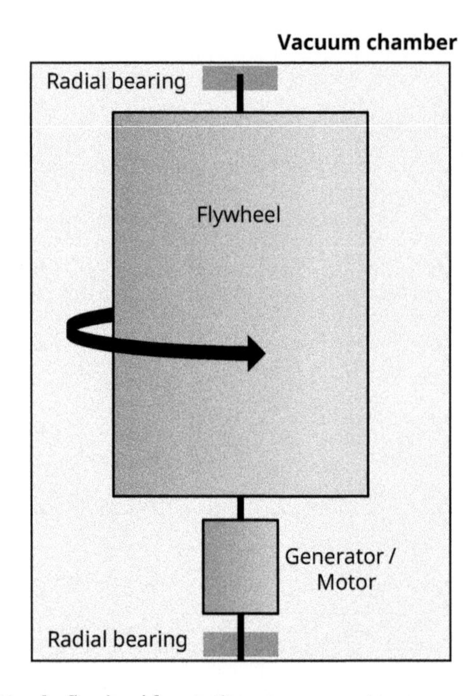

Figure 3.9 Schematic of a flywheel for stationary energy storage.

Category	Mechanical energy storage—Rotational motion
Status	Commercial stage—widely deployed in transport, but less than 100 MW for stationary applications
Application	Fast response and long cycle life make flywheels suitable for regenerative braking and acceleration in the transport sector (e.g. trains) and high frequency applications in the stationary sector, such as • Frequency regulation: Automatically correct the continuous changes in supply or demand that distort the grid frequency
Variations	Variations of the technology refer to the material used as the spinning mass (e.g. steel, aluminium, carbon fibre) and whether it rotates at high (> 10,000 rpm) or low speed (< 10,000 rpm).
Formula	$E_{Flywheel} = \frac{1}{2} I\omega^2 \eta = \frac{1}{2}(K_m m r^2)(2\pi RPS)^2 \eta = K_m m \frac{\sigma_{max}}{\rho} \eta$ E: Nominal energy capacity* [J] ω: Angular velocity [1/s] m: Rotor mass [kg] r: Radius of mass [m] σ_{max}: Maximum stress [N/m^2] I: Moment of inertia [kg m^2] η: Component efficiency [%] K_m: Shape factor of rotating mass [-] RPS: Rotations per second [1/s] ρ: Mass density [kg/m^3]
Advantages	+ High round-trip efficiency (~90%) + Rapid response time (< 1 second) + Very long lifetime (> 100,000 cycles) + High power density (1,000–5,000 kW/m^3) + Modular capacity sizing (kW–MW size)
Disadvantages	– Low energy density (~50 kWh/m^3) – High specific energy capacity cost (> 1,000 USD/kWh) – High self-discharge (up to 20% per idle hour) – Complex engineering to minimize losses and contain the spinning mass in case of a failure
Sample players (flywheel supplier)	• Beacon Power (US) • Amber Kinetics (US) • Stornetic (Germany) • Piller (UK)
Innovation focus	• Materials: High stress resistance for higher rotational speeds • Bearings: More robust bearings for lower operation and maintenance (O&M) cost • Mass: Higher rotor mass for higher energy-to-power ratio

* To convert from joules (J) to kilowatt hours (kWh), divide by 3,600,000.

3.2.4 Lead-acid battery

Lead-acid battery cells have an anode of elemental lead (Pb) in sponge-like form and a cathode of powdered lead dioxide (PbO_2) in a grid (Figure 3.10). During discharge, the aqueous sulphuric acid electrolyte (H_2SO_4) is converted to water (H_2O), while each electrode turns to lead sulphate ($PbSO_4$). When recharging by applying an external voltage, lead sulphate is converted back to sulphuric acid, leaving the layers of lead dioxide on the cathode and pure lead on the anode.[2]

Lead-acid batteries are the oldest and most widely deployed electricity storage technology. First commercialized around 1880, they are widespread as engine starter batteries for cars. In stationary applications, they are used for back-up power (e.g. hospitals) and power supply in remote locations (e.g. telecom towers).

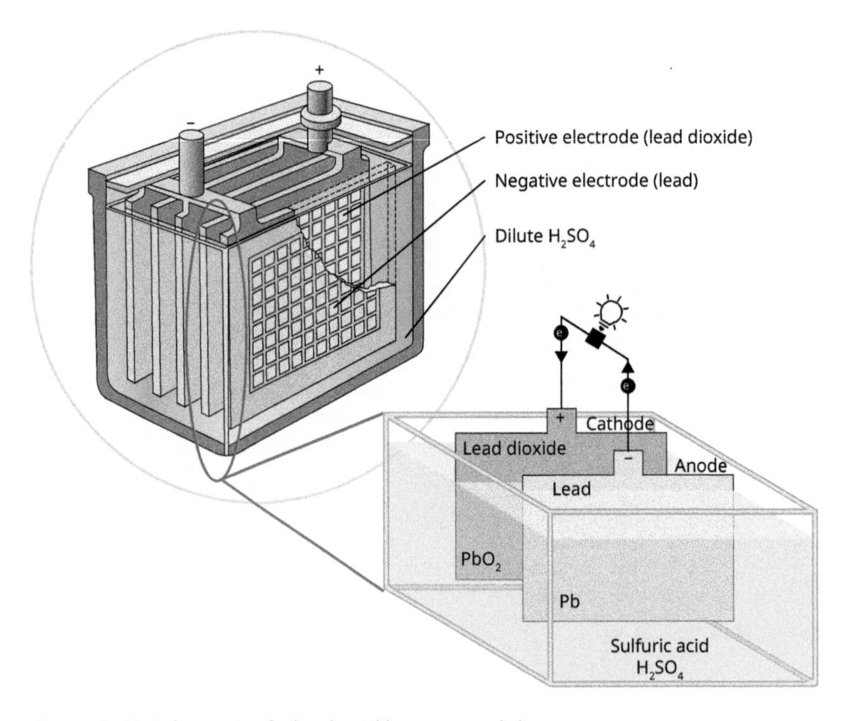

Figure 3.10 Schematic of a lead-acid battery module.

Category	Electrochemical energy storage—Sealed battery
Status	Commercial stage—widely deployed with overall annual demand at ~415 GWh per year[14] (corresponds to ~3,300 GW assuming typical 1:8 energy-to-power ratio for car engine starter batteries)
Application	In transport applications, lead-acid batteries are mostly used as engine starter batteries for internal combustion engine vehicles. In stationary applications, relatively low cost and limited cycle life make lead-acid batteries well suited for • Backup power: Fill rare, unexpected, sustained gap between supply and demand (e.g. hospitals or data centres)
Variations	Variations of the technology refer to the electrolyte, which can be a liquid mixture of water and sulphuric acid (i.e. flooded), a gel, or an acid-saturated fibreglass mat (i.e. sealed). While some systems are designed for shallow depth-of-discharge operation, others tolerate deep discharge cycles.
Formula	**Half cell reactions** (discharge): Anode: $Pb + HSO_4^- \rightarrow PbSO_4 + H^+ + 2e^-$ Cathode: $PbO_2 + 3H^+ + HSO_4^- + 2e^- \rightarrow PbSO_4 + 2\,H_2O$ **Overall reaction** (discharge): $Pb + PbO_2 + 2\,H_2SO_4 \rightarrow 2\,PbSO_4 + 2\,H_2O$ (Voltage: 2.1 V)
Advantages	+ High technical maturity (commercial since ~1880) + Relatively low cost for battery pack (< 200 USD/kWh) + Capable of high discharge rates + Wide range of sizes and specifications available
Disadvantages	– Low energy density vs other batteries (~70 kWh/m³) – Low depth of discharge for standard systems (30–50%) – Contain toxic materials (lead) – Limited lifetime (< 1,000 cycles)
Sample players (battery supplier)	• Trojan (Germany) • EnerSys (US) • Exide (US) • GS Yuasa (Japan)
Innovation focus	• Electrodes: Addition of carbon for deeper discharge and longer cycle life • System set-up: Pairing with supercapacitors to enable longer cycle life at high discharge rates

3.2.5 Lithium-ion battery

In state-of-the-art lithium-ion batteries, the cathode is made of a lithium metal oxide (e.g. $LiMO_2$) and the anode is made of graphitic carbon (C_6). Lithium ions (Li^+) move from anode to cathode when discharging and back when charging (Figure 3.11). In its charged state, the carbon anode has lithium intercalated between the carbon layers (LiC_6). The electrolyte is a non-aqueous organic liquid containing dissolved lithium salts, such as lithium hexafluorophosphate ($LiPF_6$) in ethylene carbonate.[15,16] Lithium ion is a family of technologies with multiple chemistry variations.

Lithium-ion batteries were developed in the 1990s for consumer electronics. Resulting cost reductions and modular sizing made them suitable for electric vehicles (EVs), which led to further cost reductions. As a result, they became attractive for stationary use for short-term power flexibility as well. Further cost reduction potential makes them a promising technology to integrate renewable electricity.

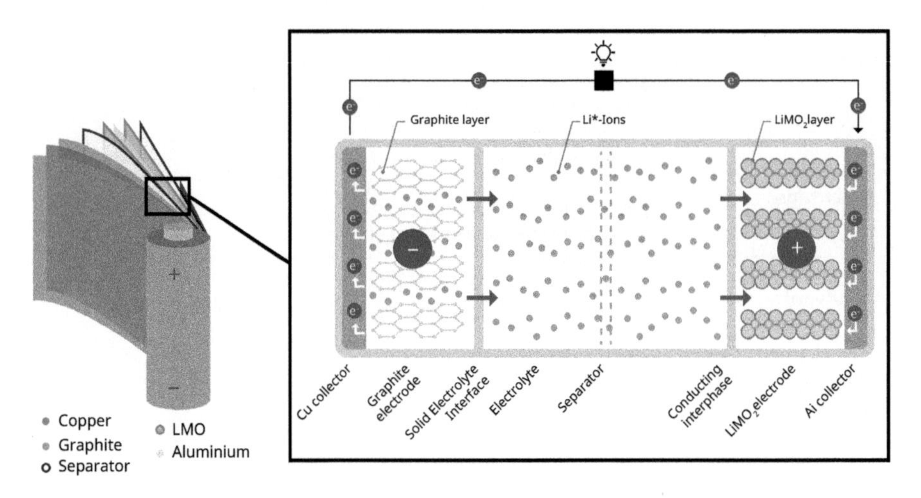

Figure 3.11 Schematic of a lithium-ion battery cell with lithium manganese oxide (LMO or $LiMO_2$) cathode and pure graphite anode.

Category	Electrochemical energy storage—Sealed battery
Status	Commercial stage—widely deployed with annual demand of ~500 GWh in 2020 and projection to increase to > 3,000 GWh by 2030.
Application	Widespread in consumer electronics, EVs, and stationary storage. In stationary storage, they are used for (among others) • Frequency response: Automatically stabilize frequency after unexpected, rare change in supply or demand • Self-consumption: Increase self-consumption of energy produced by non-dispatchable distributed generation
Variations	• Variations are based on electrode chemistry/structure and design • Cathode chemistry: lithium cobalt oxide (LCO), lithium manganese oxide (LMO), nickel manganese cobalt (NMC), nickel cobalt aluminium (NCA), lithium iron phosphate (LFP) • Anode chemistry: pure Graphite, Graphite-silicon mix, lithium titanite oxide (LTO) • Electrode structure: layered, olivine, spinel • Cell design: cylindrical, prismatic, laminate/ pouch
Formula	**Half-cell reactions** (example for LMO cathode, discharge): Anode: $LiC_6 \rightarrow C_6 + Li^+ + e^-$ Cathode: $MO_2 + Li^+ + e^- \rightarrow LiMO_2$ **Overall reaction** (discharge): $LiC_6 + MO_2 \rightarrow C_6 + LiMO_2$ (Voltage: 3.2–3.8 V)
Advantages	+ High energy/power densities (6,000 kW/m^3, 450 kWh/m^3) + High round-trip efficiency (~85%) + Modular sizing + Fast response time (< 1 second) + Strong cost-reduction potential due to several large markets
Disadvantages	– Limited cycle life (~3,500 at 80% depth of discharge) – Degradation throughout operational lifetime – Safety risks through thermal runaway – Potential resource scarcity (e.g. Lithium, Nickel, Cobalt)
Sample players (battery cell/pack supplier)	• LG Chem (South Korea) • Panasonic (Japan) • Samsung (South Korea) • CATL (China) • BYD (China) • Northvolt (Sweden)
Innovation focus	• Electrolyte: Solid material for higher energy density and cycle life • Anode: Adding silicon for higher energy density • Cathode: Pure lithium for higher energy density • Materials: Replace lithium with sodium to form sodium-ion batteries with lower material cost and lower risk of thermal runaway

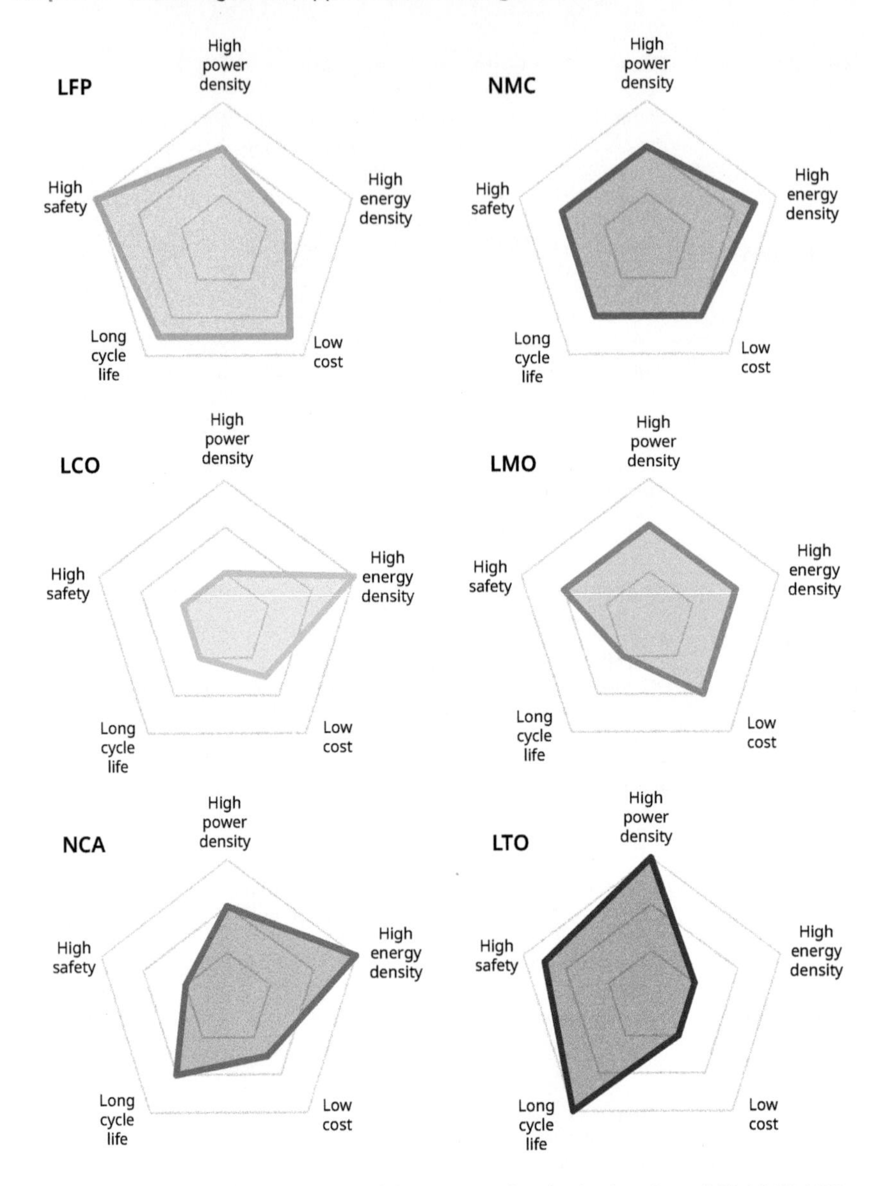

Figure 3.12 Qualitative assessment of the impact of cathode chemistry (LFP, NMC, LCO, LMO, NCA) and anode chemistry (LTO) on lithium-ion battery performance, cost, and safety. The extent of colour along each dimension indicates how well criterion is fulfilled.

Figure 3.12 highlights that lithium ion is a technology family. Different choices for cathode chemistry (LFP, NMC, LCO, LMO, NCA) and anode chemistry (LTO) have significant impacts on performance, cost, and safety parameters. The figure highlights how well these desired parameters are met by the chemistry subtypes. LFP and NMC fulfil most desired parameters reasonably well and are relatively balanced overall. These chemistries currently dominate the market among all subtypes.

Cell design represents another important dimension for technology variation. Table 3.1 depicts the three dominant designs, including their market share in 2020 as well as key advantages and disadvantages. Prismatic and laminate designs currently dominate the market as they offer better packing efficiency, which drives volumetric and gravimetric energy density, an important parameter for battery packs used in EVs.

Table 3.1 Qualitative assessment and 2020 market share of the most common cell designs for lithium-ion battery cells.

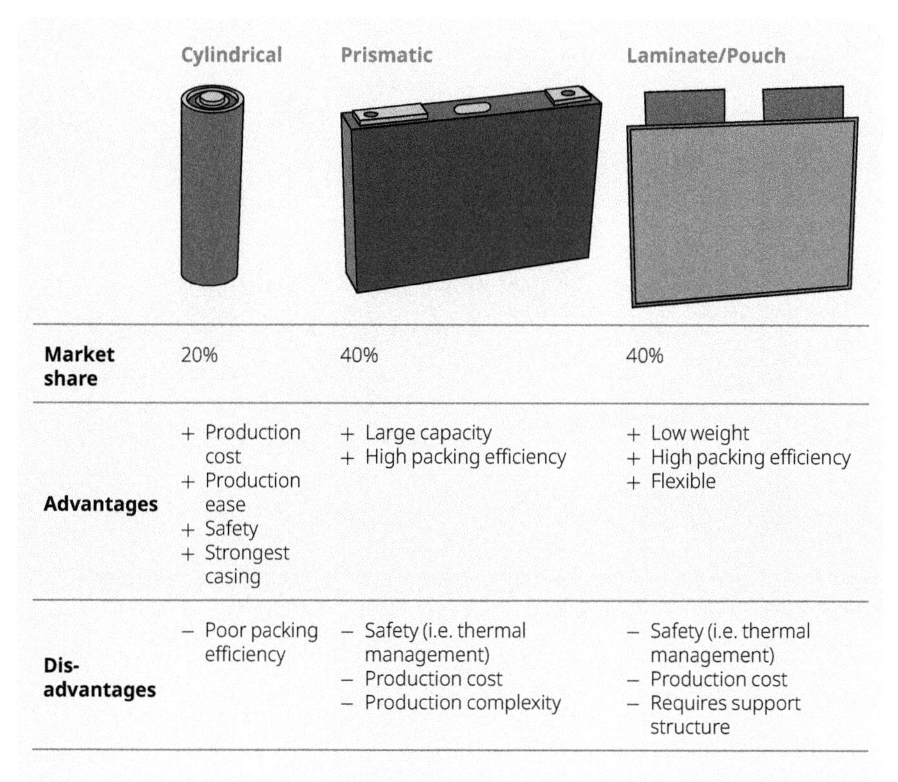

	Cylindrical	Prismatic	Laminate/Pouch
Market share	20%	40%	40%
Advantages	+ Production cost + Production ease + Safety + Strongest casing	+ Large capacity + High packing efficiency	+ Low weight + High packing efficiency + Flexible
Dis-advantages	– Poor packing efficiency	– Safety (i.e. thermal management) – Production cost – Production complexity	– Safety (i.e. thermal management) – Production cost – Requires support structure

3.2.6 Sodium-sulphur battery

Sodium sulphur is a molten salt battery containing molten sodium (Na) and sulphur (S). The sulphur is absorbed in a carbon sponge. The battery casing forms the cathode while the molten sodium core is the anode (Figure 3.13). The battery operates at high temperatures of between 300 °C and 350 °C. During charging, sodium ions are transported through the beta-alumina solid electrolyte (a ceramic of sodium polyaluminate) to the sulphur reservoir. Discharge is the reverse of this process. Once running, the heat produced by charging and discharging cycles is sufficient to maintain the operating temperature.[5]

Sodium-sulphur batteries were developed in the 1960s for transport applications and commercialized for stationary applications after 2000. Since then, deployments have been steadily increasing towards 560 MW cumulative capacity by 2020.[18] The only commercial manufacturer is the Japanese company NGK in partnership with BASF. The system is best suited for energy storage applications requiring 6–7 hours discharge duration.

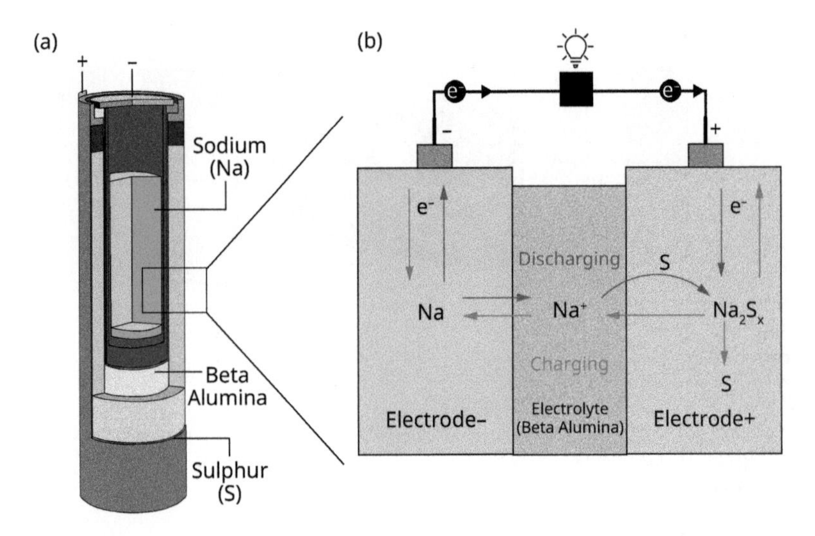

Figure 3.13 Schematic of a sodium-sulphur battery cell:[17] (a) tubular design battery, (b) during discharging and charging.

Category	Electrochemical energy storage—Sealed battery
Status	Commercial stage—cumulative deployment at ~560 MW in 2020
Application	Sodium-sulphur batteries are only used for stationary applications. Most deployed systems are used for • Renewables integration: Storing large amounts of excess renewable electricity supply to be used at a later time • Network investment deferral: Avoidance of power network upgrades when peak power flows exceed existing capacity
Variations	Sodium-nickel chloride (NaNiCl) is an alternative high-temperature battery based on sodium, nickel, and chloride. However, it is better suited for transport applications. The high nickel content makes it too costly for other applications.
Formula	**Half-cell reactions** (discharge)**:** Anode: $2\ Na \rightarrow 2\ Na^{+} + 2\ e^{-}$ Cathode: $x\ S + 2\ e^{-} \rightarrow S_x^{2-}$ **Overall reaction** (discharge)**:** $Na_2 + S_x \rightarrow Na_2S_x$ (Voltage: 2V)
Advantages	+ High energy density (~200 kWh/m^3) + Inexpensive, non-toxic raw materials (Na, S) + Wide ambient temperature range (e.g. hot climates) + Long lifetime for a sealed battery (4,500 cycles or 15 years)
Disadvantages	– High self-discharge in idle state due to need to maintain high operating temperature (~300–350 °C) – Relatively low discharge rates (systems are optimized for 6 hours discharge duration) – Safety risk due to high reactivity of sodium with water – Uncertain cost reduction potential with only one leading manufacturer
Sample players (battery supplier)	• NGK (Japan)
Innovation focus	• Electrolyte: Replace solid beta-alumina ceramic with liquid electrolyte to reduce operating temperature and cost

3.2.7 Vanadium redox-flow battery

Flow batteries use two liquid electrolytes as energy carriers. The electrochemical reaction takes place in the cell with carbon-based electrodes that are separated by an ion-selective membrane (Figure 3.14). This design allows power and energy capacity to be scaled independently as the electrolyte is stored in separate tanks and pumped into the cell when required. Energy storage capacity can be increased through more electrolyte and larger tanks. The most common electrolyte is a vanadium redox couple (anolyte: V^{2+}/V^{3+}; catholyte: VO^{2+}/VO_2^{+}), that is prepared by dissolving vanadium pentoxide (V_2O_5) in sulphuric acid (H_2SO_4).[6,7]

Flow batteries were developed in the 1980s. By 2020 less than 100 MW were installed globally.[19] The long lifetime and limited degradation make flow batteries suitable for high-throughput applications requiring many charge–discharge cycles and multiple hours of discharge duration. However, demand for these applications has been limited until now.

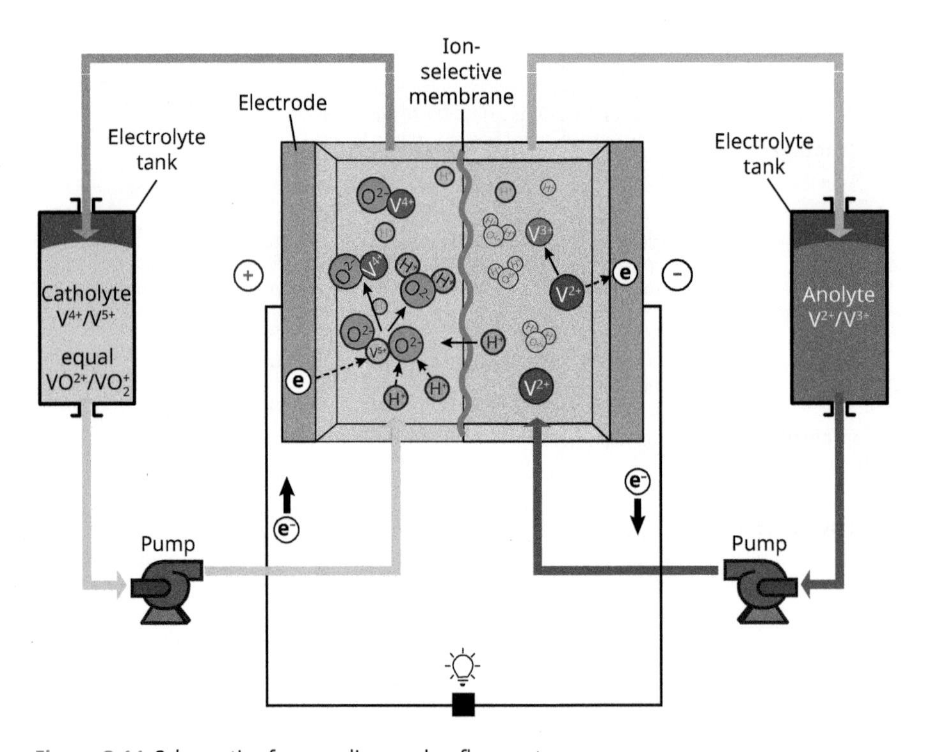

Figure 3.14 Schematic of a vanadium redox-flow system.

Category	Electrochemical energy storage—Flow battery	
Status	Commercial stage—less than 100 MW deployed by 2020.	
Application	Flow batteries are only used in stationary applications. Long cycle life and independent sizing of power and energy capacity make them suitable for high-throughput applications, such as • Energy arbitrage: Purchase power in low-price periods and sell in high-price periods • Congestion management: Avoid risk of overloading network infrastructure by consuming excess production and delivering it when network capacity is available	
Variations	There is a wide range of alternative flow battery technologies. These are categorized by type of active material (inorganic vs organic) and solvent (aqueous vs non-aqueous), which affects design (full flow, hybrid). Commercially available alternatives to vanadium are zinc-bromine and all-iron flow batteries.	
Formula	**Half cell reactions** (discharge): Anode: $V^{2+} \rightarrow V^{3+} + e^-$ Cathode: $VO_2^+ + 2H^+ + e^- \rightarrow VO^{2+} + H_2O$ **Overall reaction** (discharge): $V^{2+} + VO_2^+ + 2H^+ \rightarrow VO^{2+} + V^{3+} + H_2O$ (Voltage: 1.4V)	
Advantages	+ Independent sizing of energy capacity and power capacity + Long lifetime (~20,000 cycles) + Full depth of discharge (100%) + Limited degradation during operational life + Large operating temperature window (–20–50 °C)	
Disadvantages	– Low energy density vs other battery types (~30 kWh/m³) – Relatively immature industry with limited track record – Higher system complexity vs other battery types (e.g. pumps required, risk of electrolyte leakage) – High share of raw material cost in final product creating exposure to volatile raw material prices (e.g. vanadium)	
Sample players (battery supplier)	• Invinity (Canada/UK) • CellCube (Austria)	• Sumitomo Electric (Japan) • Rongke Power (China)
Innovation focus	• Electrolyte materials: New combinations of low-cost metals and/or gases	

3.2.8 Supercapacitor

Supercapacitors or electric double-layer capacitors utilize an electrochemical double layer of charge to store energy (Figure 3.15). As voltage is applied, charge accumulates on the electrode surfaces. Ions in the electrolyte solution diffuse across the separator into the pores of the electrode of opposite charge. The electrodes are engineered to prevent the recombination of the ions with the accumulated charge carriers, thus a double layer of charge is created at each electrode.[7] However, ions are not intercalated in the electrodes in contrast to battery technologies. As a result, energy density is typically an order of magnitude lower and power density an order of magnitude higher.

Supercapacitors were developed in the 1950s. They are ideally suited for applications where a lot of power is needed for a very short time. Typical applications are in consumer electronics (e.g. the flash in digital cameras) and transport (e.g. fast acceleration). For stationary power systems, supercapacitors are most suitable to provide voltage and frequency stability. However, their global deployment in these applications was < 100 MW by 2020.[20]

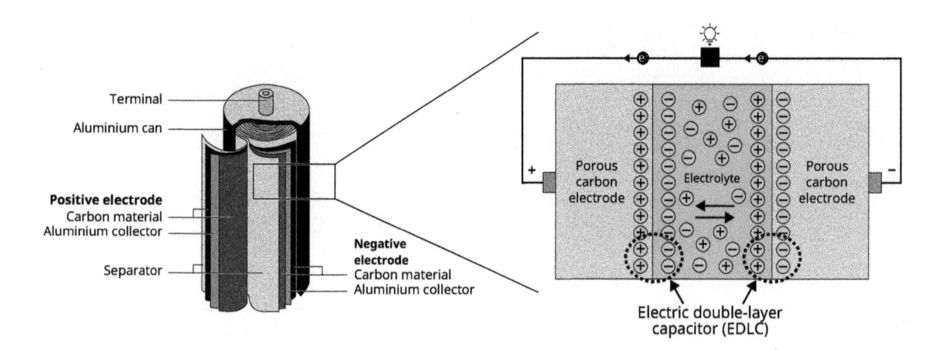

Figure 3.15 Schematic of an electric double-layer capacitor (i.e. supercapacitor).

Category	Electrical energy storage—Capacitance
Status	Commercial stage—less than 100 MW deployed in stationary storage applications by 2020.
Application	Supercapacitors can provide large amounts of power over a very short time. They are mainly used in consumer electronics and transport. The most common stationary storage application is • Frequency regulation: Automatically correct the continuous changes in supply or demand that distort the grid frequency
Variations	Alternative supercapacitor types are pseudo- and hybrid capacitors. Pseudocapacitors feature fast redox reactions at the electrode surface, which means that there is not only capacitive, but also electrochemical energy storage. Hybrid capacitors combine electric double-layer or pseudocapacitor with actual battery electrodes to form a capacitor–battery hybrid.[21]
Formula	$E = \frac{1}{2}\varepsilon A\frac{U^2}{d} = \frac{1}{2}CU^2 = \frac{1}{2}QU$ E: Nominal energy capacity* [J] A: Area of capacitor plates [m²] d: Distance between electrodes [m] Q: Charge stored in [C] ε: Permittivity of the material [F/m] U: Voltage of the electric field [V] C: Capacitance [F]
Advantages	+ High power density (~100,000 kW/m³) + High round-trip efficiency (> 90%) + High cycle life (> 100,000 cycles) + Fast response time + Very little or no maintenance required + Wide ambient temperature range (–40–70 °C)
Disadvantages	– Low energy density (~20 kWh/m³) – Short discharge duration (seconds to minutes) – High self-discharge in idle state (~20–40% per day) – Very high energy specific cost (> 10,000 USD/kWh)
Sample players (battery supplier)	• Skeleton Technologies (Estonia) • Rubycon (Japan) • Ioxus (US)
Innovation focus	• Electrode materials: Novel materials to improve performance • System set-up: Combination with batteries to leverage synergies between both technologies

* To convert from joules (J) to kilowatt hours (kWh), divide by 3,600,000.

3.2.9 Hydrogen

The enabling technology for storing electricity in chemical bonds is the conversion of electricity to hydrogen (also called 'Power to gas'). Hydrogen can be produced through electrolysis of water by imposing a voltage between two electrodes that exceeds the thermodynamic stability range of water. As a result, water splits into oxygen and positively charged hydrogen ions. These migrate through the membrane to the anode to form hydrogen gas. The hydrogen can then be stored directly as a gas in an underground reservoir and later re-electrified using a fuel cell. The overall process is called 'Power-to-gas-to-Power' and shown in Figure 3.16. Electrolyzers and fuel cells are related technologies. Improvements in one may also benefit the other.

Alternatively, the hydrogen could also be processed to ammonia, synthetic hydrocarbons, or alcohols through the addition of nitrogen or carbon dioxide. These derivatives can be stored and transported via the respective infrastructure, which often exists already. The hydrogen could also be stored in overground tanks in gaseous or liquid form or within metal hydrides. Similarly, it could be re-electrified using combustion engines or turbines. Also, instead of re-electrifying, hydrogen or its derivatives could be used as a chemical feedstock in industry or as fuel in the transport or heating sector. This is called sector coupling or 'Power to X'.

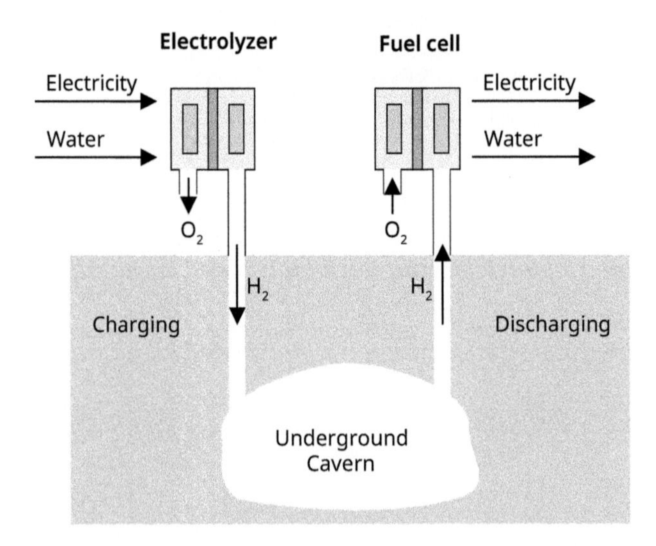

Figure 3.16 Schematic of a hydrogen-based 'power-to-gas-to-power' storage system using an electrolyzer to convert electricity into hydrogen (H_2), an underground cavern for gaseous H_2 storage, and a fuel cell for re-electrification of H_2.

Category	Chemical energy storage—Hydrogen	
Status	Commercial stage—The power-to-gas-to-power concept is not economically viable under current market conditions and less than 100 MW are deployed as of 2020.[20] However, since renewably produced hydrogen can also be used in other energy sectors and as industrial feedstock, electrolyzer capacities are expected to reach multiple GW by 2030.[22]	
Application	Due to the low cost of storing chemical compounds, hydrogen-based energy storage in the power-to-gas-to-power concept is mostly discussed for • Seasonal storage: Compensate longer-term supply disruption or seasonal variability in supply and demand	
Variations	• Electrolysis: Alkaline, PEM, Solid oxide • Storage: Cavern, tanks, conversion to solid material • Usage: PEM fuel cell, solid oxide fuel cell, gas turbine, reciprocating engine, gas burner, directly as feedstock	
Formula	**Half-cell reactions:** Anode: $2OH^- \rightarrow H_2O + ½\, O_2 + 2e^-$ Cathode: $2H_2O + 2e^- \rightarrow H_2 + 2OH^-$ **Overall reaction:** $H_2O \rightarrow H_2 + ½\, O_2$ (Voltage: 1.23V)	
Advantages	+ Fully independent power capacity and energy capacity sizing + Potential to use existing gas network capacity + High energy density (600 kWh/m³ at 200 bar) + Provision of renewable electricity to other energy sectors	
Disadvantages	– Need for compression to reach sufficient energy density – Low round-trip efficiency for re-electrification (< 40%) – Lack of a dedicated hydrogen infrastructure – High investment cost for water electrolyzers – Production of NO_x when burnt in turbine/engine/burner	
Sample players (electrolyzer supplier)	• Nel (Norway) • ITM Power (UK) • Thyssenkrupp (Germany)	• Hydrogenics (Canada) • Cockerill Jingli (China) • Shandong Saikesaisi (China)
Innovation focus	• Electrolyzer technologies: Anion exchange membrane (AEM) • Hydrogen tolerance of gas infrastructure: pipelines, turbines • Reduction of catalyst loading: Platinum, Iridium	

3.2.10 Technology overview

Table 3.2 gives an overview of the cost and performance parameters introduced in Chapter 2 for the nine most widely deployed stationary electricity storage technologies. The parameters are quantified through mean values rather than ranges to facilitate comparison between technologies; however, these values may vary significantly depending on system size, product quality, supplier, and other location- or use-case-specific conditions.

While sealed batteries (e.g. lithium ion, sodium sulphur, lead acid) and supercapacitors offer lowest power specific investment cost, pumped hydro, compressed air, and hydrogen storage offer lowest energy specific investment cost. This indicates that the former technologies are likely to be most cost effective at short discharge durations (e.g. < 8 hours) and the latter technologies are most cost effective at long discharge durations (e.g. > 8 hours).

Replacement cost refers to the substitution of mechanical components (e.g. turbine blades) or the electrochemical cells in vanadium flow batteries. For all other technologies, no replacement parts are needed within their specified lifetime. Their lifetime may be prolonged through augmentation (e.g. cell replacement in lithium-ion systems). This depends on product quality and use case and must be discussed with suppliers individually.

FREQUENTLY ASKED QUESTION

Are replacement costs considered for lithium-ion systems and how is the augmentation performed when it is due?

In merchant markets, energy storage project developers do not explicitly account for required component replacement in lithium-ion systems. They may do this indirectly by ensuring that land area is sufficient to add additional containers in future though. In markets where storage systems receive contracted revenues for > 10 years, project developers consider replacement of energy capacity components in order to maintain the contracted capacity.

In practice, systems will be overdesigned in terms of nominal energy capacity to ensure usable energy capacity meets application requirements even after 5–10 years. Once energy capacity has degraded beyond that, the most practical way is to add additional containers with their own inverters/converters to the site since these are electrically separate units. Once the warranty for the battery cells has expired, it is also an option to replace cells. However, this may involve adjusting set points of the existing inverters/converters.

End-of-life cost refers to the disposal of power electronics components for all technologies and the residual value of the electrolyte for vanadium redox-flow batteries. Other technologies may incur additional cost or have a residual value at their end of life (e.g. second life of lithium-ion batteries), but this is too uncertain to be quantified.

Temporal degradation is given instead of calendar lifetime. This is because calendar life is a theoretical concept since technologies are unlikely to remain unused for multiple years. The operational lifetime of a storage system is roughly given as cycle lifetime divided by annual cycles. Temporal degradation will further shorten this (see Chapter 8 for details).

FREQUENTLY ASKED QUESTION

What is the discharge duration for a particular energy storage technology? Why are discharge durations or energy to power ratios not shown in Table 3.2?
There is no 'correct' value for the discharge duration of an energy storage technology; this is a flexible parameter that can be varied based on the specific design. It is possible to have a pumped hydro plant with huge turbines and small reservoirs that are emptied within 10 minutes. Similarly, a huge flywheel made of a very heavy material could discharge for multiple hours. This is the reason why no limitations on the minimum discharge duration are listed in Table 3.2.

In terms of practical system design, the balance between a technology's energy-specific and power-specific investment costs will determine the discharge duration that gives optimal total cost, and this will be the configuration that developers usually design for (see Chapter 2). If turbines are expensive and additional reservoir capacity is cheap (or even free), then it makes economic sense to have pumped hydro systems with longer than 10-minute duration. Figure 3.17 shows the average discharge duration for operational projects as of 2022. This gives an indication of the discharge durations that are preferred for different energy storage technologies in real-world use cases.

For example, the median discharge duration across 480 lithium-ion battery projects is 1.3 hours (a C-rate of 0.75). Half of these projects lie in the range of 40 minutes (25th percentile) to just under 4 hours (75th percentile), but the longest systems are in excess of 8 hours (a C-rate of 0.125). This serves as a guide, but does not mean that all lithium-ion projects will continue to fit within this range. If, for example, longer-duration services become more profitable or the balance of power and energy costs change, then developers may tailor projects (across all technology groups) towards longer discharge durations.

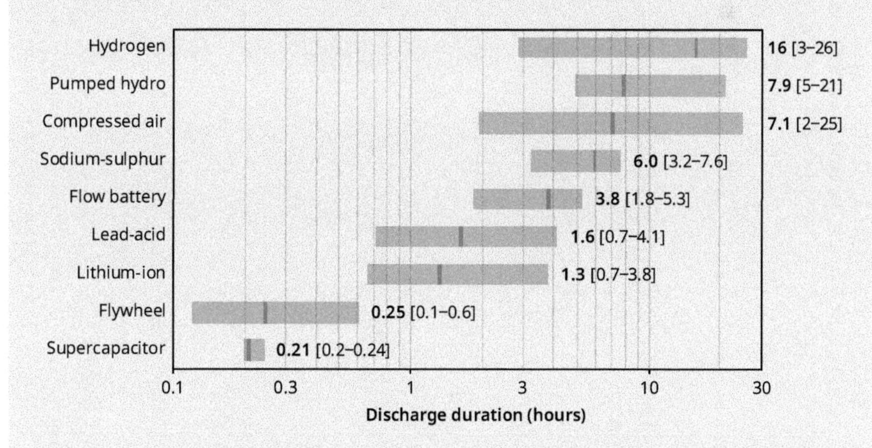

Figure 3.17 The range of minimum discharge durations (energy-to-power ratios) for operational storage projects using different energy storage technologies. Pale bars span from the 25th to 75th percentile, and darker lines show the 50th percentile (median) across projects listed as operational in the energy storage database of the US Department of Energy.[20] Infeasible values of zero hours or above 1,000 hours were excluded.

Table 3.2 Technology input parameters for 2020. This table shows specific cost and performance parameters for the nine most widely deployed stationary electricity storage technologies. Parameters are based on a review of studies by research institutes, international organizations, industry and academia, and industry interviews.[6,23–26] The values reflect mean values from within the ranges given by the reviewed studies.

				Pumped hydro	Compressed air	Flywheel	Lithium ion	Sodium sulphur	Lead acid	Vanadium redox flow	Hydrogen	Supercapacitor
Cost parameters	**Investment cost—power**	USD/kW	$C_{p,inv}$	1,100	1,300	600	250	650	300	700	5,000	300
	Investment cost—energy	USD/kWh	$C_{e,inv}$	50	40	3,000	300	450	320	450	30	10,000
	Operation cost—power	USD/kW-year	$C_{p,om}$	20	14	5	5	5	5	10	30	1
	Operation cost—energy	USD/MWh_{el}	$C_{e,om}$	0.4	2	2	0.4	0.4	0.4	2	0.4	0
	Replacement cost—power	USD/kW	$C_{p,rep}$	120	100	200	0	0	0	90	0	0
	Replacement cost—energy	USD/kWh	$C_{e,rep}$	0	0	0	0	0	0	0	0	0
	Replacement interval	cycles	Cyc_{rep}	7,300	1,500	20,000	n/a	n/a	n/a	3,500	n/a	n/a
	End-of-life cost—power	USD/kW	$C_{p,eol}$	20	20	20	20	20	20	20	20	20
	End-of-life cost—energy	USD/kWh	$C_{e,eol}$	0	0	0	0	0	0	–100	0	0
	Discount rate	%	r	Depends on technology, use-case, and investor type—sample value: 8% (mature technology, utility investor)								

(Continued)

Table 3.2 (*Continued*)

				Pumped hydro	Compressed air	Flywheel	Lithium ion	Sodium sulphur	Lead acid	Vanadium redox flow	Hydrogen	Supercapacitor
Performance parameters	**Round-trip efficiency**	%	η_{RT}	80%	45%	86%	86%	75%	72%	68%	35%	92%
	Depth-of-discharge	$\%_{cap}$	DoD	100%	100%	100%	80%	80%	80%	100%	100%	100%
	Cycle lifetime	cycles	$Life_{,cyc}$	30,000	15,000	200,000	3,500	4,000	900	20,000	10,000	300,000
	Temporal degradation	%/year	Deg_t	0%	0%	0%	1%	1%	1%	0.1%	0%	0%
	Self-discharge	$\%_{cap}$	η_{self}	0%	0%	10%	1%	5%	1%	0%	5%	15%
	Response time	seconds	T_{res}	> 10	> 10	< 10	< 10	< 10	< 10	< 10	< 10	< 10
	Construction time	years	T_{con}	3	2	1	1	1	1	1	1	1
	Power density	kW/m³	-	1	1	3,000	6,000	160	200	5	5	100,000
	Energy density	kWh/m³	-	1	4	50	450	200	70	30	600 (at 200 bar)	20

Notes: All cost parameters reflect *specific cost* for energy and power components. Cycle lifetime refers to full equivalent charge–discharge cycles at the indicated depth of discharge. Temporal degradation is given instead of calendar lifetime. Replacement interval refers to period after which selected components need to be replaced. No replacement of any energy components assumed, i.e. values of '0'. Power and energy density are given in volumetric terms as this metric is more relevant for stationary storage technologies than gravimetric energy density.

Power and energy density do not impact the economic performance of energy storage systems, but rather the geographic footprint and general suitability to certain applications. Flywheels, lithium-ion batteries, and supercapacitors can provide most power in a confined space, while lithium ion and (compressed) hydrogen can provide most energy in a confined space. This is important in transport applications as well as in densely populated areas.

Cost and performance parameters will vary from the ones given in Table 3.2. The impact of different values on the competitiveness of each technology can be explored on the companion website to this book: <www.EnergyStorage.ninja>.

FREQUENTLY ASKED QUESTION

Does investment cost not scale with energy storage system size? Why do you only give one value per technology?

Scale effects do apply to the investment cost of energy storage projects. For lithium ion it has been found that smaller commercial and industrial-scale projects have on average 20% higher investment cost than utility-scale projects of the same storage duration.[27] This is likely also to be the case for the alternative electricity storage technologies. The analyses in this book rely on only one average investment cost value for each technology for simplicity, and to keep focus on the methodologies. This is one reason why we also provide the companion website www.EnergyStorage.ninja where you can redo the analyses performed in this book with own custom data.

Energy storage technologies differ in their suitability to storage applications as shown in Figure 3.18. This is due to the different cost and performance parameters, which are ultimately defined by the underlying energy storage principles.

Direct storage as electrical energy (capacitors, coils) allows for rapid and frequent discharge due to avoiding an energy conversion process and the high cycle life of the technologies. Storing large amounts of energy is challenging though because electrically conductive materials that store electrons directly are expensive.

In contrast, chemical energy storage (hydrogen) is suitable for long discharge durations and large energy capacities due to low cost of storing chemical compounds.

The same is true for most mechanical energy storage technologies (pumped hydro, compressed air) based on the low cost of basic materials (e.g. water, air, rock). Here, better cycle life and round-trip efficiency, and lower power capacity cost, make the technology suitable for more frequent operation throughout the year. Flywheels are an extreme example for that.

Electrochemical battery technologies (lithium ion, sodium sulphur, lead acid, redox flow) sit in between electric and chemical/mechanical technologies in terms of discharge duration. Their power capacity and energy capacity costs are more balanced, making them suitable for applications that require a couple of hours of discharge duration. Heat storage is of similar size in terms of energy capacity, but suitable for longer discharge durations.

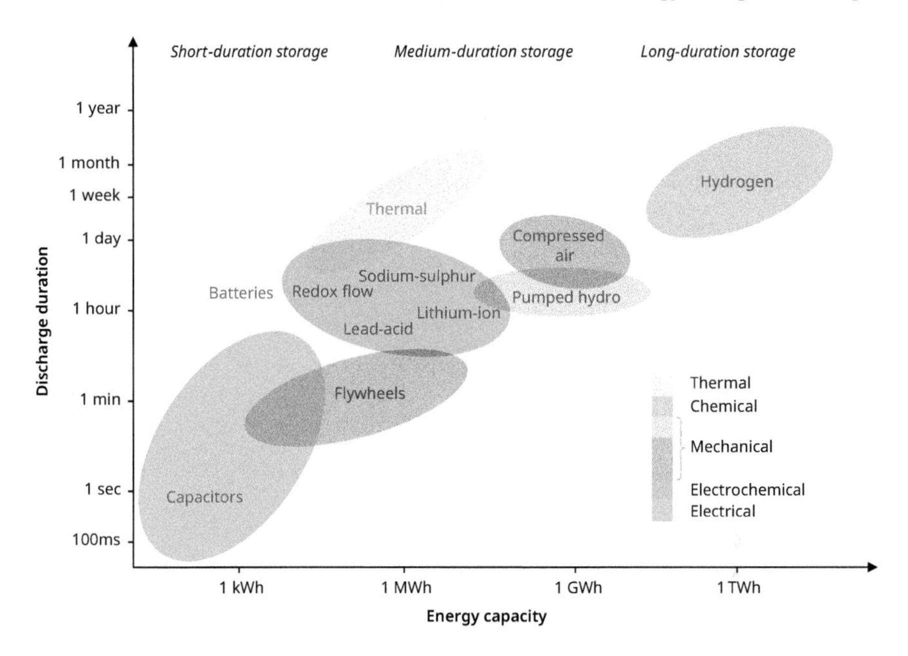

Figure 3.18 Suitability of various electricity storage technologies across the five energy storage categories to the application requirements of discharge duration and electrical energy storage capacity. Adapted from Sterner et al.[28]

KEY INSIGHT

Electricity storage technologies from same category are, roughly speaking, suitable for similar use cases. *Chemical* storage is suited to large energy capacities and long discharge durations. *Thermal*, *mechanical*, and *electrochemical* storage are suitable for medium-sized energy capacities and short to long durations. *Electrical* storage is best suited to small energy capacities and short durations.

The impact of the different types of energy storage is reflected in the power-specific and energy-specific investment costs of each technology. Figure 3.19 shows the nine technologies listed in Table 3.2 along these costs. This reveals four technology groups that can be identified, which confirm the discussed suitability to different storage applications.

3.2.11 Novel technologies

In addition to the established electricity storage technologies described in the previous sections, there is a wide range of novel technologies being developed in response to the growing need for energy storage. Their deployment depends on whether they can economically outperform the established technologies in selected use cases. One way to investigate that is to assess their levelized cost of storage (see Chapter 5). Table 3.3 introduces a selection of these novel technologies.

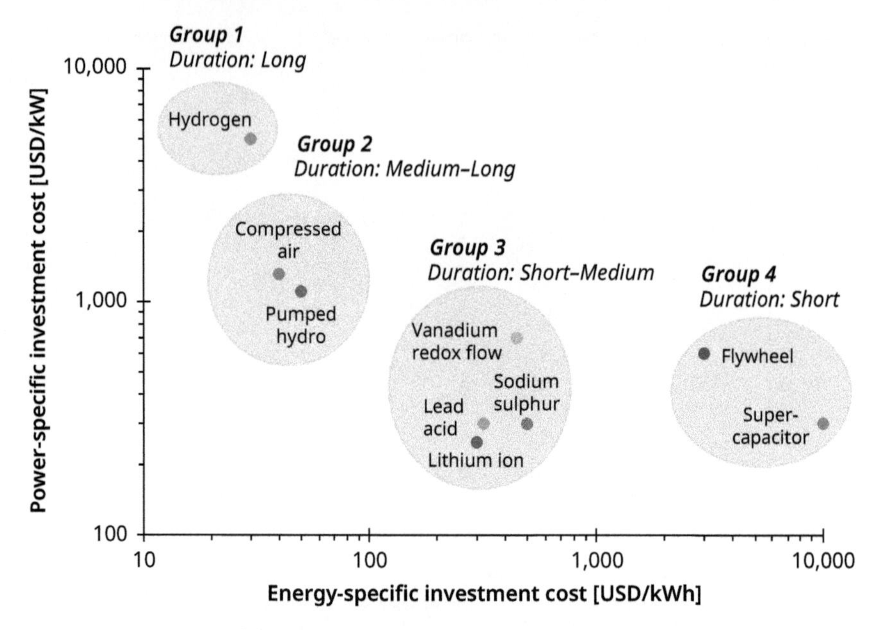

Figure 3.19 Grouping of the nine most widely deployed stationary electricity storage technologies along their power-specific and energy-specific investment costs as listed in Table 3.2

Table 3.3 Overview of a selection of novel energy storage technologies.

Gravity-based energy storage 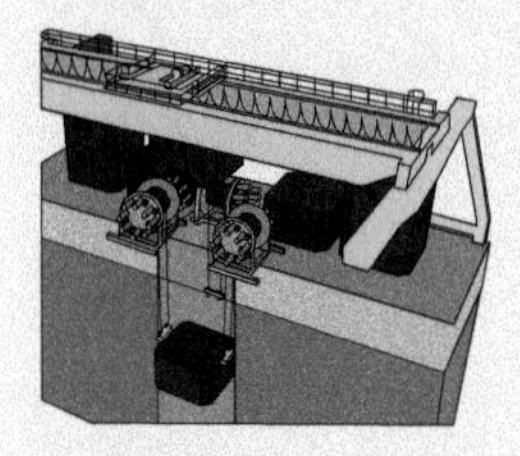	**Category:** Mechanical—Gravitation
	Concept: Electricity is stored in the form of potential energy by raising weights. It is discharged by lowering the weights, which drives a generator. Gravitricity utilizes deep shafts (e.g. mine shafts) to suspend the weight.
	Company: Gravitricity Ltd.
	Alternatives: • Lifting multiple concrete blocks with cranes (Energy Vault) • Lifting a rock cylinder by pumping water below it (New Energy Let's Go)
Liquid air energy storage 	**Category:** Mechanical—Compression
	Concept: Air is compressed and cooled until it reaches a liquid state. It is stored in cryogenic tanks at −190 °C. When power is needed, the air is pumped to high pressure and expanded to ambient temperature, which drives a turbine.
	Company: Highview Enterprises Ltd.
	Alternatives: • Compression of CO_2 (Energy Dome)

(Continued)

Table 3.3 (*Continued*)

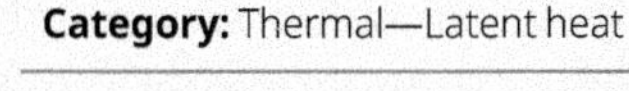

Thermal battery	**Category:** Thermal—Latent heat
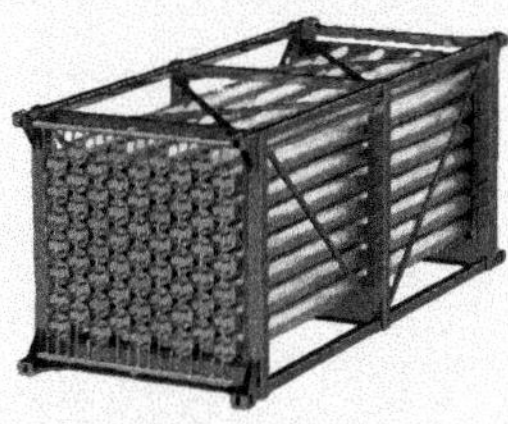	**Concept:** Electricity is used to heat a transfer fluid. This fluid transfers the heat to a storage material, for example a mixture of concrete and steel. When electricity is needed, the transfer fluid takes up the heat stored in the material and drives a turbine. Since heat is used in manufacturing, thermal batteries can be integrated with industrial processes for heat storage or to increase the efficiency of electricity storage.
	Company: EnergyNest AS
	Alternatives: • Sand (Polar Night Energy) • Volcanic rock (Siemens Gamesa) • Steel (Lumenion)
Liquid metal batteries	**Category:** Electrochemical—Sealed battery
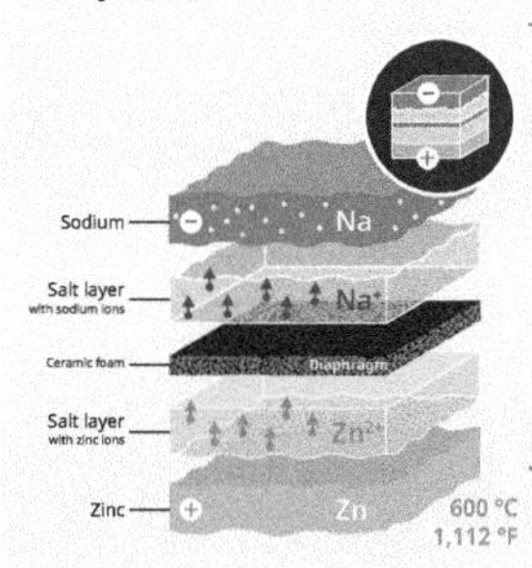	**Concept:** Liquid metal batteries are based on electrochemical reactions of metals with salt electrolyte. Both electrodes and the electrolyte are liquid and at relatively high temperatures. The liquid interfaces enable fast process kinetics for relatively abundant and cheap metals. The SOLSTICE research consortium is exploring a liquid metal battery that uses zinc and sodium as active electrode materials.
	Company: SOLSTICE research consortium
	Alternatives: • Calcium-antimony (Ambri)
Metal air batteries	**Category:** Electrochemical—Sealed battery
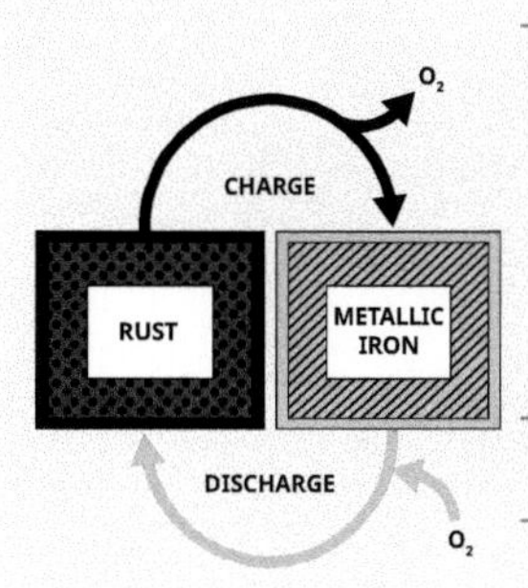	**Concept:** Metal air batteries are based on the electrochemical process of corrosion, also called 'rusting' for iron. In iron-air batteries, the battery uses oxygen from the air to convert iron metal to rust, which releases electricity (discharging). The application of an electrical current converts the rust back to iron and the battery releases oxygen (charging).
	Company: Form Energy, Inc.
	Alternatives: • Zinc-Air (Zinc8) • Aluminium-Air (Phinergy)

3.3 The energy storage industry

This section gives an overview of the energy storage industry. It covers three areas: the current and projected market size for energy storage, the structure of the industry, and the geographic concentration of the energy storage value chain.

3.3.1 Market size

In terms of installed capacity, pumped hydro is by far the most widely deployed electricity storage technology with more than 158 GW operational in 2020 (see Figure 3.20).[8] This is nearly 90% of the global stationary energy storage capacity. This capacity has grown in parallel with the deployment of nuclear power plants during the second half of the twentieth century, mostly to shift electricity supply from periods of low demand overnight to peak demand periods during the day.[29] The need for flexibility to enable the low-carbon transformation of the power system renews the interest in stationary electricity storage. As a result, an additional ~5 GW of new pumped hydro plants are projected to be commissioned every year, potentially increasing the total to ~220 GW by 2030.[30]

KEY INSIGHT

Almost all stationary electricity storage is either pumped hydro or lithium ion. These made up 89% and 10% of global capacity in 2020. Both are also growing rapidly, by at least 5 GW per year.

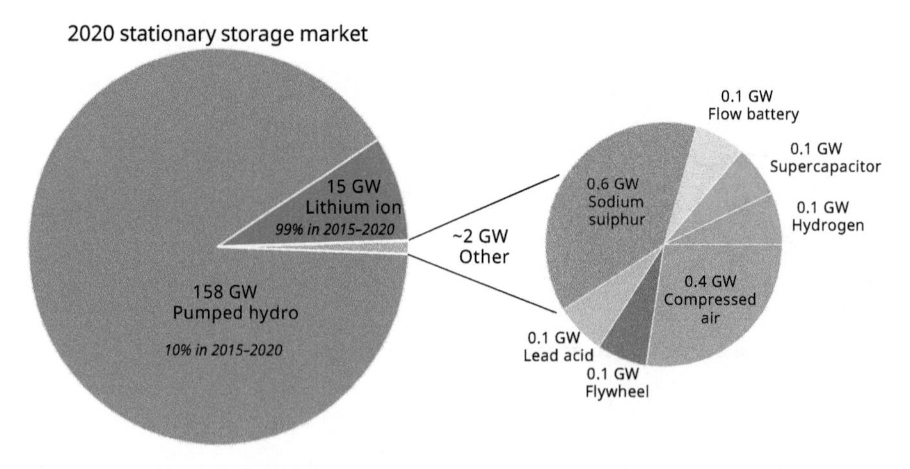

Figure 3.20 Global installed stationary electricity storage capacity by technology in 2020.[9,20,31,32] This chart includes stationary lead-acid systems connected to electricity networks and listed in the DoE energy storage database.[20] The total market for lead-acid batteries was 415 GWh in 2020, which would correspond to 3,300 GW with a 1:8 energy-to-power ratio (typical for engine starter batteries).[14] However, the vast majority is either for automotive applications or specialized stationary use cases like telecom towers and uninterruptable power supply.

The distributed and variable character of renewable electricity generation (i.e. wind and solar) has led to a rapid increase in the deployment of battery technologies. By 2020, ~15 GW of stationary lithium-ion battery systems were deployed. This was ~10% of total global stationary electricity storage capacity in 2020, and 99% was deployed in the previous 5 years.[9,32] About half of this stationary battery capacity is utility-scale size and deployed in short-term power system flexibility applications. The other half was deployed in small-scale systems at customer sites to enable self-consumption of renewable energy, bill management, or backup power. Only ~2 GW or 1% of stationary electricity storage capacity is in technologies other than pumped hydro and lithium ion. This share may increase in the future as applications that require longer discharge durations become more important to ensuring power system flexibility and stability.

By 2030, more than 400 GW of stationary battery systems are projected to be deployed in total,[33] and pumped hydro storage capacity is also projected to grow to 220 GW (see Figure 3.21a).[30] Projections for the future stationary battery market size do not give granular detail on the mix of technologies that will make up the fleet, but given current trends it can be expected these will mostly be lithium ion. The analysis developed in Chapter 5 can serve as a basis for assessing the competition between battery technologies based on their relative costs.

Figure 3.21b provides some context to the current and projected stationary storage market size by comparing them to the current and projected power capacity for the global fleet of EVs. It is anticipated that EV adoption will grow very rapidly over the current decade, so this market will grow from three times the size of stationary batteries in 2020 to at least 10 times the size in 2030.

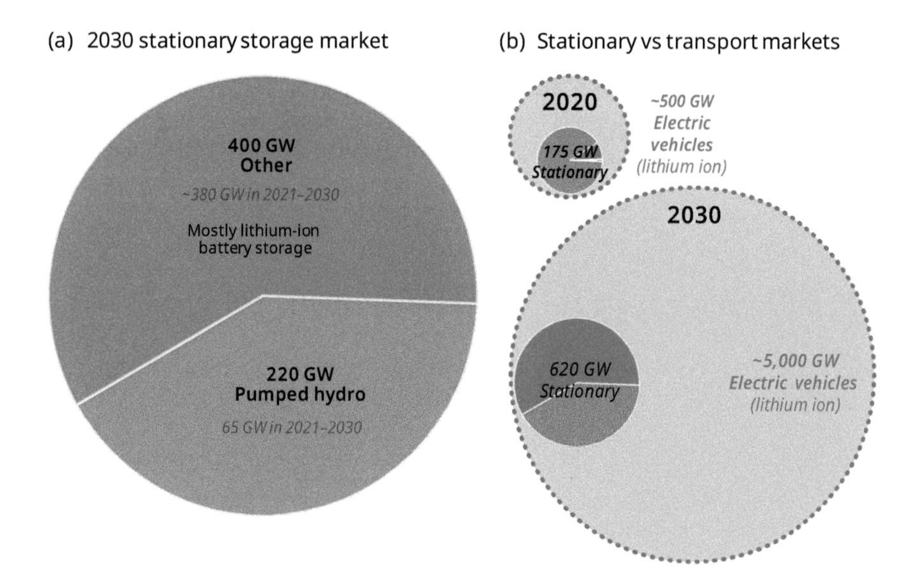

Figure 3.21 Current and future energy storage market size. (a) Projections for global installed stationary electricity storage capacity in GW in 2030.[30,33] (b) Current and future stationary storage capacity compared to the power capacity of electric vehicle (EV) batteries. Values are based on global EV stocks in 2020 (~10 million) and 2030 (~100 million, conservative assumption) and an average ~50 kW power capacity per battery.[34]

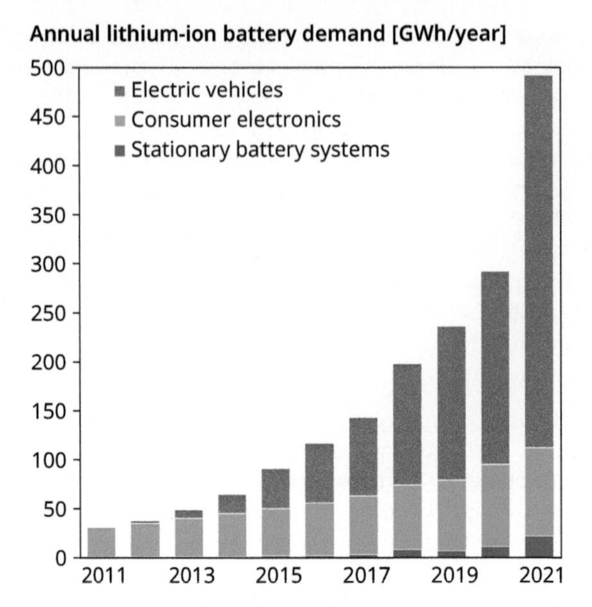

Figure 3.22 Annual lithium-ion battery demand split into deployment between electric vehicles, consumer electronics, and stationary systems.[35]

The importance of EVs for the market size of lithium-ion batteries also becomes evident when looking at the split of annual lithium-ion battery demand for EVs, consumer electronics and stationary battery systems (see Figure 3.22).[35] In 2015, lithium-ion battery demand for EVs had already overtaken the demand for consumer electronics. By 2020, the demand for EVs constituted two-thirds of total lithium-ion battery demand. Consumer electronics made up ~30% and stationary battery systems a mere 3%, despite its dominant role in stationary storage next to pumped hydro plants. The likely implications are that battery chemistries and designs are optimized for use in EVs and car manufacturers see lower prices and prioritized delivery in times of shortage due to their higher order volumes.

3.3.2 Industry structure

The energy storage industry is defined by the interplay of all the companies that ensure the deployment of energy storage technologies. Depending on the market size for each technology, individual companies may specialize in one step in the value chain or cover multiple value chain steps. Also, the number of companies active in each value chain step may differ. Figure 3.23 analyses the energy storage industry along the product value chain for stationary systems, which was introduced in Chapter 2.

Figure 3.23a lists sample companies that provide services in each value chain step. The reason for displaying individual firms is to showcase the typical ranges of value chain steps that individual firms cover across the different technologies. Naturally, there may be firms that cover different sets of value chain steps, but these do not represent the majority of firms deploying the respective technology. Overall, the overview reveals that companies are likely to cover multiple value chain steps in the energy storage industry within four overarching categories.

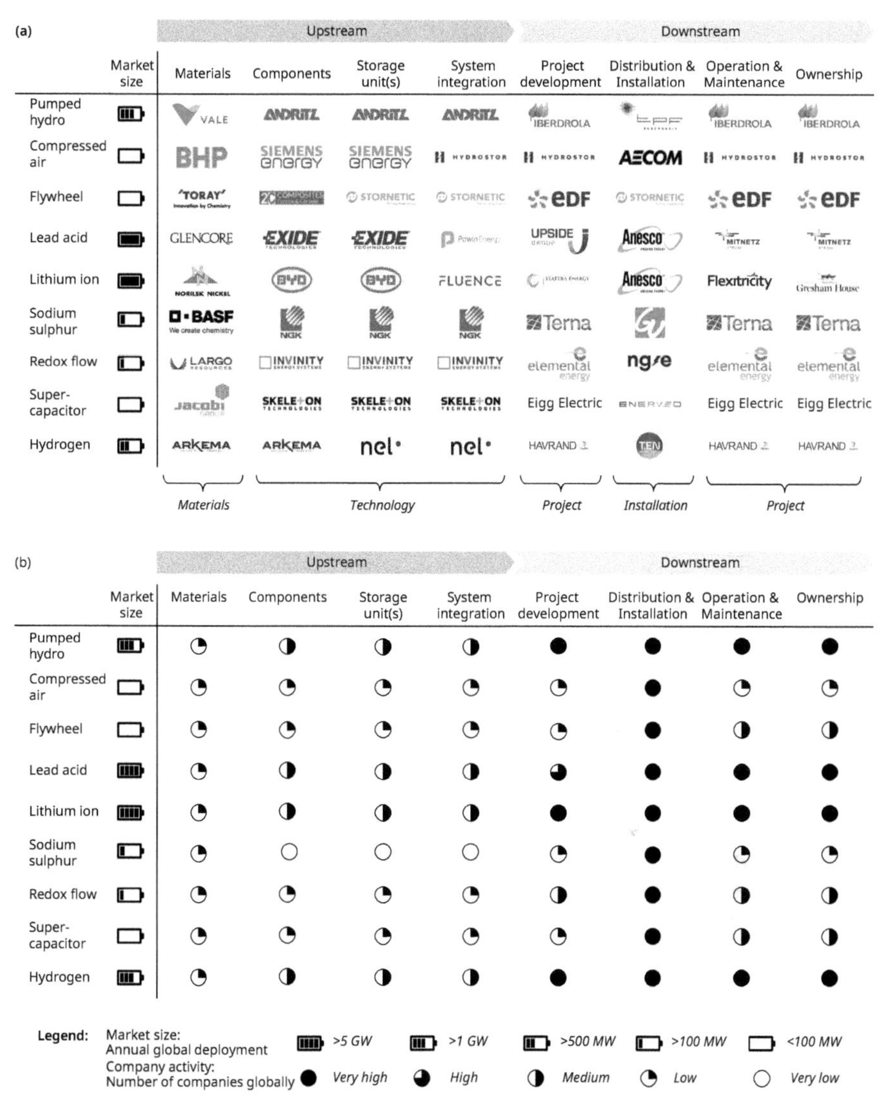

Figure 3.23 Overview of the energy storage industry. (a) Sample firms active in each value chain step and technology. The selection of firms for each technology is either taken from specific projects or industry reports on market leaders. Chapter 8 lists respective products based on which sample companies were selected. (b) Qualitative assessment of company activity for each storage technology and value chain step. Company activity is indicated relative to other technologies and value chain steps.

- **Materials:** Manufacturers of raw and processed materials (e.g. iron ore for steel turbines or polymers for membranes)
- **Technology:** Manufacturers of (sub-)components, the storage unit, and/or the integrated system (e.g. electrodes/membranes, cells/packs, and complete system integrated with power conversion and balance-of-plant components)

- **Project:** Companies that develop (e.g. land acquisition, grid connection permission, financial and technical studies), operate (e.g. dispatching, managing accounts), and/or own the system (e.g. funding)
- **Installation:** Providers of engineering, procurement, and construction services

Depending on the commercial maturity of the energy storage technology, companies may cover less or integrate beyond the steps covered by these four categories. On the one side, there are technologies with a relatively small market and only a small number of technology providers; for example compressed air or flywheels. Due to the small number of projects, technology providers may forward-integrate into the downstream value chain to also develop and/or install projects to support its deployment.

On the other side, for technologies with a large market like lithium-ion batteries, firms tend to specialize on individual value chain steps. Here, storage projects are realized through the interplay of firms that are specialized in nearly each individual value chain step. This is a case in point for the theory that bigger markets increase the 'toughness' of competition, as in the number of companies active along the entire value chain.[36] In bigger markets, firms need to specialize in order to develop competitive advantages in specific value chain steps. However, they are also more able to specialize due to the larger market that is available within each value chain step.

In terms of ownership, Figure 3.23a shows that there are four main options for storage technologies:

1. **Technology provider**: For technologies with a small market, the technology provider may act as the owner of the system and finance it, in order to push the technology into the market (e.g. Hydrostor).
2. **Integrated developer**: Project developers may not sell the project once under construction or operational but keep it to de-risk the development business with regular operational revenues (e.g. Elemental Energy).
3. **Energy utility**: Similar to integrated developers, utilities are likely to develop, operate, and own storage assets to leverage their full benefit. They are more likely to own than developers as their scale enables easier access to capital (e.g. Iberdrola).
4. **Infrastructure fund**: Similar to the solar PV business, infrastructure funds are starting to own storage projects of de-risked technologies with a long track record and large market. In this case, operation will be sub-contracted to other players (e.g. Gresham House Energy Storage Fund).

In jurisdictions with unbundled energy systems, that is, the strict separation of network operation and energy sales, network operators are explicitly precluded from operating or owning electricity storage systems (examples: Germany and the UK)[37,38]. Due to the potential significance of electricity storage to network stability, there can be exceptions, for example if storage systems only provide network system services and do not operate on electricity markets, or if no third party can be found to operate and own required storage systems.[39]

Competitive activity is the focus of Figure 3.23b. It shows that the number of active companies is higher for technologies with a larger market. The number is also higher in downstream than upstream. This is due to the knowledge required in local markets for downstream activities. Locally, there may still be a limited number of companies that provide engineering, procurement, and construction (EPC) services for a technology, however, across all markets this number will be significantly higher than the number of technology providers, which are usually active globally.

Figure 3.23b reveals the monopoly on components, storage unit, and system integration for sodium-sulphur batteries held by NGK, who operate in a joint venture with BASF.[40] There is no structural reason for this monopoly (e.g. market entry cost, economies of scale, patents), but rather the limited market size that made past competitors like General Electric leave the market.

The analysis of company activity therefore indicates that the energy storage industry for technologies with large markets is unlikely to be subject to monopolies.

KEY INSIGHT

Energy storage is a competitive industry with no structural drivers for monopolies to form. The number of active players in each step of the value chain depends on the technology's market size.

3.3.3 Geographic concentration

While individual companies may not pose a serious threat of holding monopoly power over energy storage, the upstream battery value chain is strongly concentrated in one country. China manufactured three-quarters of the world's lithium-ion batteries in 2020 and also dominates component and material production (see Figure 3.24).[41] In 2021 China built one battery 'gigafactory' per week, compared to just one every four months in the US.[42] China is also one of the largest miners of copper, lithium, graphite, and rare earth minerals, and the largest processor of these plus nickel and cobalt, covering almost the full spectrum of materials needed in lithium-ion battery storage.[43,44]

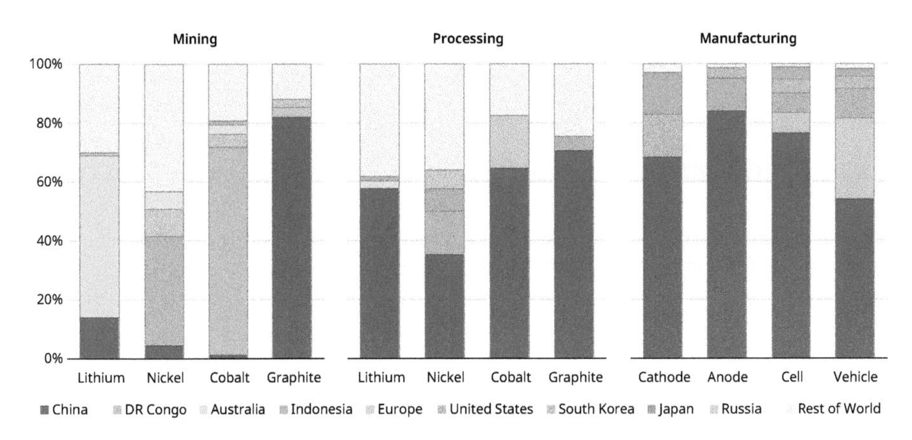

Figure 3.24 The geographical concentration within different stages of the lithium-ion battery value chain.[34] Mining and processing refer to four key materials: lithium (Li), nickel (Ni), cobalt (Co), and graphite (C), and shares are based on mass of raw material produced and refined respectively. Manufacturing of electrodes and cells is based on production capacity data, and for electric vehicles (EVs) is based on number produced. Note that Indonesian nickel production is primarily not battery grade and is used outside of the sector.

There is growing concern among Western governments that it is 'almost impossible to side-step China in a clean tech powered future'.[45] Just as the geopolitics of oil and gas rose to prominence in the 1970s,[46] the geopolitics of renewable energy is raising international tensions in the 2020s.[47,48] Geopolitical tensions, especially between the US and China, could form a roadblock to energy transition, reducing the deployment of renewables and storage.[49,50]

Batteries are thought to be more exposed to geopolitical tensions than either solar photovoltaics or wind turbines because of 'the sheer scale of their minerals requirements' and use of the 'highest risk materials': cobalt, nickel, and lithium.[51] However, a range of policy instruments in Europe and the US aim to ramp up local manufacturing capacities to reduce the share of lithium-ion batteries manufactured in China.[52,53] The 2022 US Inflation Reduction Act (IRA) puts increasingly strict requirements on battery mineral sources,[54] which may help to diversify supply chains but may also push up costs and restrict manufacturing capacity, thus slowing the rollout of US EV deployment. Goldman Sachs suggested that investments of around 20 billion USD per year into domestic companies in the US and Europe could eliminate their reliance on Chinese battery supply chains by 2030.[55]

There is little agreement among financial analysts on the near-term future for lithium production,[56,57] let alone whether the broader geopolitical tensions surrounding battery supply chains will worsen or naturally resolve themselves as the energy transition progresses. Three factors could drive the latter: diversification of mining and processing, innovation in the materials requirements of batteries, and the strong uptake of recycling (thus creating a circular economy for critical materials).[43,47,58] However, commercial pressure towards least-cost production casts doubt on whether these factors will have a material effect on reducing reliance on China.[59]

3.4 Applications of energy storage

The increasing penetration of variable or inflexible low-carbon electricity generation increases the need for power system flexibility (see Chapter 1.2). Electricity storage is one option that can provide this flexibility.

3.4.1 Archetypes

There is a wide range of applications for electricity storage that increase power system flexibility. These applications can be characterized along three qualitative categories:

1. Type of economic value
2. Location in the power system
3. Relation to variable renewable electricity (RE) generation

The type of economic value describes the actual service electricity storage provides to the power system and how it creates value. The four fundamental services are:[60]

- **Power quality**: Keeping frequency and voltage within permissible limits
- **Power reliability**: Providing electricity in case of supply reduction

- **Increased utilization**: Optimizing the use of existing assets in the power system
- **Arbitrage**: Exploiting temporal price differentials

The location in the power system describes the point within the electricity network the service is most likely to be required. This affects the voltage level at which electricity storage has to provide electricity.

- **Generation**: at generator site (front of the meter), match voltage output of generator
- **Network**: at substation of transmission or distribution network (front of the meter), 1–200 kV
- **Consumption:** at consumer site (behind the meter), < 1 kV

The relation of electricity storage applications to variable renewable electricity generation can be characterized as:

- **Direct**: Example—self-consumption increases the share of consumed variable RE and would not exist as a service without variable RE generation
- **Indirect**: Example—frequency regulation contains frequency fluctuations as a result of supply and demand imbalances; this is a common power system service, however more variable RE is likely to increase supply and demand imbalances
- **Unrelated**: Example—black start is required when parts of the electricity system fail, regardless of the reason for the failure

A comprehensive overview of electricity storage applications is provided in Figure 3.25 where the 23 most common applications are listed along these three categories.

Table 3.4 provides a more detailed description of the 23 most common electricity storage applications and alternative names that are commonly used for them. It also allocates the applications to 13 archetype applications based on similar requirements for discharge duration and annual charge–discharge cycles.

FREQUENTLY ASKED QUESTION

What's the difference between transmission and distribution (T&D) deferral and congestion management?
Applications may be defined very differently across energy markets. In this book, congestion management refers to frequent but short-lived T&D network constraints, for example during peak demand periods. Here, an electricity storage system with a relatively short discharge duration can alleviate these punctual constraints. T&D deferral refers to the next step, where constraints last for extended periods and would usually warrant the construction of a new transmission or distribution line. Here, an electricity storage system with a longer discharge duration can defer or even replace the need for a new line.

Figure 3.26 displays the 12 archetype applications with explicit discharge duration and annual cycle requirements (excluding voltage support). In addition, a currently hypothetical service at very high charge–discharge cycle frequency is displayed (named high cycle). These 13 applications and the respective annual cycle and discharge duration requirements are used in Chapter 5 to model the lifetime cost of storage. Naturally, the exact

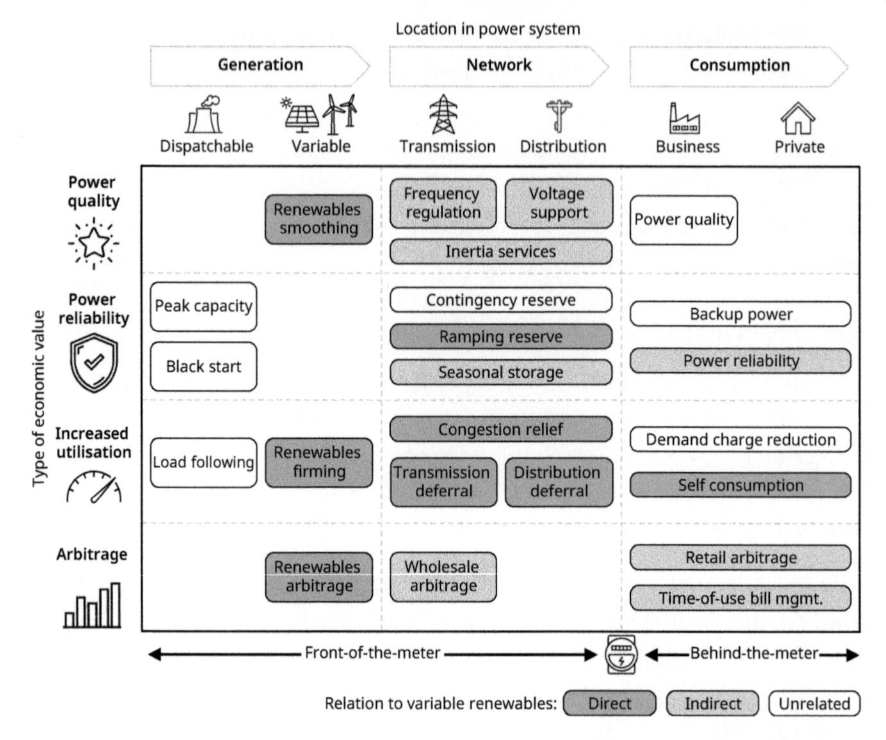

Figure 3.25 Overview of the 23 most common electricity storage applications along the dimensions of location within power system and type of economic value, with colour signifying the relation to variable renewable electricity generation. The selection of 23 applications is based on a review of reports from research institutes, international organizations, industry, and academia.[20,26,61–69] A detailed description of all 23 applications can be found in Table 3.4. Schematic inspired by Battke.[60,70]

Table 3.4 Detailed description of the 23 most common electricity storage services, including alternative names and allocation to archetypal applications based on similar technical requirements (i.e. discharge duration, annual cycles).[20,26,61–68,70]

Archetype	Application	Description	Alternative name
1. Frequency regulation (FG)	Frequency regulation	Automatically correct the continuous changes in supply or demand within the shortest market interval	Frequency control, Automatic generation control
	Renewables smoothing	Smooth output from variable supply when generation is out of line with short-term forecasts	Correcting forecast inaccuracy
	Power quality	Condition frequency and voltage of power supply for sensitive loads in unstable grids	

(*Continued*)

Table 3.4 (*Continued*)

Archetype	Application	Description	Alternative name
2. Frequency response (FS)	Contingency reserve	Automatically stabilize frequency after unexpected, rare, instantaneous change in supply or demand	Primary reserve/ response, Frequency response, Non-spinning reserve, Spinning reserve, Replacement reserve
	Ramping reserve	Automatically stabilize frequency after unexpected, rare, non-instantaneous change in supply or demand	
	Inertia services	Automatically stabilize frequency and voltage in response to changes to phase angle between voltage and current through reactive and active power response	
3. Black start (BS)	Black start	Restore power plant operations after network outage without external power supply	
4. Peak capacity (PC)	Peak capacity	Ensure availability of sufficient generation capacity at all times	Electric supply capacity, System capacity
5. Seasonal storage (ST)	Seasonal storage	Compensate longer-term supply disruption or seasonal variability in supply and demand	
6. Power reliability (PR)	Power reliability	Fill regular, sustained gap between supply and demand	Off-grid, Microgrid
	Backup power	Fill rare, unexpected, sustained gap between supply and demand (e.g. total loss of power from grid)	Home backup, Emergency supply, Resiliency, Secondary/Tertiary reserve
7. Renewables integration (RE)	Renewables firming	Storing large amounts of excess renewable electricity supply to be used at a later time	Variable resource integration, Renewable generation shifting
8. Congestion management (CM)	Congestion relief	Avoid risk of overloading existing infrastructure that could lead to re-dispatch and local price differences	Network efficiency, Virtual Power Line

(*Continued*)

Table 3.4 (*Continued*)

Archetype	Application	Description	Alternative name
	Load following	Maintain balance between supply and demand, while allowing generators to operate at optimal capacity	Balancing reserves
9. T&D deferral (TD)	Transmission upgrade deferral	Deferral, reduction or avoidance of transmission and/or distribution network upgrades when peak power flows exceed existing capacity	Transmission support, Virtual Power Lines, Distribution substation support
	Distribution upgrade deferral		
10. Demand reduction (DR)	Demand charge reduction	Reduce demand supplied by the network during periods of highest network charges	Peak reduction, Demand shifting, Retail demand charges
11. Self-consumption (SC)	Renewable energy self-consumption	Increase self-consumption of energy produced by non-dispatchable distributed generation	Prosumer
12. Energy arbitrage (EA)	Renewables arbitrage	Storing energy produced by variable renewable plants when prices are low to sell when prices are high	
	Wholesale arbitrage	Purchase power in low-price periods and sell in high-price periods on the wholesale market	Electric energy time-shift, Price arbitrage, Time-shifting
	Retail arbitrage	Purchase power in low-price periods and sell in high price periods on the retail market	End-consumer arbitrage
	Time-of-use (ToU) bill management	Purchase power in low-price periods and use during high-price periods	Retail energy time-shift, Energy management, ToU charges
13. Voltage support (VS)	Voltage support	Maintain voltage levels across networks via supplying or absorbing reactive power. *Note that as no active power is drawn from the device, there are no cycle or discharge duration requirements.*	VAR support

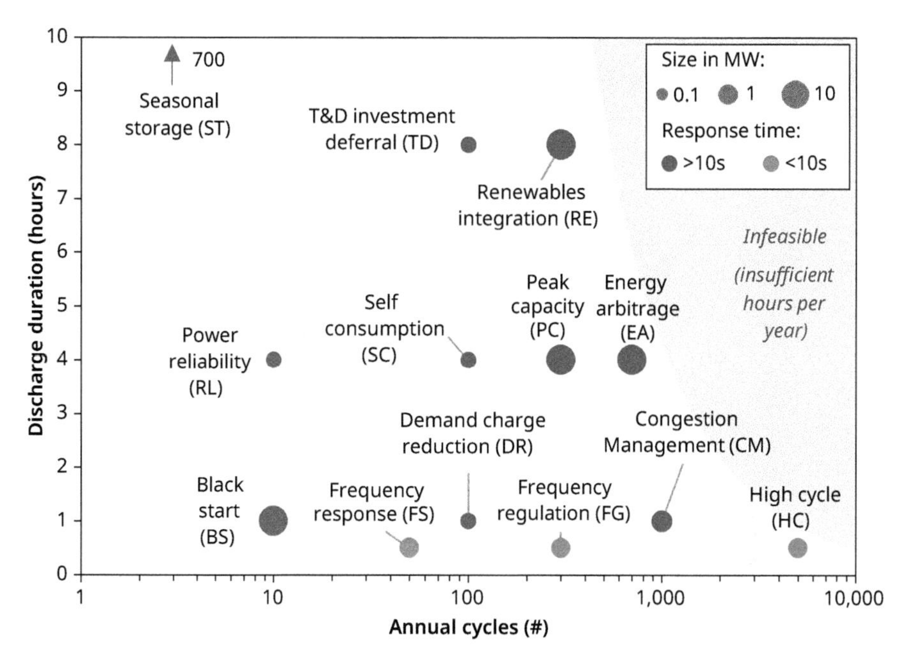

Figure 3.26 Thirteen electricity storage applications with illustrative requirements. The annual cycle and discharge duration requirements are chosen from within the common range of each application, such that the entire spectrum for these parameter combinations is represented. Annual cycles refer to full equivalent charge–discharge cycles. Size in MW refers to the nominal power capacity of a typical electricity storage system serving the respective application. The quantitative values for discharge duration and annual cycles are derived in Chapter 8.

requirements vary by market. For example, resilient power systems may have a black start event per decade while more fragile systems see multiple a year. Here, annual cycle and discharge duration requirements are chosen from within common ranges observed in various markets to cover the entire spectrum of cycle and discharge duration requirements.

The high cycle service (HC) is fictive and does not represent a currently common electricity storage service. However, given the increasing annual cycle requirements of frequency regulation services and the increasing need to compensate reducing physical inertia in the system with instantaneous response, it is conceivable that such services will be required in future.[71]

KEY INSIGHT

There are 23 unique electricity storage applications that create economic value through increasing power quality, power reliability, asset utilization, or arbitrage at different locations in the power system. When focusing on discharge duration and annual cycle frequency requirements alone, these can be reduced to 13 archetype applications.

Table 3.5 Key characteristics of variable renewable electricity generators that trigger the need for flexibility.

Characteristic	Description	Applications
Uncertainty	Availability of resource cannot be predicted with absolute certainty	Renewables smoothing, Ramping reserve, Inertia services
Variability	Power generation fluctuates with availability of renewable resource	Renewables firming, Frequency regulation, Seasonal storage, Power reliability
Low short-run cost	Once built, electricity is generated at very low operating cost	Renewables arbitrage, Wholesale arbitrage, Retail arbitrage, Time-of-use bill management
Location-constraint	Resource is not equal in all locations and cannot be transported	Transmission upgrade deferral, Congestion management, Voltage support
Modularity	Scale of individual generators is relatively small	Self consumption, Distribution upgrade deferral

The electricity storage applications that are directly or indirectly driven by the increasing share of variable RE (see Figure 3.25) can be classified along the key characteristics of RE generators in Table 3.5. These characteristics drive the need for flexibility and are the reason the described electricity storage applications are directly or indirectly related to increasing shares of variable RE.[72]

Inflexible low-carbon generation technologies like nuclear have similar characteristics that also require respective electricity storage applications: uncertainty of an outage, output inflexibility instead of variability, low short-run cost, and location-constraint for cooling and areas of low population.[73]

3.4.2 Deployment potential

Table 3.6 provides a conceptual framework to understand the deployment potential and timeline for the various electricity storage applications.[74] It highlights that storage deployment can be described as function of discharge duration with shorter durations required earlier at low renewables and nuclear penetrations, while longer durations are required at higher penetrations.

Historically, an integrated energy market and low short-run cost of nuclear generation drove the deployment of pumped hydro storage plants for a range of applications, mostly peak capacity. This came to a halt due to the restructuring of energy markets and a slowdown in nuclear deployment.

Since 2010, the increasing share of variable renewable generators combined with the restructured energy market that offers dedicated short-term flexibility services drove the deployment of battery technologies for regulation and reserve applications in many markets

Table 3.6 Phases of electricity storage deployment in centralized power systems. Conceptual framework adapted from NREL.[74] Battery symbols indicate the potential scale of deployment within each phase. RE—Renewable electricity.

RE / nuclear energy share	Phase	Description	Archetype application	Deployment potential	Discharge duration	Response time
<20%	Pre-2010	Integrated energy market & low-cost nuclear power	Various	low	Mostly 8-12 hours	Minutes
	1	Restructured energy market & reducing system inertia	Frequency regulation	low	< 1 hour	Milliseconds to seconds
	2	Narrowing of peak periods & reducing RE+storage cost	Peak capacity	mid	2-6 hours	Seconds to minutes
	3	RE+storage cost lower than other generators	Renewables integration	high	4-12 hours	Minutes
>80%	4	No fossil fuel generators & very low storage cost	Seasonal storage	uncertain	>12 hours	Minutes to hours

low mid high uncertain

(Phase 1). The increasing share of variable renewable plants reduces the share of conventional plants, which passively provided regulation services through generator inertia.

Phase 2 sees the deployment of longer discharge duration storage technologies for peak capacity. This is driven by narrowing daytime peak demand periods as a result of larger shares of solar PV and reducing renewables plus storage cost. Conventional, thermal 'peakers' are less economic at shorter peak demand periods.

Phase 3 could be described as the 'holy grail' for electricity storage. It is when the levelized cost of electricity from renewables plus storage falls below that of fossil-fuel based alternatives and their wide-scale deployment serves the majority of electricity demand. While for Phase 1 and 2 battery technologies look most promising as of today, the longer discharge duration required in Phase 3 may mean that alternative technologies play a stronger role.

Phase 4 is reached when variable and inflexible low-carbon electricity supply reaches > 80%. In that case, very few conventional thermal generators remain and long-duration storage is required to ensure supply and demand are matched at all times. Deployment of electricity storage in this phase is highly uncertain because it is unclear whether costs will reach levels that make it competitive. Alternatives are flexible low-carbon generators (e.g. hydropower) or flexible conventional generators that pay a high carbon price, are equipped with carbon capture, or run on low-carbon fuels. In addition, electricity network expansion and increased demand-side response can offer alternatives to provide flexibility (see Chapter 1). Practical examples for these phases are:

KEY INSIGHT

Electricity storage deployment can be clustered into four phases. These are characterized by the share of low-carbon power generation in a given market, and influence the required storage discharge duration. Archetype applications are frequency regulation in Phase 1, peaking capacity in Phase 2, renewables integration in Phase 3, and seasonal storage in Phase 4.

- **Pre-2010**:
  - The growth of pumped hydro and nuclear plants in the US from 1960 to 2010 (Figure 3.27)
- **Phase 1**:
  - Germany: In 2020 more than 50% of frequency regulation services or ~300 MW were provided by batteries (up from 0% in 2012)[75]
  - Great Britain: In 2016 only batteries succeeded in the new Enhanced Frequency Response tenders of ~200 MW[76]
- **Phase 2**:
  - In South Australia multiple large-scale battery projects are developed (e.g. 300 MW/800 MWh—Neoen, 225 MW/450 MWh—Maoneng) following analyses that 2- to 4-hour battery storage facilities are 30% cheaper than peaking gas plants in 2021[77-79]

Figure 3.28 shows the share of variable wind and solar PV electricity generation for the countries or states with the highest penetration, confirming that multiple European countries like Germany and the UK are in Phase 1.

South Australia has reached an average wind and solar penetration of 64% in 2022.[80] This confirms it is deep in Phase 2, approaching Phase 3. In contrast, Denmark has relatively few battery installations despite also having a high share of variable renewables. Its high interconnection with other European markets (Nordics, Germany) provides the required flexibility. This highlights how the actual deployment of electricity storage technologies is not only dependent on variable renewables or nuclear penetration, but also the availability of other flexibility options (e.g. interconnection, flexible generation).

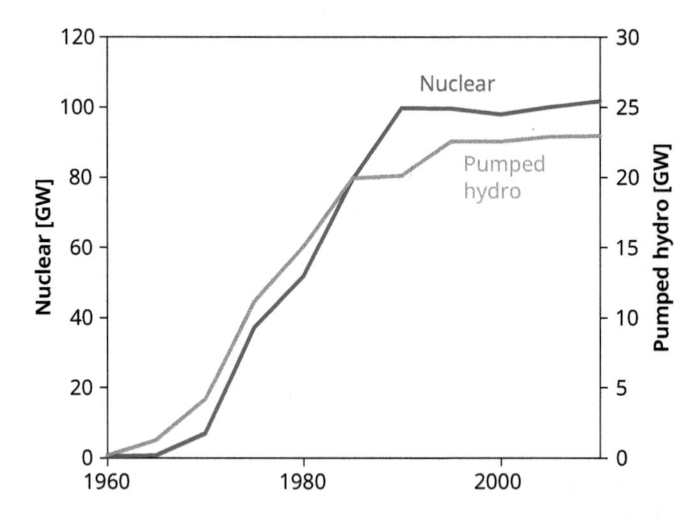

Figure 3.27 Deployment of nuclear and pumped hydro power capacity in the US between 1960 and 2010. Both displayed on separate axes to highlight the similar growth rates.[81,82]

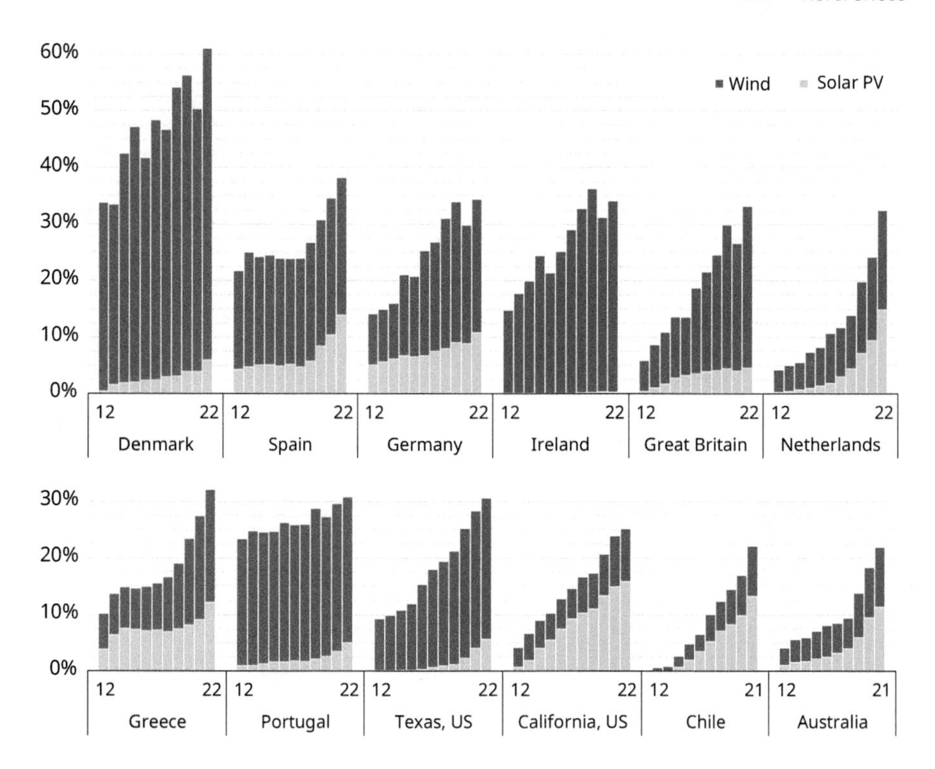

Figure 3.28 Share of wind and solar electricity generation by country/state.

3.5 References

1. Specht M, Sterner M, Stuermer B, Frick V et al. Renewable Power Methane—Stromspeicherung durch Kopplung von Strom- und Gasnetz—Wind/PV-to-SNG (2009).
2. Huggins R. *Energy Storage* (New York: Springer International Publishing, 2016).
3. Hauer A. *Technology Brief 4: Thermal Storage* (IEA-ETSAP and IRENA, 2013).
4. Zhang H, Li X, and Zhang J. *Redox Flow Batteries: Fundamentals and Applications* (Boca Raton: CRC Press, 2017).
5. Arup. *Five Minute Guide—Electricity Storage Technologies* (London: Arup, 2012).
6. Luo X, Wang J, Dooner M, and Clarke J. Overview of Current Development in Electrical Energy Storage Technologies and the Application Potential in Power System Operation (2015) 137 *Applied Energy* 511–36.
7. Decourt B and Debarre R. *Electricity Storage—Factbook* (The Hague: SBC Energy Institute, 2013).
8. International Hydropower Association (2022). Available at: <https://www.hydropower.org/> accessed 5 November 2022.
9. International Energy Agency. *Energy Storage* (Paris: IEA, 2022).
10. Energy Storage Association. Compressed Air Energy Storage (CAES) (2022). Available at: <https://energystorage.org/why-energy-storage/technologies/compressed-air-energy-storage-caes/> accessed 5 November 2022.
11. Hydrostor (2022). Available at: <https://www.hydrostor.ca/> accessed 5 November 2022.

12. Renewable Energy News. Planning withdrawn for Gaelectric storage (2019). Available at: <https://renews.biz/54683/solar> accessed 5 November 2022.
13. Ibrahim H, Belmokhtar K, and Ghandour M. 'Investigation of Usage of Compressed Air Energy Storage for Power Generation System Improving—Application in a Microgrid Integrating Wind Energy' (2015) 73 *Energy Procedia* 305–16.
14. Consortium for Battery Innovation. Lead battery market data (2021). Available at: <https://batteryinnovation.org/resources/lead-battery-market-data/> accessed 5 November 2022.
15. Schmuch R, Wagner R, Hörpel G, Placke T, *et al.* 'Performance and Cost of Materials for Lithium-based Rechargeable Automotive Batteries' (2018) 3 *Nature Energy* 267–78.
16. Liu C, Neale ZG, and Cao G. 'Understanding Electrochemical Potentials of Cathode Materials in Rechargeable Batteries' (2016) 19 *Materials Today* 109–23.
17. Hannan MA, Hoque MM, Mohamed A, and Ayob A. 'Review of Energy Storage Systems for Electric Vehicle Applications: Issues and Challenges' (2017) 69 *Renewable and Sustainable Energy Reviews* 771–89.
18. Tamakoshi T. 'Sodium-Sulfur (NAS) Battery' in DOE Long Duration Energy Storage Workshop (Nagoya: NGK Insulators Ltd, 2021), 1–11.
19. Sánchez-Díez E, Ventosa E, Guarnieri M, Trovò A, *et al.* 'Redox Flow Batteries: Status and Perspective Towards Sustainable Stationary Energy Storage' (2021) 481 *Journal of Power Sources* 228804.
20. Sandia National Laboratories. DOE Global Energy Storage Database (2022). Available at: <https://sandia.gov/ess-ssl/gesdb/public/index.html> accessed 5 November 2022.
21. Lund PD, Lindgren J, Mikkola J, and Salpakari J. 'Review of Energy System Flexibility Measures to Enable High Levels of Variable Renewable Electricity' (2015) 45 *Renewable and Sustainable Energy Reviews* 785–807.
22. International Energy Agency. *Global Hydrogen Review* (Paris: IEA, 2021).
23. Augustine C and Blair N. *Storage Futures Study: Storage Technology Modeling Input Data Report* (Washington, DC: U.S. Department of Energy, 2021).
24. Mongird K, Viswanathan V, Alam J, Vartanian C, *et al. 2020 Grid Energy Storage Technology Cost and Performance Assessment* (Washington, DC: U.S. Department of Energy 2020).
25. Frith J. *Lithium-Ion Batteries: The Incumbent Technology Platform for Coal Regions in Transition* (BloombergNEF, 2019).
26. Akhil A, Huff G, Currier AB, Kaun BC, *et al. DOE/EPRI Electricity Storage Handbook* (Albuquerque: Sandia National Laboratories, 2013).
27. BloombergNEF. *1H 2022 LCOE Update* (2022).
28. Sterner M and Stadler I. *Energiespeicher—Bedarf, Technologien, Integration* (Berlin: Springer, 2014).
29. Barbour E, Wilson IAG, Radcliffe J, Ding Y, *et al.* 'A Review of Pumped Hydro Energy Storage Development in Significant International Electricity Markets' (2016) 61 *Renewable and Sustainable Energy Reviews* 421–32.
30. International Energy Agency. *Hydropower Special Market Report* (Paris: IEA, 2021).
31. NGK Insulator Ltd. Energy Storage System Products (2022). Available at: <https://www.ngk-insulators.com/en/product/search-business/battery/> accessed 5 November 2022.
32. Wood Mackenzie. *Global Energy Storage Outlook H1 2021* (Edinburgh: Wood Mackenzie, 2021).
33. BloombergNEF. Global Energy Storage Market to Grow 15-fold by 2030 (2022). Available at: <https://about.bnef.com/blog/global-energy-storage-market-to-grow-15-fold-by-2030/> accessed 5 November 2022.

34. International Energy Agency. *Global EV Outlook* (Paris: IEA, 2022).
35. BloombergNEF. *Electric Vehicle Outlook* (London: BloombergNEF, 2022).
36. Melitz MJ and Ottaviano GIP. 'Market Size, Trade, and Productivity' (2008) 75 *The Review of Economic Studies* 295–316.
37. *§ 8 EnWG—Eigentumsrechtliche Entflechtung* (German Federal Ministry of Justice).
38. Ofgem. Enabling the competitive deployment of storage in a flexible energy system: decision on changes to the electricity distribution licence (London: Office of Gas and Electricity Markets, 2018).
39. *§ 11b EnWG—Ausnahme für Energiespeicheranlagen, Festlegungskompetenz* (German Federal Ministry of Justice).
40. BASF. *BASF and NGK Enter into Sales Partnership Agreement for NAS® Battery* (2019). <https://tinyurl.com/basf-19-241> accessed 5 November 2022.
41. Grid News. China is Owning the Global Battery Race. That Could Be a Problem for the U.S. (2022). Available at: <https://www.grid.news/story/global/2022/01/18/china-is-owning-the-global-battery-race-that-could-be-a-problem-for-the-us> accessed 5 November 2022.
42. Moores S. 'The Global Battery Arms Race: Lithium-Ion Battery Gigafactories and Their Supply Chain' (2021) 126 *Oxford Energy Forum* 26–30.
43. IRENA. *A New World: The Geopolitics of the Energy Transformation* (Masdar City: IRENA, 2019).
44. International Energy Agency. *The Role of Critical Minerals in Clean Energy Transitions* (Paris: IEA, 2022).
45. Scholten D. *The Geopolitics of Renewables* (Cham: Springer, 2018).
46. Yergin D. *The Quest: Energy, Security and the Remaking of the Modern World* (London: Penguin, 2012).
47. Øverland I. 'The Geopolitics of Renewable Energy: Debunking Four Emerging Myths' (2019) 49 *Energy Research & Social Science* 36–40.
48. Vakulchuk R, Øverland I, and Scholten D. 'Renewable Energy and Geopolitics: A Review' (2020) 122 *Renewable and Sustainable Energy Reviews* 109547.
49. Bazilian M, Bradshaw M, Gabriel J, Goldthau A, *et al.* 'Four Scenarios of the Energy Transition: Drivers, Consequences, and Implications for Geopolitics' (2020) 11 *Wiley Interdisciplinary Reviews: Climate Change* e625.
50. Levi M, Economy EC, O'Neill S, and Segal A. 'Globalizing the Energy Revolution: How to Really Win the Clean-Energy Race' (2010) 89 *Foreign Affairs* 111–21.
51. Senior J. *Expert Views Regarding China–U.S. Tensions over Production of Solar PV, Wind and Battery Technologies, and Their Potential to Reduce Deployment of Those Technologies During the Transition* (London: Imperial College London, 2022).
52. European Battery Alliance. Building a European Battery Industry (2022). Available at: <https://www.eba250.com/> accessed 5 November 2022.
53. Argonne National Laboratory. Li-Bridge (2022). Available at: <https://www.anl.gov/li-bridge> accessed 5 November 2022.
54. US Senate. *Inflation Reduction Act* (Washington, DC: US Senate, 2022).
55. White E and Lewis L. West Could End Reliance on Chinese Batteries by 2030, Says Goldman Sachs. *Financial Times* (2022). Available at: <https://www.ft.com/content/458ebaf3-c1ee-499c-b7f3-2e5d7f1bb6df> accessed 25 November 2022.
56. Home A. Goldman Sparks Bear–Bull Battle in the Lithium Market (2022). Available at: <https://www.reuters.com/markets/commodities/goldman-sparks-bear-bull-battle-lithium-market-2022-06-15> accessed 25 November 2022.

57. Archer C. Lithium Price 2023: Are Goldman Sachs and Credit Suisse Misguided? (2022). Available at: <https://www.ig.com/en-ch/news-and-trade-ideas/lithium-price-2023--are-goldman-sachs-and-credit-suisse-misguide-221123> accessed 25 November 2022.
58. Golev A, Scott M, Erskine PD, Ali SH, et al. 'Rare Earths Supply Chains: Current Status, Constraints and Opportunities' (2014) 41 *Resources Policy* 52–59.
59. Smith Stegen K. 'Heavy Rare Earths, Permanent Magnets, and Renewable Energies: An Imminent Crisis' (2015) 79 *Energy Policy* 1–8.
60. Battke B and Schmidt TS. 'Cost-efficient Demand-pull Policies for Multi-Purpose Technologies—The Case of Stationary Electricity Storage' (2015) 155 *Applied Energy* 334–48.
61. Stephan A, Battke B, Beuse MD, Clausdeinken JH, *et al.* 'Limiting the public cost of stationary battery deployment by combining applications' (2016) 1 *Nature Energy* 16079.
62. Ela E, Milligan M, and Kirby B. *Operating Reserves and Variable Generation* (Golden: National Renewable Energy Laboratory, 2011).
63. Rastler DM. *Electric Energy Storage Technology Options: A White Paper Primer on Applications, Costs and Benefits* (Palo Alto: Electric Power Research Institute, 2010).
64. Fitzgerald G, Mandel J, Morris J, and Touati H. *The Economics of Battery Energy Storage: How Multi-use, Customer-sited Batteries Deliver the Most Services and Value to Customers and the Grid* (Basalt, CO: Rocky Mountain Institute, 2015).
65. International Electrotechnical Commission. *Electrical Energy Storage White Paper* (Geneva: International Electrotechnical Commission, 2011).
66. International Energy Agency. *Technology Roadmap—Energy Storage* (Paris: IEA, 2014).
67. Lazard. *Lazard's Levelized Cost of Storage Analysis—Version 2.0* (New York: Lazard Ltd, 2016).
68. Everoze Partners. *Cracking the Code: A Guide to Energy Storage Revenue Streams and How to Derisk Them* (Bristol: Everoze Partners Ltd, 2016).
69. AEMO. *Application of Advanced Grid-scale Inverters in the NEM* (Melbourne: Australian Energy Market Operator, 2021).
70. Malhotraa A, Battke B, Beuse M, Stephan A, *et al.* 'Use Cases for Stationary Battery Technologies: A Review of the Literature and Existing Projects' (2016) 56 *Renewable and Sustainable Energy Reviews* 705–21.
71. Bosch J. A New Frontier in Dynamic Frequency—Does It Stack Up?—Open Energi (2022). Available at: <https://openenergi.com/articles/a-new-frontier-in-dynamic-frequency> accessed 5 November 2022.
72. International Energy Agency. *The Power of Transformation* (Paris: IEA, 2014).
73. Lokhov A. *Technical and Economic Aspects of Load Following with Nuclear Power Plants* (Paris: OECD & NEA, 2011).
74. Denholm P, Cole W, Frazier AW, Podkaminer K, *et al. The Four Phases of Storage Deployment: A Framework for the Expanding Role of Storage in the U.S. Power System* (Golden: National Renewable Energy Laboratory, 2021).
75. Schäfer C. Batteriespeicher dominieren den PRL-Markt. *Regelleistung-Online* (2022). Available at: <https://www.regelleistung-online.de/batteriespeicher-dominieren-den-prl-markt/> accessed 5 November 2022.
76. National Grid. *Enhanced Frequency Response—Market Information Report* (London: National Grid plc, 2016).
77. Clean Energy Council. Battery storage—The New, Clean Peaker (2021). Available at: <https://www.cleanenergycouncil.org.au/resources/resources-hub/battery-storage-the-new-clean-peaker> accessed 5 November 2022.

78. RenewEconomy. Neoen Reveals Plans for Another 300MW Big Battery in South Australia (2021). Available at: <https://reneweconomy.com.au/neoen-reveals-plans-for-another-300mw-big-battery-in-south-australia/> accessed 5 November 2022.

79. Energy Storage News. 450MWh Battery Storage Project Granted South Australia Government Development Approval (2021). Available at: <https://www.energy-storage.news/450mwh-battery-storage-project-granted-south-australia-government-development-approval/> accessed 5 November 2022.

80. Parkinson G. South Australia set to become first big grid to run on 100% renewables. Available at: <https://www.climatechangenews.com/2022/09/16/south-australia-set-to-become-first-big-grid-to-run-on-100-renewables/> accessed 5 November 2022.

81. U.S. Energy Information Administration (EIA). Most Pumped Storage Electricity Generators in the U.S. Were Built in the 1970s (2019). Available at: <https://www.eia.gov/todayinenergy/detail.php?id=41833> accessed 5 November 2022.

82. U.S. Energy Information Administration (EIA). Nuclear Energy Overview (2022). Available at: <https://www.eia.gov/totalenergy/data/monthly/pdf/sec8_3.pdf> accessed 5 November 2022.

PART II: Cost and Value of Energy Storage

4 Investment cost
Projecting cost developments

KEY INSIGHT	WHAT IT MEANS
Experience curves match historical product price data to cumulative installed capacity. This allows the development of investment cost to be quantified by an *experience rate*, which is the change in product price for each doubling of cumulative installed capacity.	Experience curves are a simple framework to structure the historical development of investment cost for a technology. This framework can be used to compare technologies and project potential future investment cost.
The cost reduction across electricity storage technologies reveals three maturity categories based on cumulative installed capacity and product price: • Emerging (< 1 GWh, > 1,000 USD/kWh) • Growing (1–100 GWh, > 300 USD/kWh) • Mature (> 100 GWh, < 300 USD/kWh)	This can provide a useful shorthand to assess the maturity of electricity storage technologies and identify if investment cost potentially exceeds or undercuts comparable technologies at the same maturity level.
Experience rates for electricity storage technologies range from –3% to 30%. The highest rates belong to lithium-ion cells (30%), packs (24%), and utility-scale systems (19%); the lowest to pumped hydro plants (–3%).	Among all major electricity storage technologies, lithium ion has the strongest potential for cost reductions. In contrast, investment cost for pumped hydro plants has increased as more capacity is deployed.
A technology's future investment cost can be projected by extending its experience curve forwards to future amounts of cumulative installed capacity.	Experience curves are a tool to project future investment cost in a structured way. The projections rely on the continuation of the historical cost reduction trend.
All major electricity storage technologies are on a cost reduction trajectory towards 100–500 USD/kWh once 1 TWh of energy capacity of the respective technology has been installed.	Given that most technologies follow a similar trend, the technology that brings most capacity to market is likely to reach lowest investment cost first. This gives an advantage to modular technologies that can be used in multiple applications.

(*Continued*)

Monetizing Energy Storage. Oliver Schmidt & Iain Staffell, Oxford University Press. © Oliver Schmidt & Iain Staffell (2023).
DOI: 10.1093/oso/9780192888174.003.0004

KEY INSIGHT	WHAT IT MEANS
Experience curves can be used to make long-term projections of future product prices for electricity storage technologies. They are ill-suited to provide accurate short-term price forecasts.	Short-term prices may fluctuate due to imbalances in supply and demand (e.g. manufacturing capacities, raw materials, logistics). Therefore, experience curves should only be used to identify overarching cost reduction trends.
Raw material availability is unlikely to become a limiting factor in electricity storage deployment for any of the major technologies. It is also unlikely to affect the product price projections based on experience curves in the long-term.	Concerns about raw material availability and cost are usually based on supply and demand imbalances which can be resolved with new raw material mining and processing capacity. There is no fundamental shortage of the key raw materials required for any major electricity storage technology.
Combining experience curves with market growth projections allows future investment costs to be projected as a function of time. This suggests that by 2030, lithium-ion packs for EVs could fall to 60 USD/kWh and large-scale 4-hour utility-scale systems to 100 USD/kWh.	Widespread commercialization of electric vehicles and renewables-plus-storage for baseload power generation can be expected by 2030. The projections represent industry averages. Variations are likely to be based on design, system size, or geography of deployment.
Experience curves are based on historical trends and are therefore ill-suited to capture potential step-change effects like radical innovations.	Cost reduction potentials should be vetted against technology roadmaps for radical innovations that diverge from the current predominant design.

4.1 Role of investment cost

The investment cost, or upfront capital cost, is a key determinant of a technology's competitiveness. It determines whether a novel technology takes off and enters the mainstream, or remains forever stuck in the laboratory and demonstration phase. However, investment costs for new low-carbon technologies are typically higher than those of the incumbents they seek to replace. It is therefore essential to understand the cost reduction potential of novel technologies to assess their mass-market potential, even though their current market is small.

This is very much the case for energy storage technologies. Uncertainty around their potential future investment cost is still a key barrier to unlocking their potential.[1]

Therefore, this chapter compiles an extensive dataset of product prices (i.e. investment cost) and cumulative deployed capacity for 11 electricity storage technologies. It derives cost reduction curves, so-called experience curves, and shows how to conduct a comparative appraisal of the observed cost reduction trends. It demonstrates how future prices can be projected based on increased cumulative capacity and how their feasibility can be tested against possible cost floors set by raw material costs. Using market growth models, feasible timescales for realizing these prices and the required cumulative investments in deployment are determined. Key implications are discussed with two stylized examples to show how the derived cost reduction rates can be used to assess uncertainty around future competitiveness of storage. The impact of extraordinary raw material price fluctuations on short- and long-term technology cost development is analysed. Finally, cost reduction insights are compared to an alternative method for future investment cost projections. As such, this chapter presents a step-by-step walk-through for a comprehensive assessment of future cost reduction potentials using the example of electricity storage technologies.

KEY INSIGHT

Experience curves match historical product price data to cumulative installed capacity. This allows the development of investment costs to be quantified by an *experience rate*, which is the change in product price for each doubling of cumulative installed capacity.

The cost reduction dataset and analyses presented here are publicly available from our companion website <www.EnergyStorage.ninja>, and are regularly updated to stay abreast of continuing developments in the industry.

4.2 Experience curves for storage technologies

Figure 4.1 shows product prices per unit of energy capacity for the most common electricity storage technologies as a function of increasing cumulative installed energy capacity. Prices for storage technologies differ by product scope and application. Therefore, product price and installed capacity data in Figure 4.1 are differentiated by product scope: cell, pack, and installed system, and application category: portable (consumer electronics), transport (electric vehicles and hybrid electric vehicles), and stationary (residential, utility).

Analysing the cost reduction across all electricity storage technologies facilitates the categorization into *mature*, *growing*, and *emerging* technologies:

- Mature technologies have above 100 GWh installed and exhibit prices below 300 USD/kWh. Examples include pumped hydro (system), lead acid (pack), alkaline electrolysis (pack), and lithium ion for consumer electronics (cell) and electric vehicles (pack).
- All remaining technologies are growing with above 1 GWh installed capacity and prices above 300 USD/kWh.
- When technologies were emerging (i.e. below 1 GWh installed capacity), prices were generally above 1,000 USD/kWh.

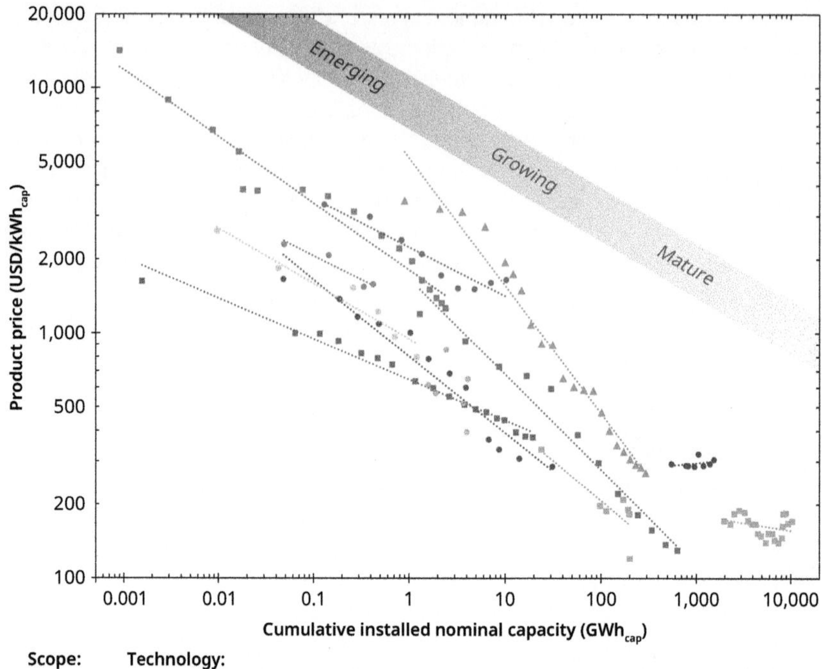

Figure 4.1 Experience curves for storage technologies, measured by energy capacity. Product prices measured in real 2020 USD per nominal energy capacity as a function of cumulative installed nominal energy capacity. Dotted lines show the experience curves resulting from linear regression of the data. The legend indicates product scope (installed system, pack, cell) and technology (including application, experience rate with uncertainty, and years covered by the dataset). Experience rate uncertainty is quantified with the 95% standard error confidence interval. The grey stripe indicates the overarching trend in cost reduction for technologies relative to technology maturity as a function of cumulative installed capacity: Emerging (< 1 GWh), Growing (< 100 GWh), and Mature (> 100 GWh). Fuel cells and electrolysis must be considered in combination to form a hydrogen-based storage technology (with an assumed energy to power ratio of 10). Data for lead acid (pack) refer to multiple applications, including uninterruptable power supply and heavy-duty transportation. kWh$_{cap}$—nominal energy storage capacity.

KEY INSIGHT

The cost reduction across electricity storage technologies reveals three maturity categories based on cumulative installed capacity and product price:

- Emerging (< 1 GWh, > 1,000 USD/kWh)
- Growing (1–100 GWh, > 300 USD/kWh)
- Mature (> 100 GWh, < 300 USD/kWh)

Experience rates are derived from the slope of experience curves and quantify the percentage change in product price with each doubling of cumulative installed capacity.

All experience rates for the analysed electricity storage technologies are between 10% and 30%, except for pumped hydro systems and lead-acid packs. Negative experience rates observed for pumped hydro systems could be due to good sites becoming more difficult to access, increased development time due to public hearings and licensing, and stronger environmental protections.[2,3] Lead-acid pack product prices are already relatively close to raw material cost of lead with respective fluctuations visible in the product price development. This dilutes cost reductions in the residual cost contributors, which were determined at 19%, in line with the other technologies displayed in Figure 4.1.[4] This observation suggests a potential 'flattening' of experience curves once products are sufficiently commercialized and raw material cost constitute a significant portion of total product cost.[5]

The highest experience rates can be observed for lithium-ion cells (consumer electronics) and battery packs (EVs). This explains the dominance of lithium-ion technology in these applications. The strong experience rates in combination with significant deployment levels enabled competitive price levels in the respective applications. As a result, residential lead-acid systems have not been deployed in significant amounts since 2016. Data points and experience rate are still displayed for reference.

Nickel-metal hydride batteries are also displayed in Figure 4.1 with an experience rate of 11%. As with lithium ion, this electricity storage technology is used in hybrid electric vehicles (HEV), but has little other major applications.

When analysing experience rates, it is also essential to assess uncertainty through a regression analysis, for example with a 95% standard error-based confidence interval. The experience rates shown in Figure 4.1 have uncertainties equal to or below 6%, which indicates relatively high confidence in the identified cost reduction trend. The only exceptions are sodium-sulphur systems and alkaline electrolysis packs. The uncertainty for sodium sulphur is so high that no experience rate can be derived. For electrolysis, a rate of 20% can be derived, albeit with an uncertainty of ±11%.

Electricity storage technologies with insufficient price data are not shown, but these may still hold promise in the future, for example flywheels, compressed air, gravity storage, new electrochemical battery types, or thermal storage.

KEY INSIGHT

Experience rates for electricity storage technologies range from –3% to 30%. The highest rates belong to lithium-ion cells (30%), packs (24%), and utility-scale systems (19%); the lowest to pumped hydro plants (–3%).

Figure 4.2 displays experience curves as a function of power capacity. These were determined by applying the average power-to-energy ratio to product price and cumulative installed capacity in energy terms. These are fixed for each technology, therefore experience rates stay the same as in Figure 4.1. However, it can be observed that the range of prices is much wider, clearly differentiated along the power-to-energy ratio of the different

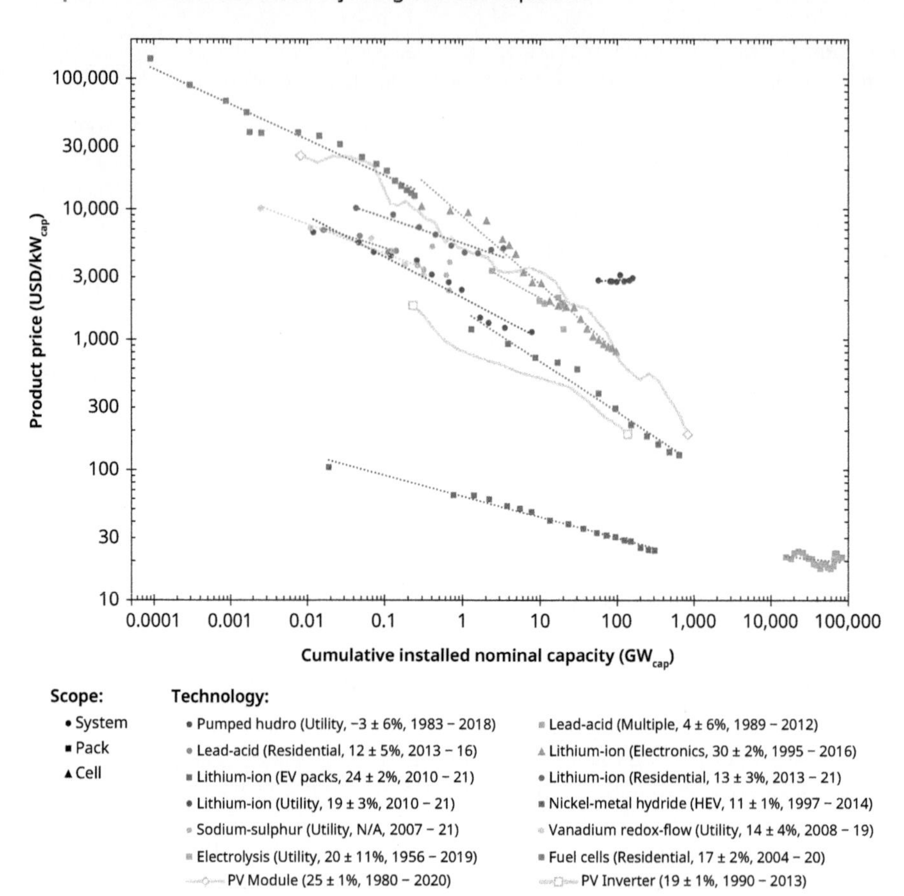

Figure 4.2 Experience curves for storage technologies, measured by power capacity. Product prices measured in real 2020 USD per nominal power capacity as a function of cumulative installed nominal power capacity. Dotted lines show the experience curves resulting from linear regression of the data. The legend indicates product scope (installed system, pack, cell) and technology (including application, experience rate with uncertainty, and years covered by the dataset). Experience rate uncertainty is quantified with the 95% standard error confidence interval. Indicative power to energy ratios are used to convert prices and capacity from energy to power terms. Experience curves for solar PV modules and PV inverters are shown in grey for context.

technologies. Nickel-metal hydride battery packs that are used in HEVs have a power-to-energy ratio of 15 and cost 20 USD/kW. Lithium-ion packs with a ratio of 1 cost 130 USD/kW. All other technologies with ratios below 0.5 cost more than 500 USD/kW. The experience rates of solar PV modules and inverters are displayed for comparison, indicating that experience rates of storage technologies are within the same range.

In addition, it can be observed that experience rates for lithium-ion technologies decrease with increasing product scope (see Figure 4.3). Overall, higher experience rates for cells than for packs and higher rates for packs than for systems imply that cost reductions are likely driven by increased production experience in cell manufacturing, followed by pack manufacturing.

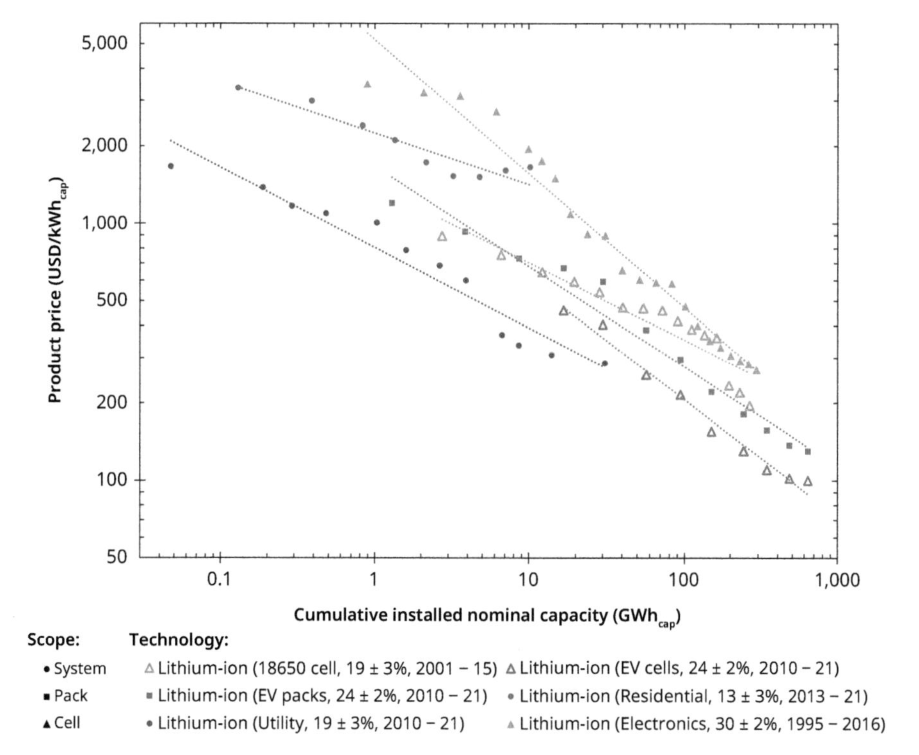

Figure 4.3 Experience curves for lithium-ion technologies, measured in energy terms. Product prices measured in real 2020 USD per nominal energy capacity as a function of cumulative installed nominal energy capacity. Dotted lines show the experience curves resulting from linear regression of the data. The legend indicates product scope (installed system, pack, cell) and technology (including application, experience rate with uncertainty, and years covered by the dataset). Experience rate uncertainty is quantified with the 95% standard error confidence interval.

FREQUENTLY ASKED QUESTION

Don't pumped hydro plants only cost ~20 USD/kWh?

Some reports indicate that average installed system cost for pumped hydro plants is 20 USD/kWh.[7] While there is indeed a wide variation in pumped hydro cost based on the given geology, this is an order of magnitude less than the ~300 USD/kWh reported in this book and elsewhere.[8] What has happened?

There is a frequent mix-up between the total system cost expressed per unit of total energy capacity (USD_{total}/kWh) and specific energy cost expressed per marginal unit of capacity ($USD_{specific}$/kWh). While the former includes all cost components of the plant divided by the total energy capacity, the latter reflects the cost of adding a kWh of storage capacity and thereby only refers to the energy-related cost of the plant. Simply put, the former includes turbine and reservoir cost, while the latter only refers to the cost of the reservoir. See Figure 2.3 in Chapter 2 for a detailed example.

Stronger cost reduction for consumer electronics cells compared to 18,650 cells could reflect the ongoing shift from cylindrical 18,650 to more cost-competitive prismatic and laminate cells in consumer electronics.[6] The difference in experience rate for residential and utility systems could be the result of lower cost reduction pressure in the less competitive residential business and reduced participation in cell and pack cost reductions due to lower procurement volumes of residential system manufacturers.

4.3 Future investment cost

Experience rates reveal the underlying trend in how historical prices have fallen as a function of increasing cumulative deployed capacity. It is possible to project these experience curves forwards to potential future deployment levels. Of course, there is no guarantee that prices will continue to fall at the same rate as they have done in the past; however, this approach does give an objective, evidence-based view on how costs might develop.

KEY INSIGHT

A technology's future investment cost can be projected by extending its experience curves forwards to future amounts of cumulative installed capacity.

Figure 4.4 shows the experience curves projected forwards to 1 TWh of cumulative capacity. This allows a common cost reduction trajectory to be derived across all the analysed electricity storage technologies. This implies that all commercially successful technologies have the potential to cost between 100 and 500 USD/kWh once 1 TWh has been deployed. As this applies to all technologies, it means that the technology that manages to bring most capacity to market is likely to reach this range first and be most cost competitive. Thus, modular technologies that can be used in multiple applications have an advantage (e.g. lithium ion).

KEY INSIGHT

All major electricity storage technologies are on a cost reduction trajectory towards 100–500 USD/kWh once 1 TWh of energy capacity of the respective technology has been installed.

When accounting for uncertainty of the underlying price and capacity data, the price range at 1 TWh expands to 80–750 USD/kWh. Also, it is important to note that experience curves have an empirical, rather than analytical, nature. Extrapolation into the future is not only subject to uncertainty of the derived rates but also uncertainty associated with unforeseeable future changes. This can include technology breakthroughs, knowledge spill-overs, and commodity price shifts, which may all fundamentally change the cost reduction rate.[9,10]

The lower bound of the range is set by stationary utility-scale lithium-ion systems. This appears counterintuitive as these systems contain more components than lithium-ion cells or packs, which should be cheaper as a result. However, stationary lithium-ion systems will

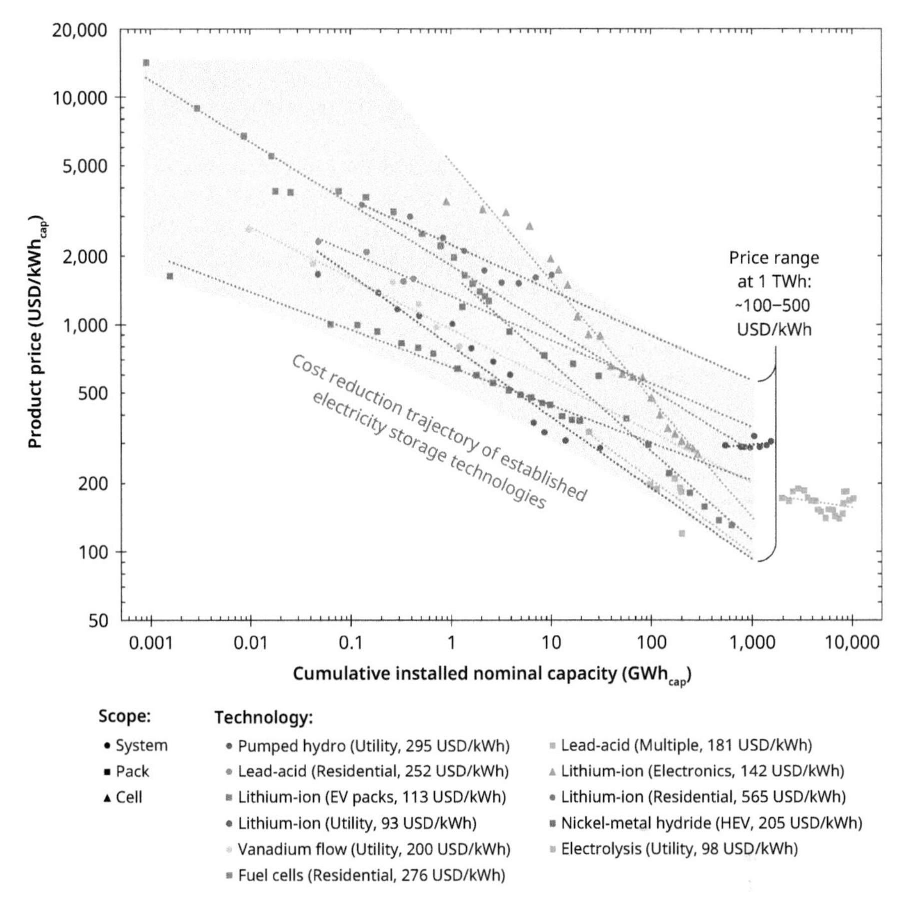

Figure 4.4 Future cost of storage technologies at 1 TWh of capacity deployed. Experience curves (dotted lines) are projected forwards to 1 TWh cumulative installed energy capacity to analyse potential future product prices. The legend indicates product scope (installed system, pack, cell) and technology (including projected product price at 1 TWh cumulative installed capacity). The shaded region indicates the cost reduction trajectory across all technologies, which narrows to the price range given on the right of the figure.

reach the 1 TWh deployment threshold later than cells and packs, which are also deployed in consumer electronics and electric vehicles. Thus, when lithium-ion systems reach the 1 TWh threshold, cells and packs will already be at several TWh of deployment, and significantly cheaper as a result. This insight highlights the importance of adding the temporal element to experience curve analysis, which is considered in section 4.5.

The analysis of the data points for utility-scale lithium-ion systems also triggers an important discussion on experience curves for novel technologies. Experience rates use product prices as proxy for production cost. Therefore, they are subject to pricing dynamics, which occur in particular during development and market introduction of a product. Figure 4.5 depicts four phases:[12]

FREQUENTLY ASKED QUESTION

Can the price target of 100 USD/kWh for lithium-ion systems be achieved in the next few years?

'The cost of the battery is only ~10–20% higher than the bill of materials – suggesting a potential long-term competitive price for lithium-ion batteries could approach ~USD 100 per kWh. Tesla currently pays Panasonic 180 USD/kWh for their batteries, although conventional systems are still selling for 500–700 USD/kWh. But Navigant says that the broader marketplace will reach the levels Tesla is paying in the next two to three years.'[11]

Such claims mix references to the cost of battery cells and installed stationary battery systems. The cost of a battery **cell** is only ~10–20% higher than the bill of materials and Tesla probably paid 180 USD/kWh for Panasonic lithium-ion **cells** in 2015. That's only ~35% of the cost for an entire **installed battery system** though. Hence, the article is flawed by comparing Panasonic cell costs to the costs of 'conventional systems' and then suggesting that 100 USD/kWh will be reached in the next 2 to 3 years. It infers that this is the price for battery systems. Table 4.1 below gives an indication of the relative cost contribution of key components in a battery system (see also section 2.2).

Table 4.1 Components of a battery storage systems and indicative cost contributions.

Technology scope	Indicative contribution
Cell (electrodes, electrolyte, collectors, housing, etc.)	35%
Pack/Rack (structure, electronics, BMS, etc.)	15%
Power conversion system (inverter, etc.)	10%
Balance-of-system (container, fire suppression, etc.)	10%
System integration	5%
Project development	10%
Engineering, procurement, commissioning	15%
Total	**100%**

- Development: Product prices are below production cost for the new product to be tried in the market
- Price umbrella: A dominant player can command a price premium for the newly introduced product
- Shakeout: As the market grows, new players enter and increase competition so that prices reduce strongly back to production cost levels
- Stability: When the market environment has stabilized, product prices reflect production cost development.

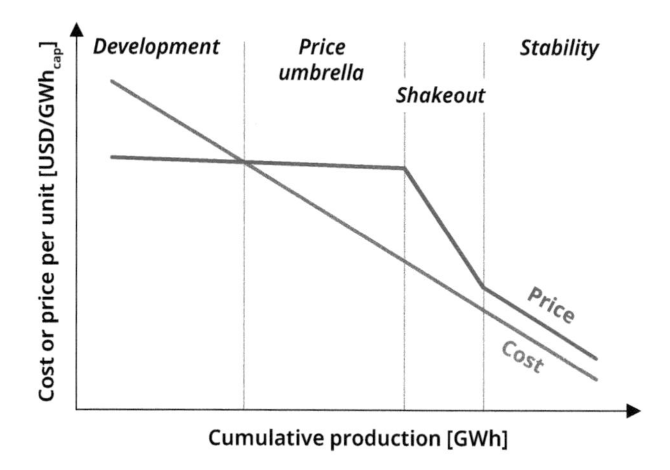

Figure 4.5 The relationship between cost and price during market introduction of a new product.[12]

The data points for lithium-ion systems (residential- and utility-scale) mirror this development. Figure 4.6 shows a price umbrella before 2016/17 when a relatively small number of players served a fast-growing market. These companies may have been able to enforce high prices, which would attract competitors. In 2015, Tesla announced the residential Powerwall and utility-scale Powerpack product for prices significantly below market average.[13] From 2017, these products were mass-produced. This is an indicator for the increased competition that took place from 2017, explaining the shakeout-like price reductions for lithium-ion battery systems thereafter.

One could argue that only the data points for 2018 to 2021 should be considered to ensure that price development mirrors cost development. This would yield an experience rate of 10% for utility-scale systems. However, the aggressive pricing strategy by Tesla may indicate that selected entrants were pricing below cost to buy market share. Also, some industry experts suggest that the high prices before 2017 were caused by a shortage in production capacity and resulting competition with the EV battery market. Product prices will always be subject to pricing dynamics and product supply/demand imbalances, so there is an argument for including these data in the experience curve.

FREQUENTLY ASKED QUESTION

I read that lithium-ion packs became more expensive in 2022. Does that mean the identified experience curves are wrong?

Prices for lithium-ion battery packs increased in 2022.[14] This is attributed to higher raw material prices. However, it does not mean that the identified cost reduction trend is wrong. Rather, annual price points vary due to imbalances in demand and supply for raw materials, manufacturing capacities and/or logistics capacities. The broader cost reduction drivers are not affected (e.g. production scale-up, product innovation) and unlock further reduction potential that materializes once demand and supply are in balance again.

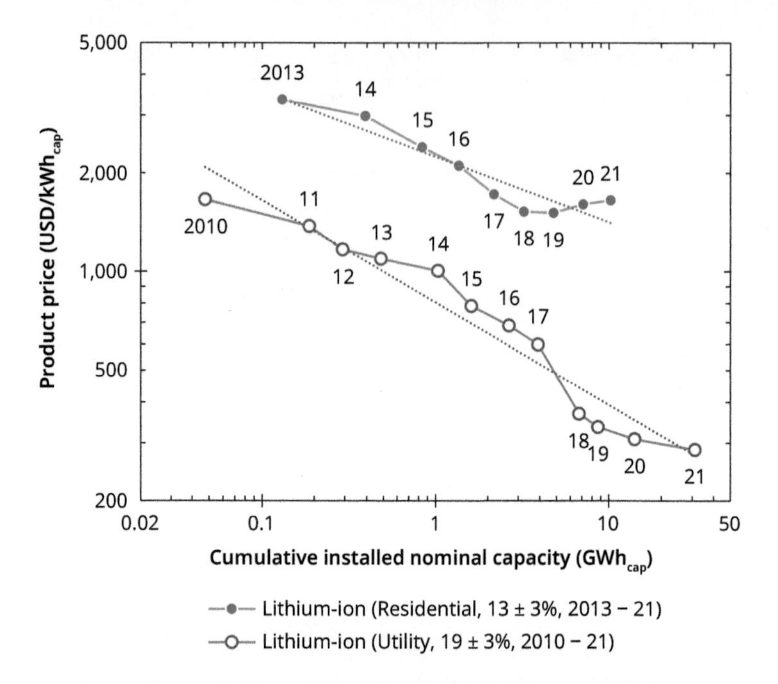

Figure 4.6 Historical price development of residential and utility-scale lithium-ion battery systems. Product prices measured in real 2020 USD per nominal energy capacity as a function of cumulative installed nominal energy capacity. Dotted lines show the experience curves resulting from linear regression of the data. The legend indicates technology (including application, experience rate with uncertainty, and years covered by the dataset). Experience rate uncertainty is quantified with the 95% standard error confidence interval.

KEY INSIGHT

Experience curves can be used to make long-term projections of future product prices for electricity storage technologies. They are ill-suited to provide accurate short-term price forecasts.

4.4 Raw material cost

Experience curve analyses with extrapolated forecasts should include cost floors to avoid excessively low cost estimates.[9,10] As highlighted before, experience curves may flatten once products are sufficiently commercialized and raw material cost constitute a significant cost share. Therefore, raw material cost can be used as a proxy for cost floors in experience curve analyses.

Figure 4.7 compiles raw material cost for each technology based on their material inventories and commodity prices from 2010 to 2020. It also compares these to the product price projections for 1 TWh cumulative deployed capacity from Figure 4.4. The raw material cost

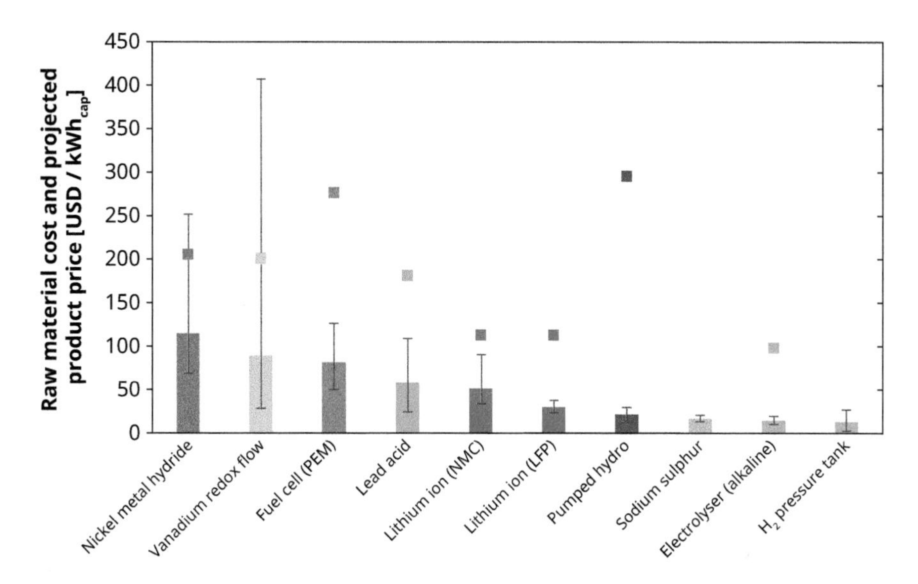

Figure 4.7 Combination of raw material cost (bars) and product price projections (squares) at 1 TWh cumulative deployed capacity for electricity storage technologies. Product prices are given for system level for pumped hydro and pack level for all other technologies. Error bars account for variation in each technology's material inventory and commodity prices from 2010 to 2020. Commodity prices are taken as monthly averages instead of daily spot prices as these better reflect the contract prices battery manufacturers will see.[15] NMC refers to NMC111 with equal shares of nickel, manganese, and cobalt.

represent cost floors below which product prices are unlikely to go with current technology and no innovation in raw material extraction.

The figure reveals that the average raw material cost for most technologies is significantly below the experience curve-based product price projections at 1 TWh for most technologies. This confirms that the identified cost reduction potentials to 100–500 USD/kWh are feasible. Note that raw materials prices increased sharply in the first years of the 2020s beyond the range shown in this figure. This is potentially a short-lived commodity cycle though, as discussed further in section 4.7.4, and so may not accurately reflect the ultimate lower-bound for technology costs.

Nickel-metal hydride batteries exhibit relatively high raw material cost due to large amounts of nickel. These batteries are also subject to commodity price fluctuations of nickel, leading to pack cost of 250 USD/kWh in the worst case. Similarly, vanadium price fluctuations mean that raw material cost for vanadium-flow batteries could be as high as 400 USD/kWh in the worst case. For both technologies, these worst-case raw material costs would be above projected product prices at 1 TWh cumulative deployed capacity. This indicates that product prices could temporarily be higher than projected by experience curves, depending on raw material cost fluctuations.

The material costs of other electrochemical storage technologies are also driven by their active materials like platinum, lithium, and lead. Lithium ion is a family of technologies with different options for materials used in the cathode. Taking average raw material costs, NMC is 66% more expensive than LFP. Mechanical storage technologies have the lowest material cost below 20 USD/kWh due to the low-cost materials employed.

The use of monthly average price data for 10 years from 2010 to 2020 covers the typical cyclical fluctuations of supply and demand as described by the cobweb theorem or 'pork cycle'.[16] However, there is still uncertainty on the future development of commodity prices, and extraordinary circumstances could lead to price fluctuations outside of the ranges that are explained by the cobweb theorem. This will be covered in section 4.7.

In order to assess the impact of raw material price increases on product prices, it is important to understand the raw material composition of electricity storage technologies.

Figure 4.8 illustrates this for lithium-ion battery packs by displaying weight and cost contribution of the key raw materials for the two most common chemistries, LFP and NMC. It shows that aluminium constitutes 22% of the total weight due to its use in the current conductor and cell and pack housing. This is followed by 12–13% graphite, which is used in the anode and 12–13% copper, which is also used as a current collector.

The individual active materials of the cathode and lithium usually constitute less than 10% of the raw material weight in lithium-ion battery packs. However, the cost contributions of lithium in LFP or nickel and cobalt in NMC batteries are larger than 20% each.

This analysis shows that each material only contributes a minor share to total raw material cost. In addition, total raw materials cost only constitutes a share of total product price. The cost increase of one raw material will therefore only have a limited impact on lithium-ion battery pack prices.

Figure 4.9 illustrates this through a sensitivity analysis for the key raw material cost contributors for LFP and NMC battery packs. It shows that a doubling of copper prices (a 100% increase), will increase the product price of LFP or NMC battery packs by only 5.4% or 5.0% respectively. A quadrupling of copper prices (a 300% increase) would increase LFP or NMC pack prices by 16.2% or 15.1% respectively.

LFP battery pack prices are most sensitive to copper, aluminium, and lithium hydroxide cost. A quadrupling of all three would increase pack prices by ~35%. In contrast, NMC battery pack prices are most sensitive to the cathode materials, nickel and cobalt. A quadrupling of the cost for both would increase NMC battery pack prices by more than 50%. This suggests that LFP battery pack prices are more robust to raw material cost changes than NMC battery packs because the cost contribution of individual materials to total raw material cost is lower.

Price spikes for raw materials usually trigger discussions on whether resource availability may be a general concern for the future deployment potential of selected electricity storage technologies like lithium ion. In fact, all active materials of the investigated storage technologies have a reserve base sufficient for the production of beyond 10 TWh storage capacity with current technology.[18]

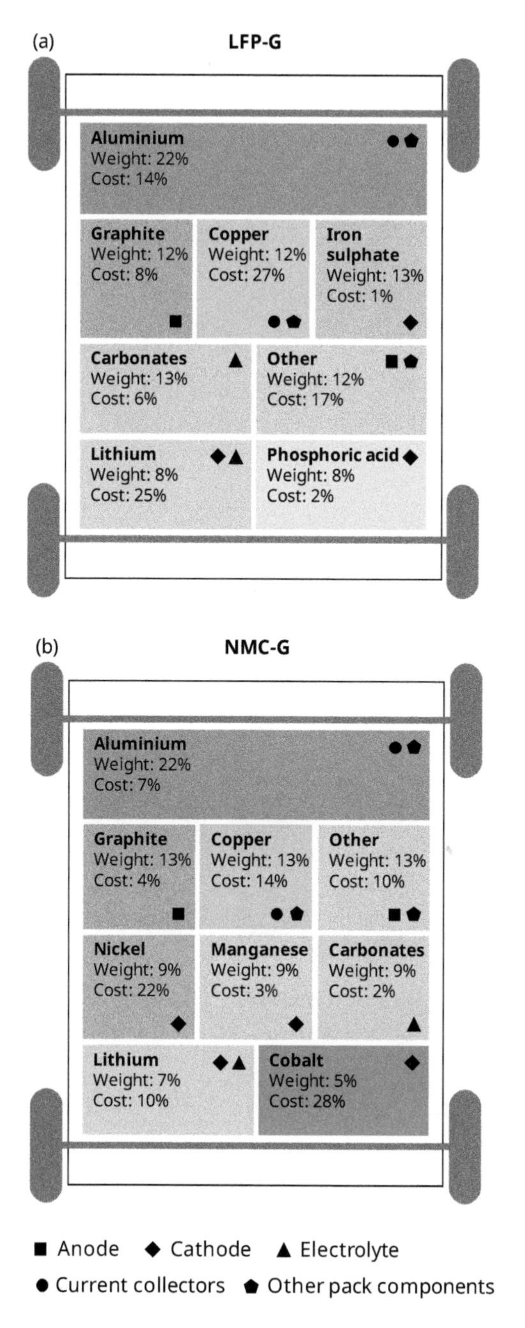

Figure 4.8 Weight and cost contribution of key raw materials in lithium-ion batteries with (a) LFP cathode and graphite anode and (b) NMC cathode and graphite anode. NMC111 is assumed with equal shares of nickel, manganese and cobalt. Symbols indicate usage of raw materials in battery cell and pack components. Raw material analysis is based on ANL's BatPac model v3.0 and the Bloomberg terminal.[15,17]

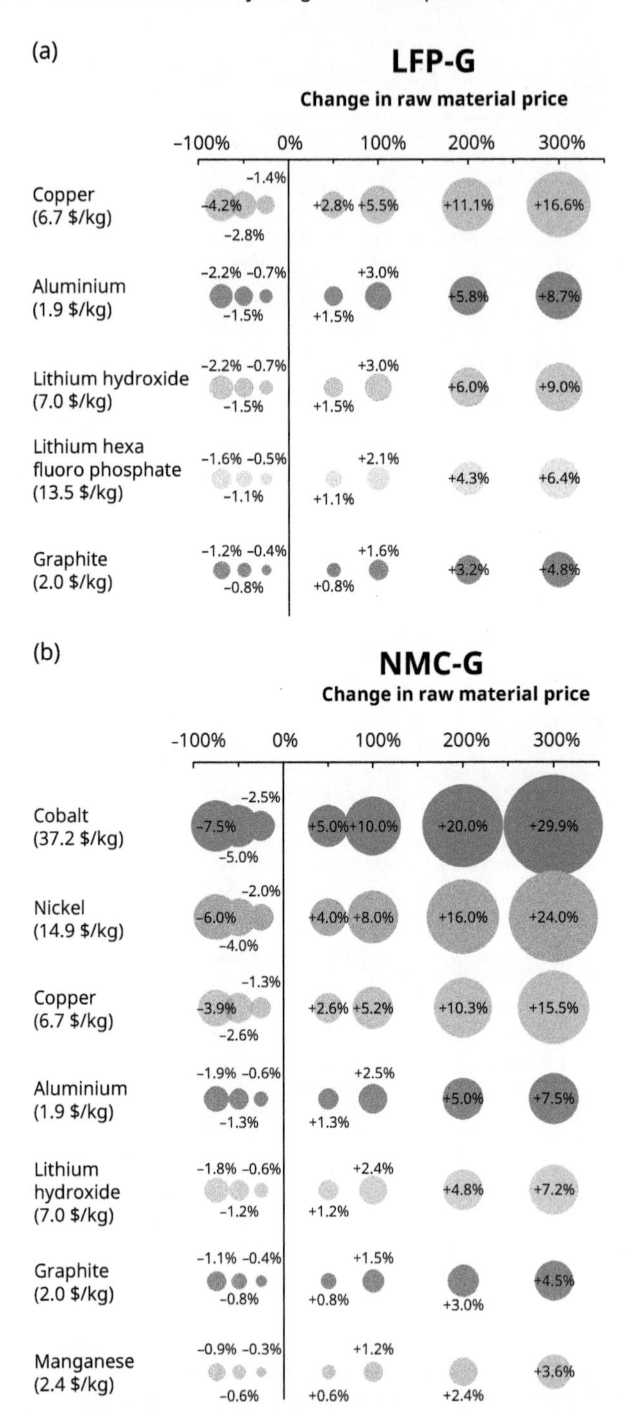

Figure 4.9 The impact of raw material cost changes on lithium-ion battery pack price for (a) LFP cathode and graphite anode and (b) NMC cathode and graphite anode. NMC111 is assumed with equal shares of nickel, manganese, and cobalt. A battery pack price of 130 USD/kWh is assumed. Values in brackets show baseline raw material cost assumptions based on monthly average prices from 2010 to 2020.

KEY INSIGHT

Raw material availability is unlikely to become a limiting factor in electricity storage deployment for any of the major technologies. It is also unlikely to affect the product price projections based on experience curves in the long-term.

Table 4.2 provides a more detailed analysis for lithium ion and vanadium redox flow. It translates the raw material availability for the active materials in both technologies to the production potential of these batteries if all currently known reserves are used. This cursory analysis reveals that raw material availability is sufficient to deploy more than 1 billion electric vehicles (global car stock in 2020 was ~1.3 billion)[19] or at least 19× the stationary battery storage capacity projected for 2030.[20]

On the one side, a clear limitation of this analysis is that it assumes full consumption of all resources for the production of batteries. On the other side, resources may increase and the material intensity in battery manufacturing may decrease in future. Alternative NMC811 lithium-ion batteries (80% nickel, 10% lithium and cobalt respectively) only use a quarter of the cobalt indicated Table 4.2 and LFP lithium-ion batteries use no cobalt at all. Thus, raw material availability is unlikely to pose a significant barrier to the deployment of lithium-ion and vanadium redox-flow batteries in the foreseeable future.

Table 4.2 Raw material availability and resulting production potential for lithium-ion NMC111 batteries with equal shares of nickel, manganese, and cobalt, and vanadium redox-flow batteries.[21] Resources refer to the amount of a geological commodity that exists in both discovered and undiscovered deposits. Reserves refer to the amount of a geological commodity that has been discovered, has a known size, and can be extracted at a profit. Average capacity of lithium-ion battery pack in electric vehicle is assumed to be 50 kWh. Projection for 2030 stationary electrical energy storage capacity in batteries is 1 TWh.[20] n/a given where a technology is not viable for the use case.

Raw material availability	Unit	Lithium	Cobalt	Nickel	Vanadium
World annual production	Mt	82	140	2500	86
World reserves	Mt	21,000	7,100	94,000	22,000
World resources	Mt	86,000	25,000	300,000	63,000
Production potential (based on resources)	**Unit**	**Lithium**	**Cobalt**	**Nickel**	**Vanadium**
Material intensity in battery	kg/kWh	0.139	0.394	0.392	3.4
Potential electrical energy storage capacity	TWh	619	63	765	19
Number of electric vehicles	bn	12.4	1.3	15.3	n/a
Multiples of stationary capacity projected for 2030	#	619	63	765	19

FREQUENTLY ASKED QUESTION

How do experience curves account for raw material constraints? If lithium becomes scarce further cost reductions appear unlikely, don't they?
Experience curves do not directly account for raw material constraints. What can be done is to consider raw material cost as potential 'cost floors' (see Figure 4.7). These cost floors could increase if raw materials become scarce or decrease if extraction costs reduce.

However, one needs to consider how likely prolonged raw material constraints are and whether there are alternatives once a certain material is too expensive or not available anymore.

For lithium-ion batteries, for example, concerns usually refer to the demand and resulting price development for nickel, cobalt, and lithium, the key materials used in the cathode. However, only a minor percentage of lithium-ion batteries is made of these materials, for example only ~5% (in terms of weight) is made up of lithium. The global reserve base for these materials (i.e. identified resources that have been discovered, have a known size and can be extracted at a profit) are sufficient to produce at least 1 billion electric vehicles (see Table 4.2). This base continuously increases as new resources are discovered and/or extraction costs fall. Thus, resource constraints appear unlikely in the foreseeable future. Recent price spikes for individual materials are rather the result of a mismatch in demand and supply due to the time lag between increased demand and new supplies (i.e. new mines coming online).

It is also important to understand that lithium-ion batteries are a technology group. The cathode can be based on a range of raw materials (nickel, aluminium, cobalt, manganese, iron phosphate, etc.). Price spikes in one material can result in battery manufacturers shifting towards alternative materials. When cobalt prices spiked in 2018, manufacturers shifted to nickel-rich cathodes. The risk of high nickel prices strongly impacting nickel-rich cathodes in turn contributed to the recent increase of the production of iron-phosphate based cathodes.[22] This flexibility reduces the risk of running into resource constraints for lithium-ion batteries. In short, high prices cure high prices.

4.5 Time frame for cost reductions

Future cost reductions will materialize in line with experience curves as new capacity is deployed. To map future cost reductions to time, the market diffusion process of new technologies must be modelled, which adds another layer of uncertainty though. A common modelling approach is the archetypal sigmoid function (known as the S-curve) which has been observed for the deployment of novel technologies.[23] Figure 4.10 shows annual market deployment and cumulative deployed capacity for lithium-ion battery storage used in utility-scale systems. Chapter 8 contains these market forecasts for all applications, including the parameters to model each respective sigmoid function.

Combining the projected capacity in a specific year with the product price at that capacity (as projected by the experience curve) returns a projected product price in that specific year. This approach is displayed in Figure 4.11.

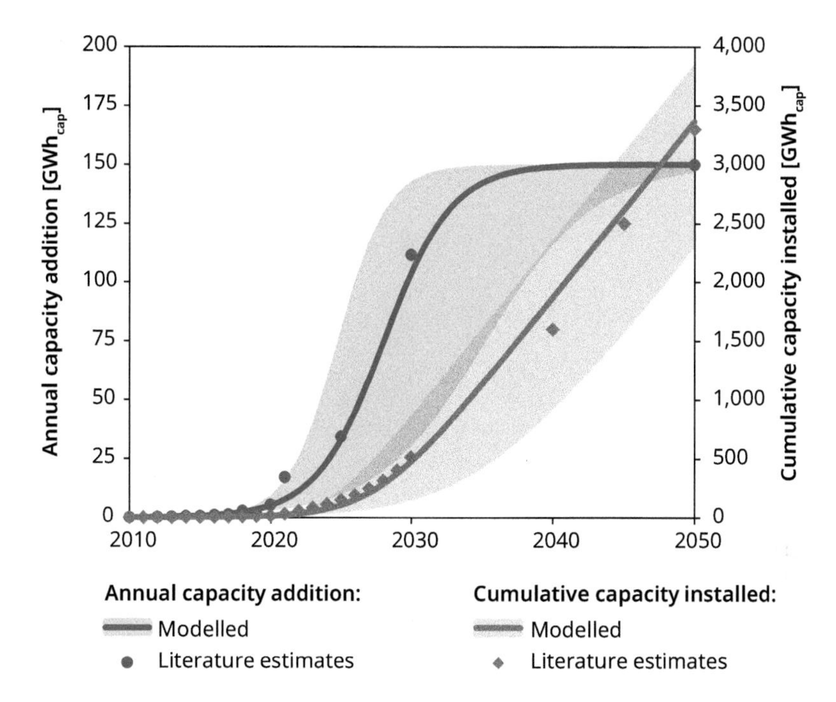

Figure 4.10 Market growth projection for stationary utility-scale lithium-ion systems.[24,25]

For the ~20 GWh of utility-scale lithium-ion systems that were deployed by 2020 (black circle) a product price of ~350 USD/kWh is projected, which is roughly in line with the 310 USD/kWh actual cost determined for a large-scale project in 2020.[26]

For the ~350 GWh that could be deployed in the high growth scenario by 2026 (blue circle), a price of ~100 USD/kWh can be projected using the high experience curve (i.e. the best case scenario for both growth and innovation).

It is found that 1 TWh of cumulative capacity could be installed for most new technology types within 5 to 20 years based on current market growth expectations (central growth scenario). Figure 4.12 shows that, as a result, by 2030 stationary systems cost between 120 USD/kWh (utility-scale lithium ion) and 700 USD/kWh (residential lithium ion). When accounting for experience rate uncertainty, the price range expands to 80–730 USD/kWh (shaded area in Figure 4.12).

Lithium-ion EV pack price reduces to 60 USD/kWh by 2030 if the high experience rate of 24% is combined with the high rate of doubling of deployed capacity if 25 million EVs are sold annually by 2030.[27] This equates to more than 1 TWh annual capacity additions, compared to 30 GWh for residential storage, explaining the wide difference in product prices by 2030 from a theoretical perspective. However, in practice, residential lithium-ion systems are likely to benefit more strongly in the cost reduction of EV packs due to spill-over effects. This effect appears to not be fully reflected in the experience curve of residential systems yet and may increase their experience rate in the future.

Lithium-ion battery cells for consumer electronics are projected to cost 140 USD/kWh by 2030. This higher cost than EV packs (which also include cells) is possibly due to the higher difficulty of standardizing across the consumer electronics industry. In contrast, cells for

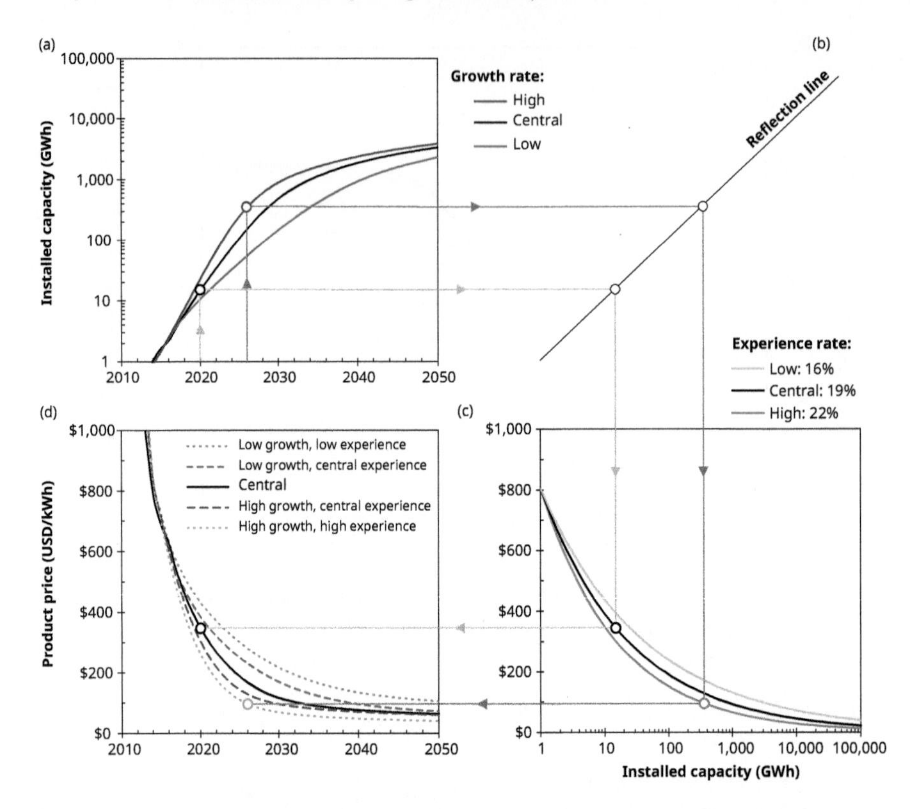

Figure 4.11 The step-by-step derivation of cost reductions over time. The convolution of (a) projection of cumulative capacity over time with (c) product price projection as a function of cumulative capacity yields (d) product price projection over time. Panel (a) is the sigmoid growth curve with logarithmic y-axis. Panel (c) is the experience curve, including uncertainty, with linear y-axis.

EV packs and stationary systems will benefit from better design standardization and larger economies of scale.

KEY INSIGHT

Combining experience curves with market growth projections allows future investment costs to be projected as a function of time. This suggests that by 2030, lithium-ion packs for EVs could fall to 60 USD/kWh and large-scale 4-hour utility-scale systems to 100 USD/kWh.

Price projections for lithium-ion battery packs reduce to below 60 USD/kWh after 2030. This is within the raw material cost range identified in Figure 4.7. So, these projections rely upon reductions in raw material cost (e.g. increased mining capacity and improved processes), improvements in energy density (e.g. silicon anode, solid electrolyte), or changes in raw material composition of lithium-ion batteries (e.g. no nickel or cobalt). The latter two developments are focus topics for the industry, which is reflected in lithium-ion innovation roadmaps.[28,29]

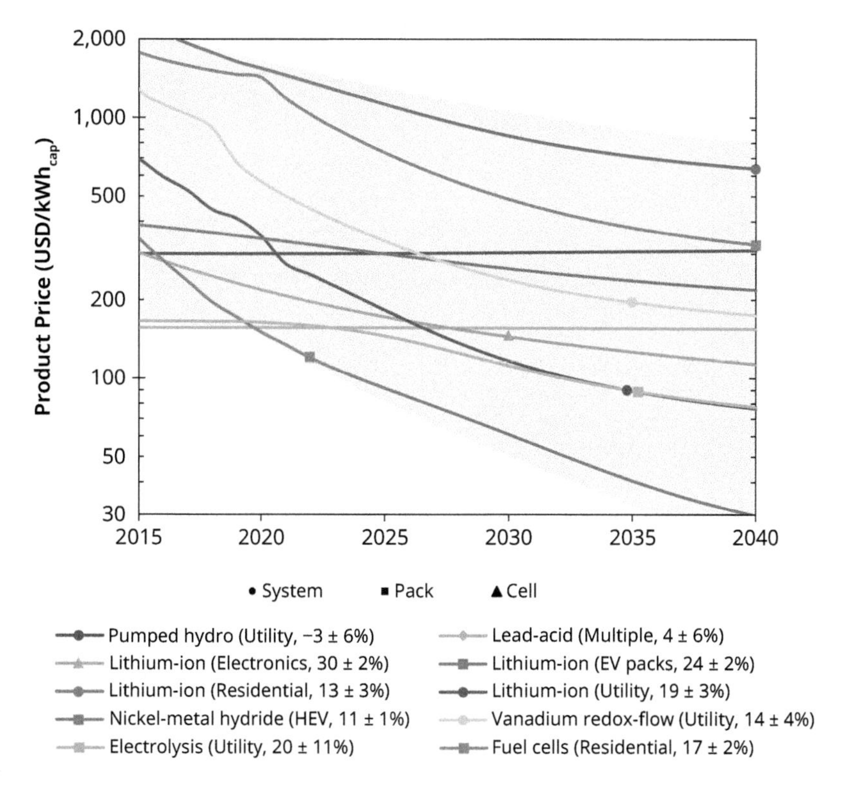

Figure 4.12 The future cost of storage technologies over time. Cost projections are based on experience rates and S-curve market growth assumptions for consumer electronics, hybrid electric vehicles, electric vehicles, residential and utility-scale storage. Market growth in different applications is mutually exclusive, but technology penetration is not (i.e. 100% market share assumed for each technology). Symbols indicate when 1 TWh cumulative installed capacity could be achieved for each technology under this condition. No symbol means 1 TWh cumulative capacity is not achieved within the given time frame (pumped hydro: 2000, lead-acid—modules: pre-1989, NiMH: 2046). Residential lead-acid systems are excluded from the analysis due to limited competitiveness with lithium ion (see section 4.2.) The shaded area marks the impact of experience rate uncertainty. The legend indicates product scope (installed system, pack, cell) and technology (including application and experience rate with uncertainty). Fuel cells and electrolysis must be considered in combination to form a hydrogen-based storage technology.

4.6 Cumulative investment needs

By linking product prices to cumulative capacity, experience curves offer the opportunity to quantify the cumulative investment required to deploy energy storage solutions. This is of great interest to academics, policy-makers, and industry.[9,10,30–32]

Global investment in renewable power generation and electricity networks was around USD 300 billion (bn) for 2018–2020.[33,34] Investment in electricity storage technologies was

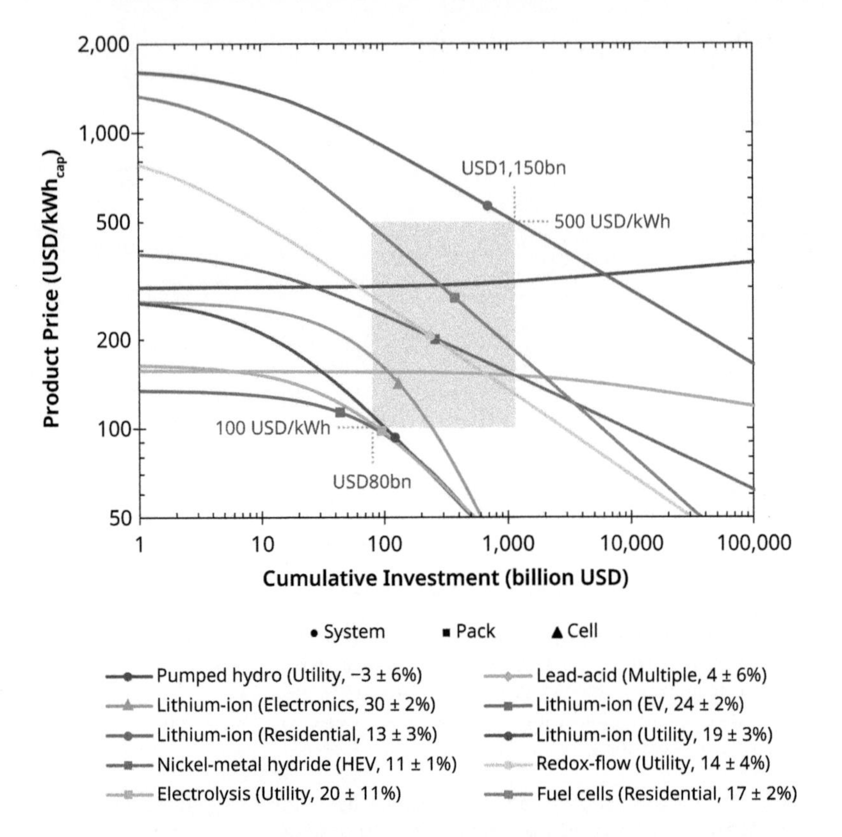

Figure 4.13 The impact of cumulative investment in storage deployment on the future cost of storage. Graph shows the investment in storage deployment required to 'pull' technologies along individual experience curves. This investment could be consumer capital, industry capital, government subsidy, or a mix of all. The shaded rectangle indicates the investment required to reach prices of 100–500 USD/kWh. Symbols mark the amount of investment required to deploy 1 TWh cumulative capacity for each technology. No symbol means 1 TWh cumulative capacity is already deployed (pumped hydro, lead-acid modules). Residential lead-acid systems are excluded from analysis due to limited competitiveness with lithium ion (see section 4.2.). The legend indicates product scope (installed system, pack, cell) and technology (including application and experience rate with uncertainty). Fuel cell and electrolysis must be considered in combination to form a form a hydrogen-based storage technology (together these require USD 240bn to reach < 500 USD/kWh).

around USD 3–4 bn, around 1% of these more established technology groups.[35] Figure 4.13 shows that investments worth USD 80 bn (lithium ion, EV pack) to USD 1,150 bn (lithium ion, residential system) would be required for the deployment of each storage technology to reach the price range of 100–500 USD/kWh that was identified for each technology at ~1 TWh cumulative installed capacity. For context, this means between 1% and 35% of the investments made into electricity networks and renewables need to be spent on each storage technology to reach this price range by 2030. Accounting for experience rate uncertainty, the investment range could be USD 70–400 bn with high experience rates or USD 100–6,800 bn in the low experience rate case.

For example, if end-users would be willing to pay 500 USD/kWh for a residential lithium-ion system, then USD 650 bn of the USD 1,150 bn total investment would be required to subsidize deployment until this price level is reached.

Such insights can inform policy-makers and industry on appropriate deployment policies and investment requirements. As the largest country-specific investments into the renewable energy industry range from USD 5 bn (Germany) to USD 80 bn (China) in 2019,[36] global cumulative investment of USD 30 bn to USD 100 bn for storage technologies by 2030 does appear conceivable.

4.7 Discussion

This analysis comes with three key implications: insights on future competitiveness of energy storage technologies, related uncertainties, and a basis for comparison to other cost projection methods.

4.7.1 Key implications

First, the common cost trajectory identified for storage technologies enables practitioners to assess proposed technologies against existing ones, with cost trajectories lying above or below signalling that the technology may remain uncompetitive or become disruptive, respectively. But such conclusions are limited to investment cost, and a complete assessment of competitiveness must include additional factors (such as lifetime and efficiency) that affect application-specific lifetime cost (see Chapter 5).[37]

Second, the future projections made for storage technology prices enable simple assessment of price targets and investment requirements established for storage to become competitive. For example, a study identified that electricity storage coupled to solar and wind power could meet baseload power demand for 95% of the time most cost effectively if storage investment cost was below 150 USD/kWh.[38] The present analysis indicates that this price threshold could be achieved once an additional 185 GWh of utility-scale lithium-ion systems have been deployed (central experience rate), which according to market growth assumptions could be achieved by 2027 if all future utility-scale deployments were lithium-ion systems. This would correspond to USD 35 bn invested in deployment. Such quantification enables informed discussions about the scale of, and split between, private, and public sector investments.[31] However, such analysis is incomplete without considering alternatives to storage, such as network expansion, demand-side management, and flexible low-carbon generation.

The third implication of this analysis is that the provision of a comprehensive experience curve dataset can remove a significant barrier to analysing the future competitiveness of storage in distinct applications. Figure 4.14 shows two stylized examples for EV transportation and residential storage, which are deliberately simplified to showcase the potential insights that can be gained from such data. The myriad of applications, technologies and location-specific contexts that are absent from this cursory analysis can now be more readily explored in future studies.

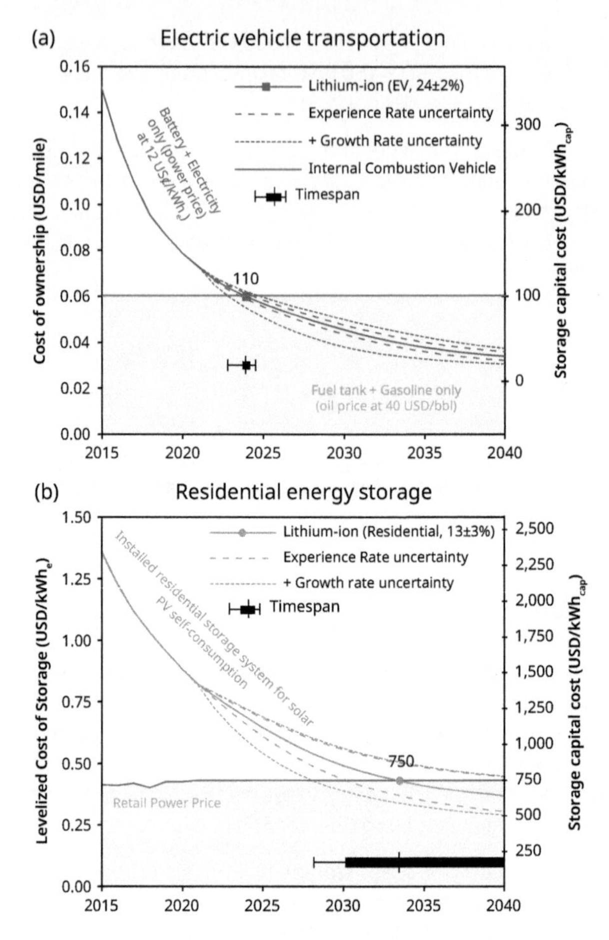

Figure 4.14 The application-specific lifetime cost of lithium ion in two stylized examples. (a): Cost of ownership for personal transport in the United States, comparing lithium-ion battery pack plus cost of electricity (blue) with a fuel tank plus cost of gasoline at an oil price of USD 40 per barrel (red). (b): Levelized cost of storage (LCOS) in Germany for solar PV coupled residential lithium-ion system (blue) compared to retail power price (red). Retail price is fixed at 2020 levels. Dashed and dotted lines in both panels represent the impact of experience rate uncertainty alone and combined with market growth uncertainty, respectively. Black bars indicate possible timespan for costs to equalize with the conventional technology based on these uncertainties (vertical line: equalization at central experience rate; thick bar: with experience rate uncertainty; thin bar: with experience and growth rate uncertainty). kWh$_e$—unit of electricity, kWh$_{cap}$—nominal energy storage capacity. All relevant parameters for the calculations can be found in Chapter 8.

4.7.2 Storage competitiveness

A prominent study suggested EVs are suitable to replace the majority of vehicles in the US based on daily driving requirements.[39] To assess their economic competitiveness, experience rate analysis can be used to project cost of ownership (in USD per mile travelled) for the energy inputs and storage components of EVs and conventional cars (see Chapter 8 for detailed methods). In this simplified example (see Figure 4.14a), EVs become competitive against internal combustion engine vehicles at a gasoline price of 2.10 USD/gallon, which corresponds to an oil price of 40 USD per barrel (bbl), in 2024 at lithium-ion pack costs of 100 USD/kWh. When taking the average oil price over the 2010s (~50 USD/bbl), EVs would be competitive at lithium-ion pack costs of ~120 USD/kWh. In 2021 pack prices stood at 132 USD/kWh on average.[14] The combined uncertainty in experience rate and growth rates could alter the date at which EVs become competitive at 40 USD/bbl oil price by up to 3 years (2022 to 2024). The required cumulative production is 1,300 GWh of battery packs or 24 million EVs (at 55 kWh per pack).[40] Note that this is a simplified example, neglecting any differences in vehicle performance or tax impacts on gasoline or electricity cost. However, the impact of experience rate uncertainty would carry through into more detailed analyses.

Integrated solar photovoltaic (PV) and storage systems are considered an effective means to increase self-consumption by residential producers in light of decreasing feed-in tariffs.[41] In the second stylized example (see Figure 4.14b) the levelized cost of storage (LCOS) for such a system in Germany is modelled in comparison to the retail power price. There appears to be much greater uncertainty regarding the future competitiveness than in the previous example. Already the spread between central experience rate projection and high experience combined with high growth rate projection is 6 years (2027 to 2033), translating into required cumulative capacities of 130 or 240 GWh respectively. Regardless of simplifications, this highlights the emerging state of the residential storage market. The rate at which costs reduce through the early phase will be a significant determinant of whether solar plus lithium-ion batteries will become competitive before 2040 at all.

FREQUENTLY ASKED QUESTION

For EVs to be competitive, lithium-ion cell cost of 100 USD/kWh is suggested as the tipping point. Does such a point exist for stationary systems?

It does, but this is dependent on the storage technology, application, and market:

- **Technology**: A wider range of technologies are viable for stationary storage. These have different performance parameters. While two technologies may have the same investment cost, technology A with a longer lifetime may be competitive already, while technology B is not.
- **Application**: Each application has different requirements and a different incumbent technology solution. At a certain investment cost a storage technology may be competitive in application A, but not yet in application B. Although we see battery storage solutions compete in frequency response and regulation applications already, this does not mean that energy storage is competitive to balance the mismatch in seasonal supply and demand.
- **Market**: Markets differ in terms of electricity price (for charging) and remuneration for specific services. So, while market A may be an island with very high remuneration for a specific application, market B may not offer the same level of remuneration and storage solutions may have a lower investment cost 'tipping point'.

Investment cost 'tipping points' can at best serve as approximation for competitiveness of specific storage technologies in a certain application and market. A direct assessment of competitiveness can be conducted by modelling application-specific lifetime cost (see Chapter 5).

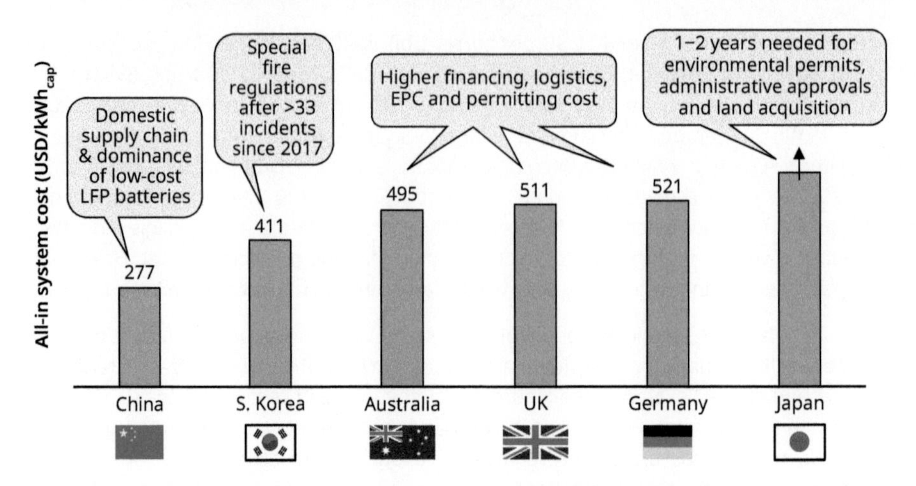

Figure 4.15 All-in cost for stationary lithium-ion battery systems with 2-hour discharge across various countries in 2020.[25] Value for Japan not explicitly determined.

4.7.3 Regional differences

An uncertainty not explicitly covered in the analysis of this book is the variation of all-in electricity storage system prices by geography. Price differences exist mainly because of differences in:

- Dominant technology—LFP vs NMC lithium-ion chemistries
- Project development costs—land acquisition, permits
- Distribution and installation costs—procurement, construction
- Other soft costs—financing, taxes

Figure 4.15 illustrates the variation of all-in lithium-ion system cost across six different markets. They are lowest in China due to the dominance of low-cost LFP technology and lower procurement cost due to access to a domestic supply chain. All other markets see higher all-in system cost, driven by the dominance of higher performance, but also more expensive NMC technology, complex procurement from non-domestic supply chains, higher labour cost driving construction cost, more complex permitting procedures, and higher financing cost. These regional differences in electricity storage system cost must be taken into account when assessing specific projects.

4.7.4 Supply and demand imbalances

For lithium-ion technology, overall raw material cost soared by 300% to 700% in the 12 months from March 2021 to March 2022 (see Figure 4.16). This is outside of the uncertainty ranges given in the section on raw material cost (see section 4.4) and driven by a combination of extraordinary circumstances, mainly:

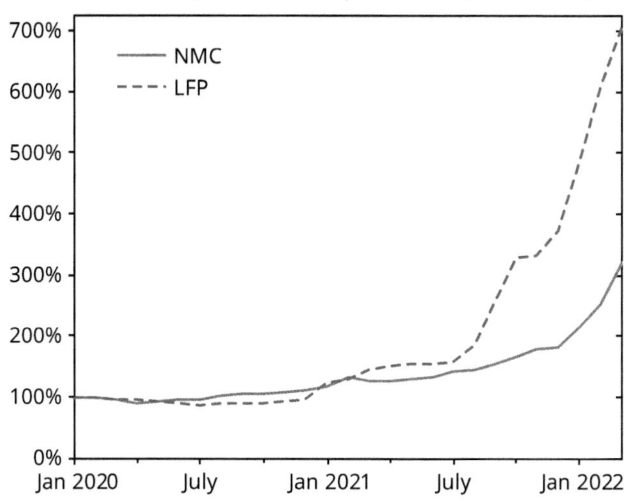

Figure 4.16 Lithium-ion battery raw material price index from January 2020 to March 2022.[43]

- the supply chain disruptions and delayed expansion of mining and raw material processing capacity due to the COVID pandemic that started in 2020
- the firm increase in lithium-ion battery production due to a strong uptake in electric vehicles and stationary lithium-ion storage in 2020 and 2021
- the economic uncertainty caused by the Russian invasion of Ukraine in February 2022
- the foreseeable increase in interest rates by central banks for 2022 to counter high levels of inflation

As a result, the cost of nearly all raw materials used in lithium-ion batteries increased significantly.[42] In particular, the cost of lithium carbonate (used in LFP) increased, which also affected the cost of lithium hydroxide as derivative (used in NMC and LFP). The raw material cost increase of 300% and 700% would increase battery pack prices by 100% for NMC and 150% for LFP respectively.

However, battery manufacturers are unlikely to be directly subject to these elevated raw material costs because raw materials are largely purchased via long-term contracts. These elevated price levels are unlikely to sustain long enough to be reflected in future long-term contracts.

Figure 4.17 illustrates this with the example of high-purity polysilicon used to manufacture solar PV modules. Three important insights can be derived from this chart.

First, the continuous alternation between shortage and oversupply of polysilicon is a great illustration of the cobweb theorem,[16] which also applies for lithium-ion raw materials and production capacities. The impact on polysilicon contract price (see Figure 4.17 a) can be seen in the product price for solar PV modules (see Figure 4.17 b). However, it does not materially distort the overall product price development from the underlying experience curve. It thereby shows that the uncertainties for raw material cost determined in Figure 4.7 are

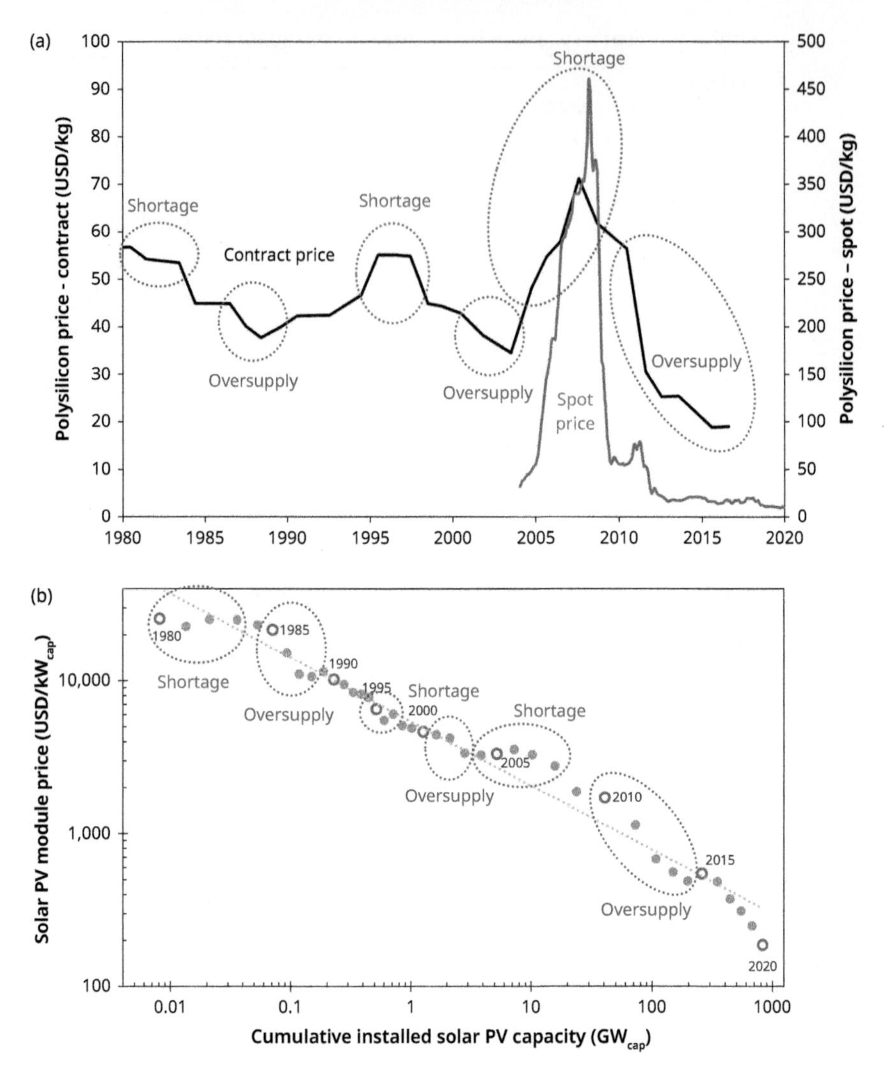

Figure 4.17 The impact of high-purity polysilicon availability on (a) polysilicon contract and spot market prices and (b) product prices of solar PV modules. Dotted circles indicate times of polysilicon shortage or oversupply. In panel (a), the solid black line refers to polysilicon contract prices (left axis) and the grey line refers to polysilicon spot prices (right axis).[44]

reasonable and unlikely to significantly affect the experience curve-based price projections for storage technologies.

Second, there was an extraordinary polysilicon price increase between 2005 and 2008. This was due to the combination of an increase in feed-in-tariffs for electricity from solar PV in Germany, which quadrupled solar PV module demand in Germany in 2004, and a high growth rate for semiconductors in 2004 (23% per year), which also use high purity polysilicon.[44] However, this raw material price increase was time-limited because it triggered innovation (e.g. the amount of polysilicon in PV modules dropped from 13 grams (g) to 10 g per Watt) and additional polysilicon processing capacity. As a result, prices collapsed in 2008 exacerbated by the financial crisis and scrapping of lucrative feed-in-tariffs in Spain. It thereby shows that the extraordinary raw material cost increases seen in 2021/22 are unlikely to sustain for more than 2 to 3 years at maximum.

Lastly, while the spot price for polysilicon increased by 1,500% from ~30 USD/kg (2005) to ~460 USD/kg (2008), contract prices only increased by 100% from ~35 USD/kg to 70 USD/kg. This shows that extraordinary increases in spot prices do not directly feed through to the long-term raw material contracts that battery manufacturers are likely to have with raw material producers.

Overall, these observations show that the experience curve-based price projections for electricity storage technologies are unlikely to be significantly distorted by raw material price fluctuations. As for 2022, raw material prices for lithium-ion storage technologies are expected to increase, which leads to a temporary increase in product prices.[14] However, these prices are likely to return to their reduction trend once raw material costs return to previous levels. This can be the result of additional mining and processing capacities, product innovation, and/or a reduction in the rate of demand increase.

However, this discussion of raw material costs shows that experience curves are ill-suited for short-term projections of product prices as they do not capture commodity price fluctuations and market developments.[9] Their purpose is to uncover underlying cost development trends and make long-term product price projections.

4.7.5 Remaining uncertainties and limitations

Uncertainty affecting all data stems from the use of product prices (i.e. investment cost) instead of production cost. The theory of learning applies to production cost.[45] So, product price-based experience curves can only be used as proxy to mirror the development of production cost. Product prices include additional cost factors (e.g. cost of sales, margin) and any price projections that rely on the theory of learning in production are subject to these additional cost factors remaining unchanged.[9] In practice, however, experience curves are widely used to project future product prices and have proven useful for multiple technologies.[46,47]

An additional uncertainty regarding the use of product prices as a proxy for production cost development relates to early-stage markets, in which these quantities might deviate more strongly.[9,12] The datasets for residential lithium-ion, lead-acid, utility-scale vanadium flow systems, and electrolysis do not cover the recommended two orders of magnitude in data length (i.e. cumulative installed capacity).

KEY INSIGHT

Experience curves are based on historical trends and are therefore ill-suited to capture potential step-change effects like radical innovations.

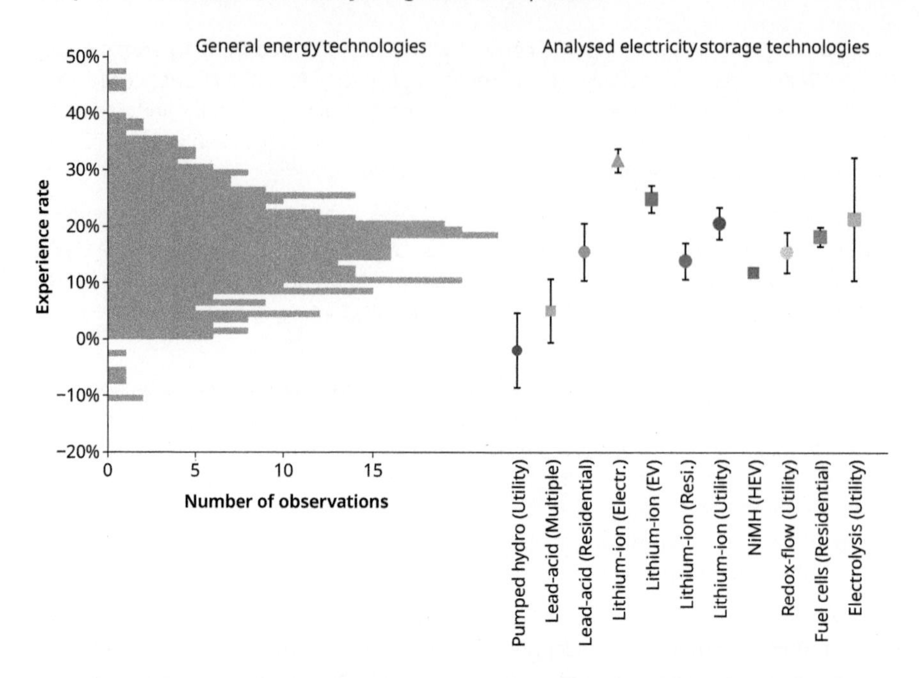

Figure 4.18 Comparison of experience rates for electricity storage technologies with experience rates observed for other energy technologies in the last 35 years.[48–51] Grey bars refer to number of observations of distinct experience rates for energy technologies. Coloured points show experience rates for electricity storage technologies. Error bars reflect the identified uncertainty ranges. NiMH - Nickel-metal hydride.

Multiple studies identified experience rates for energy technologies in the last 35 years.[48–51] These range from –11% (nuclear, CCGT) up to 47% (solar PV) with nearly two-thirds of observations between 10% and 25%. The experience rates identified in this study for electricity storage technologies, including their uncertainty, are well within these extrema (see Figure 4.18). Mirroring the distribution of experience rate observations for energy technologies, 8 of the 11 rates (i.e. around two-thirds) are also within 10% and 25%.

Another limitation regarding the identified time frames of cost reductions and cumulative investment needs is the assumption that technologies will capture 100% market share. In practice, it is likely that a portfolio of technologies will be deployed. The composition of that portfolio depends on which applications will mainly be served by electricity storage (see Chapter 5). The determined timelines for cost reductions and required investment needs therefore represent optimistic scenarios for each technology. On the other hand, the assumption of 100% market share is not completely irrational. There is a strong element of endogeneity in electricity storage deployment, in other words, which technology becomes cheapest is influenced by how much of each gets built, and that is influenced by which becomes cheapest. Technologies that see high deployment will be subject to these reinforcement loops and capture a disproportionately large market share. Lithium ion could be a case in point. Stationary systems benefit from the high deployment of EV battery packs, which facilitates initial deployment in stationary applications. This drives cost reductions for stationary systems and makes additional deployment more cost efficient. In practice, this could mean technologies that would have been cheaper than lithium ion, had they seen similar deployment, remain more expensive.

Finally, experience or learning rates are ideally derived for specific components of industry goods.[9,52] By reporting prices on pack- or system-level, there is uncertainty due to aggregation of potentially different experience rates of individual components (e.g. cells, power electronics) and due to inclusion of non-manufacturing related cost (e.g. installation, commissioning), which might follow cost dynamics not captured with the experience rates derived in this book.

4.7.6 Comparison to other cost projection methods

Due to the uncertainties and limitations that come with experience curve analysis, it can be useful to compare the price projections to those from alternative cost projection methods.

Expert elicitations use structured discussions to elicit scientific and technical judgments in the form of subjective probability distributions around uncertain variables from experts in a particular field.[53] They are a valuable tool to support investment and policy decision-making in conditions of uncertainty and limited data availability.[53,54]

In a study conducted in 2016/17, 10 experts were asked to give cost estimates for lithium-ion battery packs in 2020 and 2030.[55] Figure 4.19 compares these estimates to an experience curve-based projection as well as the actual price development. It shows the:

- Reference price range from 2005–2014 that was available as briefing material[56]
- Experience rate price projection up to 2030 based on historic price data up to 2015
- Expert estimates for 2020 and 2030 including uncertainty
- Actual average prices up to 2022 based on an industry survey[14]

Expert estimates are in line with the experience rate projection. While the 2020 estimate is slightly below, the 2030 estimate is slightly above the central projection. Studies on water electrolysis and wind turbines have shown that stakeholder expectations and expert elicitations yield cost estimates that are lower than would be indicated by historical trends and experience curves.[57,58] This is confirmed by the 2020 estimate and could show that experts tend to give overly optimisitic projections due to the limited account of historical trends and possible relation to cumulative produced capacity. The observation that 2030 estimates are higher than the experience curve confirms the hypothesis that experience curves are more sensitive to future technology deployment than experts' estimates. Deployments are a key driver for the cost reduction of lithium-ion battery packs and their impact appears to be underestimated by experts in the long-term.

However, when considering the actual price development, it becomes obvious that both methods perform badly at projecting future prices. The median experts' estimate for 2030 was realized already in 2018. The same is true for the 16% experience rate projection.

The main drivers for the sharp decline in pack prices between 2015 and 2018 were economies of scale (e.g. increase in average factory production capacity to above 5 GWh/year), chemistry changes for higher energy density (e.g. nickel-based chemistries, higher nickel

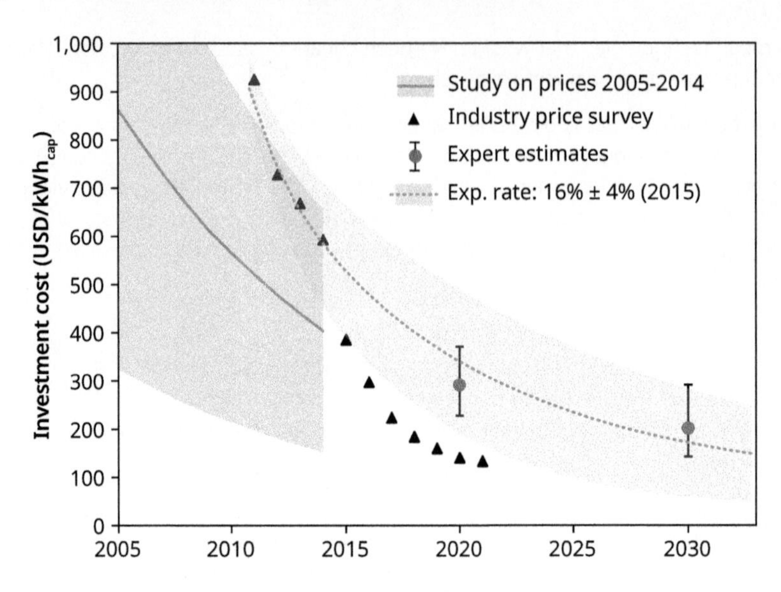

Figure 4.19 Expert estimates compared with an experience rate projection for lithium-ion battery pack investment cost. Median expert estimates (grey circles) for 2020 and 2030 investment cost are compared with projections based on an experience rate (blue trajectory). Error bars represent the median 90th and 10th percentile of expert estimates. The blue shaded area accounts for both experience rate and market size uncertainty. The experience rate is based on price survey data from 2010 to 2015 to reflect the information status when expert interviews were conducted. The grey shaded area shows the range for lithium-ion pack prices of market leading manufacturers from 2005 to 2014.[56] Actual average prices from an industry price survey are displayed for reference (black triangles).

content) and engineering improvements to cells and packs (e.g. cell size, pack design).[59,60] From expert responses to underlying innovations for future estimates,[55] it appears that experts highlighted improvements in cell chemistry but underestimated the impact of economies of scale and engineering improvements. It could therefore be argued that experts may be subject to a conservatism bias for technologies that are about to be widely deployed, because they underestimate cost efficiency potentials when moving from small-scale to mass manufacturing. This could be related to the higher complexity of this change compared to a specific scientific or technical advance. Similar observations have been made for the projection of solar PV penetration in total electricity generation, which experts systematically underestimated for more than 20 years.[61]

The poor performance of experience curves is likely to be related to the few historical data points that were available in 2015 spanning only 1.5 orders of magnitude in cumulative deployment data (1 GWh to 30 GWh). This is a common risk associated with experience curves for novel technologies, highlighting the need to continuously update respective datasets with latest data points. The majority of experience curves derived in this chapter span at least two orders of magnitude in cumulative deployment data, a requirement that has been identified for experience curves to give reliable future cost estimates.[9]

4.8 Worked example

This chapter has given you the dataset and methods for understanding the future development of electricity storage investment cost. Simple demonstrations were made of how much investment will be required to achieve a target price, and when storage could become competitive against incumbent applications.

Such analyses are more useful if they can be customized to specific situations or updated as new data arrive. This section explains how this book's companion website can be used for such purposes.

A question:
The SunShot initiative by the US Department of Energy claims that storage system investment cost of 100 USD/kWh are required to make solar PV plants coupled with storage competitive for baseload power supply.[62] How does this target align with reality, and what is needed for it to be hit by 2030?

How to answer:
This worked example walks you through how to use experience curves with real-world data to qualify this target. We will use the free online tool <www.EnergyStorage.ninja> to guide you through this example.

1. Open <www.EnergyStorage.ninja> and go to the 'Investment cost' tab
2. You will see three charts, as below.

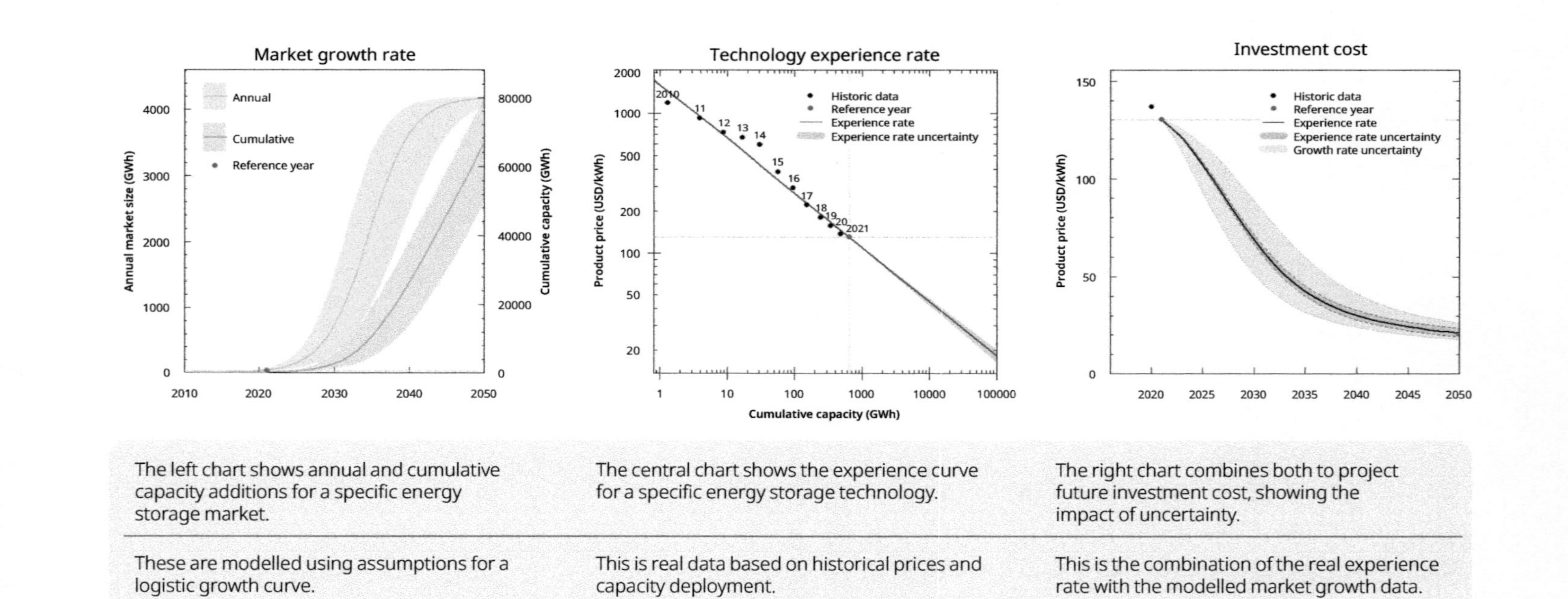

The left chart shows annual and cumulative capacity additions for a specific energy storage market.	The central chart shows the experience curve for a specific energy storage technology.	The right chart combines both to project future investment cost, showing the impact of uncertainty.
These are modelled using assumptions for a logistic growth curve.	This is real data based on historical prices and capacity deployment.	This is the combination of the real experience rate with the modelled market growth data.

Experience curves can project future investment cost based on future capacity additions (centre). Hence, the combination with market growth assumptions on future capacity additions as a function of time (left) allows investment cost projections as a function of time (right).

However, there are two important notes to make

- The default market growth assumptions represent capacity additions for entire storage markets, whereas the experience curves show cost reductions as a function of capacity deployment for distinct storage technologies.
- Historical deployment for electricity storage technologies may differ from the modelled market growth in those early years.

3. Let's first choose the market and technology that are of interest here. Go to the drop-down menu of the market growth rate and technology experience rate sections and choose 'Utility-scale storage' and 'Lithium-ion systems (utility)' and click '2020 values'.

   The charts now show the projected annual and cumulative capacity additions for utility-scale energy storage, the experience curve for utility-scale lithium-ion systems, and the resulting cost reduction trajectory.
4. Before analysing the data, we now have to ensure that modelled deployment data do not reflect the utility-scale storage market as a whole, but utility-scale lithium-ion systems only. This could be done by reducing the growth rate to 0.35 (35% per year), for example.

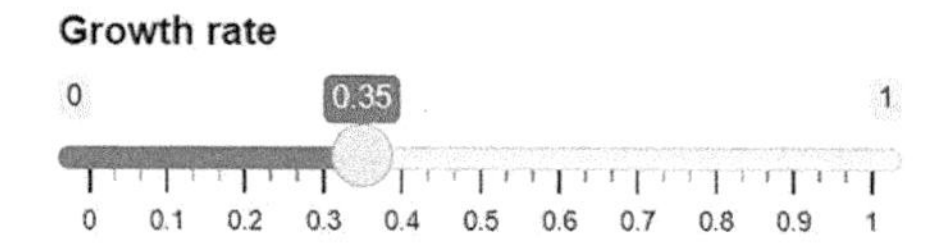

5. The investment cost chart now shows that utility-scale lithium-ion battery systems will reduce from just below 300 USD/kWh in 2021 to 100–200 USD/kWh by 2030, including experience rate and growth rate uncertainty (green area).

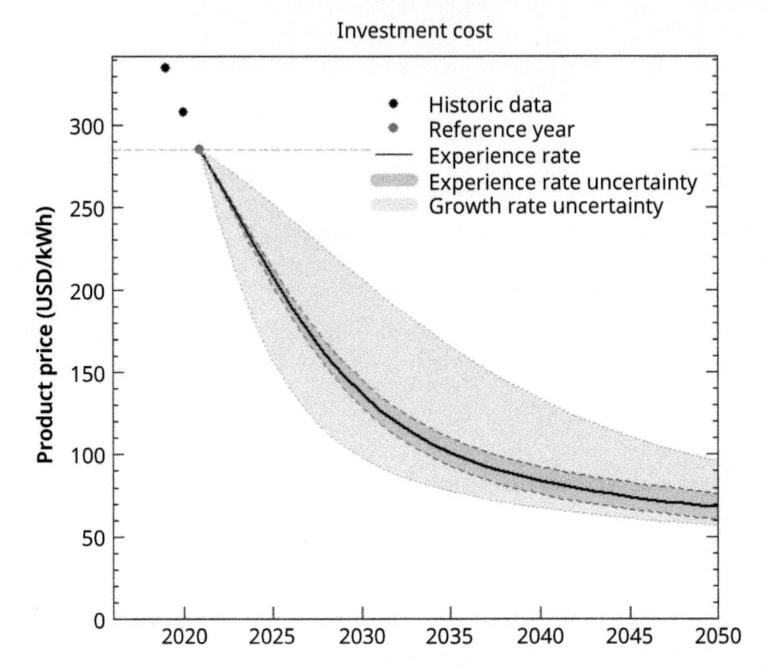

Now, let's use the market growth and experience curve levers to explore under which conditions utility-scale lithium-ion battery systems could reach ~100 USD/kWh in the central scenario by 2030.

6. The first lever is the experience rate of the technology. If the technology reduces in cost faster with additional capacity deployment, respective cost reductions will be achieved faster at a given future deployment trajectory. So, let's increase the experience rate by seven percentage points to 26.5%.

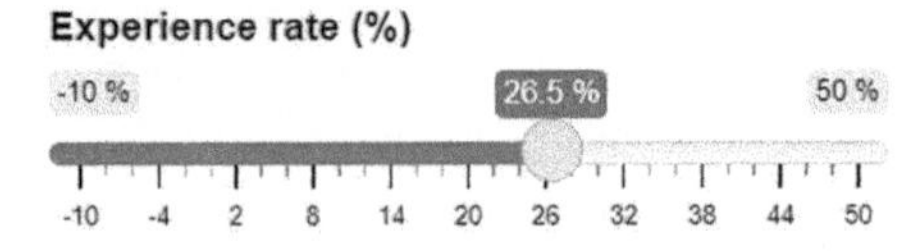

The experience curve becomes much steeper and, as a result, the annual cost reduction curve bends downwards (in the right chart) reaching ~100 USD/kWh by 2030.

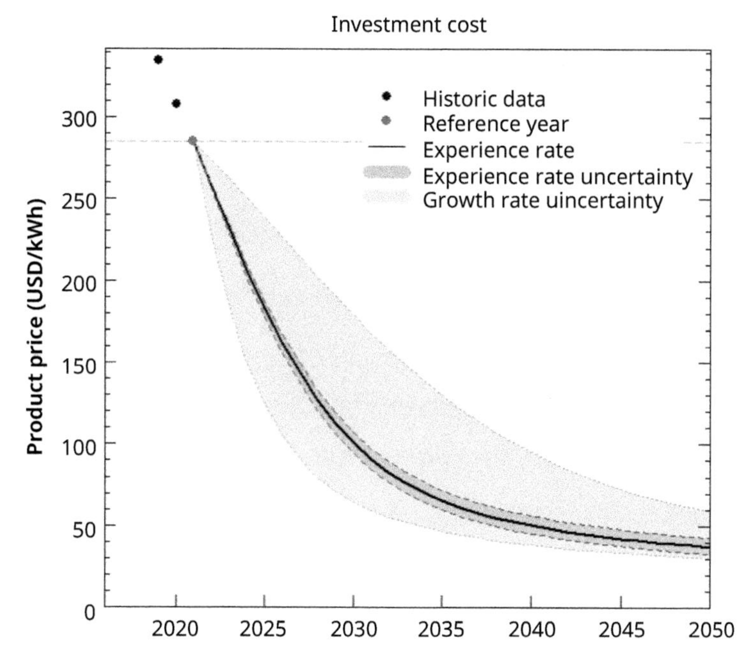

While the experience curve is steeper than cost reductions since 2010 suggest, it seems in line with the cost reductions from 2016–2019. The experience rate for lithium-ion batteries in consumer electronics has been 30% based on data between the 1990s and today. Hence, there is a possibility that an experience rate of 26.5% for utility-scale lithium-ion systems is feasible.

It is important to note that this cost reduction is based on the assumption of deploying ~100 GWh utility-scale lithium-ion systems per year by 2030.

7. Let's revert to our current assumption of the utility-scale lithium-ion experience rate (click '2020 values' next to the 'Technology' drop-down menu) and look at the market growth assumptions. Now, increase the growth rate to 0.6 (60% per year).

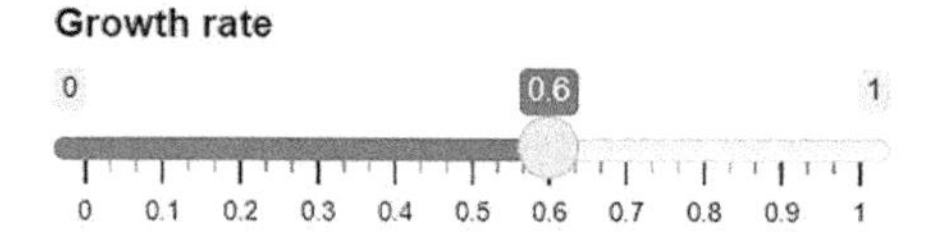

The market growth chart now indicates that by 2030 market saturation is reached already with nearly 150 GWh utility-scale lithium-ion battery system installations each year.

This significant increase in capacity additions also leads to a cost reduction to ~100 USD/kWh by 2030 as the experience rate based cost reductions materialize faster.

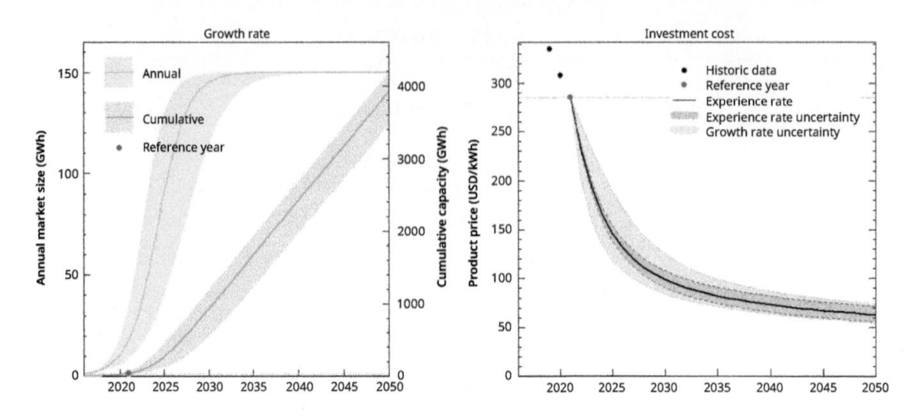

You've done it!

You have successfully explored the impact of growth rate and experience rate on the temporal cost reduction of a prominent storage technology. Now, feel free to:

- play with growth and experience rate uncertainty to explore the impact the possible cost reduction range
- change product price if you feel in your chosen reference year; for example to match the cost reductions of a particular manufacturer.
- play with start year, start capacity, saturation capacity, and growth rate if you would like to assume a specified growth trajectory for utility-scale lithium ion
- change market or technology overall to assess different markets and/or technologies
- Choose 'New technology' under Growth rate and Experience rate to get rid of historical data points and project future investment cost fully independently.

4.9 References

1. World Energy Council. *World Energy Issues Monitor* (London: World Energy Council, 2019).
2. Schmidt O, Hawkes A, Gambhir A, and Staffell I. 'The Future Cost of Electrical Energy Storage Based on Experience Curves' (2017) 2 *Nature Energy* 17110.
3. Gordon JL. 'Recent Developments in Hydropower' (1983) 35 *Water Power & Dam Construction* 21–3.
4. Matteson S and Williams E. 'Residual Learning Rates in Lead-Acid Batteries: Effects on Emerging Technologies' (2015) 85 *Energy Policy* 71–9.
5. Junginger M, Faaij A, and Turkenburg WC. 'Global Experience Curves for Wind Farms' (2005) 33 *Energy Policy* 133–50.
6. Pillot C. 'The Rechargeable Battery Market and Main Trends 2014–2025'. *32nd International Battery Seminar & Exhibit* (Fort Lauderdale: 2015), 1–69.
7. IRENA. *Electricity Storage and Renewables: Costs and Markets to 2030* (Abu Dhabi: International Renewable Energy Agency, 2017).

8. International Energy Agency. *Projected Costs of Generating Electricity* (Paris: IEA, 2015).
9. Junginger M, van Sark W, and Faaij A. *Technological in the Energy Sector: Lessons for Policy, Industry and Science* (Cheltenham: Edward Elgar Publishing, 2010).
10. Gross R, Heptonstall P, Greenacre P, Candelise C, *et al. Presenting the Future—An Assessment of Future Costs Estimation Methodologies in the Electricity Generation Sector* (London: UK Energy Research Centre, 2013).
11. Wang B. Lithium Ion Batteries Scaling Up and Costs Could Drop to $100 per kWh (2015). Available at: <https://www.nextbigfuture.com/2015/08/lithium-ion-batteries-scaling-up-and.html> accessed 5 November 2022.
12. The Boston Consulting Group. *Perspectives on Experience* (Boston: Boston Consulting Group, 1968).
13. Chappell B. Tesla CEO Elon Musk Unveils Home Battery; Is $3,000 Cheap Enough? Available at: <https://www.npr.org/sections/thetwo-way/2015/05/01/403529202/tesla-ceo-elon-musk-unveils-home-battery-is-3-000-cheap-enough> accessed 5 November 2022.
14. BloombergNEF. Battery Pack Prices Fall to an Average of $132/kWh, but Rising Commodity Prices Start to Bite (2021). Available at: <https://about.bnef.com/blog/battery-pack-prices-fall-to-an-average-of-132-kwh-but-rising-commodity-prices-start-to-bite/> accessed 5 November 2022.
15. Bloomberg Professional [Online]. Subscription Service. Available at: <https://www.bloomberg.com/professional/solution/bloomberg-terminal/> accessed 5 November 2022.
16. Mordecai E. 'The Cobweb Theorem' (1938) 52 *The Quarterly Journal of Economics* 255–80.
17. Argonne National Laboratory. BatPac Version 3.0. Excel Spreadsheet (2015). Available at: <https://www.anl.gov/cse/batpac-model-software> accessed 5 November 2022.
18. Wadia C, Albertus P, and Srinivasan V. 'Resource Constraints on the Battery Energy Storage Potential for Grid and Transportation Applications' (2011) 196 *Journal of Power Sources* 1593–8.
19. U.S. Energy Information Administration (EIA). *International Energy Outlook* (Paris, IEA, 2021).
20. BloombergNEF. Global Energy Storage Market Set to Hit One Terawatt-Hour by 2030 (2021). Available at: <https://about.bnef.com/blog/global-energy-storage-market-set-to-hit-one-terawatt-hour-by-2030/> accessed 5 November 2022.
21. Armstrong RC, Barbar M, Brown P, Brushett F, *et al. The Future of Energy Storage* (Boston, MA: MIT, 2022).
22. Hanley S. Tesla to Buy 45 GWh Of LFP Batteries from CATL. Available at: <https://cleantechnica.com/2021/10/30/tesla-to-buy-45-gwh-of-lfp-batteries-from-catl/> accessed 5 November 2022.
23. Rogers E. *Diffusion of Innovations* (New York: Free Press, 1995).
24. BloombergNEF. *New Energy Outlook* (London: BloombergNEF, 2021).
25. Wood Mackenzie. *Global Energy Storage Outlook H1 2021* (Edinburgh: Wood Mackenzie, 2021).
26. Mayr F. Storage at 20 USD/MWh? Breaking Down the Low-Cost Solar-Plus-Storage PPAs in the USA. *Apricum.* Available at: <https://apricum-group.com/storage-at-20-usd-mwh-breaking-down-the-low-cost-solar-plus-storage-ppas-in-the-usa/> accessed 5 November 2022.
27. BloombergNEF. *Electric Vehicle Outlook 2020* (London: BloombergNEF, 2020).
28. International Energy Agency. *Global EV Outlook 2018* (Paris: IEA, 2018).
29. Thielmann A, Sauer A, and Wietschel M. *Gesamt-Roadmap Energiespeicher für die Elektromobilität 2030* (2015).

30. McCrone A, Moslener U, D'Estais F, Usher E, *et al. Global Trends In Renewable Energy Investment 2016* (Frankfurt am Main: Frankfurt School-UNEP Centre/BloombergNEF, 2016).
31. Stephan A, Battke B, Beuse MD, Clausdeinken JH, *et al.* 'Limiting the Public Cost Of Stationary Battery Deployment By Combining Applications' (2016) 1 *Nature Energy* 16079.
32. International Energy Agency. *Technology Roadmap—Energy Storage* (Paris: IEA, 2014).
33. Ajadi T, Cuming V, Boyle R, Strahan D, *et al. Global Trends in Renewable Energy Investment 2020* (Frankfurt am Main: Frankfurt School-UNEP Centre/BloombergNEF, 2020).
34. International Energy Agency. *World Energy Investment* (2020).
35. Hancock E. BloombergNEF: Energy Storage Investment Levels 'Steady' at US$3.6 Billion Worldwide in 2020 (2021). Available at: <https://www.energy-storage.news/bloombergnef-energy-storage-investment-levels-steady-at-us3-6-billion-worldwide-in-2020/> accessed 5 November 2022.
36. McCrone A, Moslener U, D'Estais F, Grüning C, *et al. Global Trends in Renewable Energy Investment 2018* (Frankfurt am Main: Frankfurt School-UNEP Centre/BloombergNEF, 2018).
37. Lazard. *Lazard's Levelized Cost of Storage Analysis—Version 2.0* (New York: Lazard, 2016). Available at: <https://web.archive.org/web/20170216094455/https://www.lazard.com/perspective/levelized-cost-of-storage-analysis-20/.
38. Ziegler MS, Mueller JM, Pereira GD, Song J, *et al.* 'Storage Requirements and Costs of Shaping Renewable Energy Toward Grid Decarbonization' (2019) 3 *Joule* 2134–53.
39. Needell ZA, McNerney J, Chang MT, and Trancik JE. 'Potential for Widespread Electrification of Personal Vehicle Travel in the United States' (2016) 1 *Nature Energy* 16112.
40. International Energy Agency. *Global EV Outlook 2021* (Paris: IEA, 2021).
41. Hoppmann J, Volland J, Schmidt TS, and Hoffmann VH. 'The Economic Viability of Battery Storage for Residential Solar Photovoltaic Systems—A Review and a Simulation Model' (2014) 39 *Renewable and Sustainable Energy Reviews* 1101–18.
42. Krishna M. 'LFP Batteries: The Electric Vehicle Battery Chemistry Debate Just Got More Complicated' (2022). Available at: <https://www.fastmarkets.com/insights/the-ev-battery-chemistry-debate-just-got-more-complicated> accessed 5 November 2022.
43. Benchmark Mineral Intelligence. Benchmark Launches Lithium Ion Battery Raw Material Price Index (2022). Available at: <https://www.benchmarkminerals.com/membership/benchmark-launches-lithium-ion-battery-raw-material-price-index/> accessed 5 November 2022.
44. Bernreuter Research. Polysilicon Price: Chart, Forecast, History (2022). Available at: <https://www.bernreuter.com/polysilicon/price-trend/> accessed 5 November 2022.
45. Wright TP. 'Factors Affecting the Cost of Airplanes' (1936) 3 *Journal of the Aeronautical Sciences* 122–8.
46. Mayer JN, Philipps S, Hussein NS, Schlegl T, *et al. Current and Future Cost of Photovoltaics* (Freiburg: Fraunhofer ISE/Agora Energiewende, 2015).
47. Liebreich M. 'In Search of the Miraculous' Bloomberg New Energy Finance Summit 2016 (New York: 2016). Available at: <https://www.bbhub.io/bnef/sites/4/2016/04/BNEF-Summit-Keynote-2016.pdf>.
48. Dutton JM and Thomas A. 'Treating Progress Functions as a Managerial Opportunity' (1984) 9 *The Academy of Management Review* 235.
49. McDonald A and Schrattenholzer L. 'Learning Rates for Energy Technologies' (2001) 29 *Energy Policy* 255–61.

50. Neuhoff K. 'Learning by Doing with Constrained Growth Rates: An Application to Energy Technology Policy' (2008) 29 *The Energy Journal* 165–82.

51. Rubin ES, Azevedo IML, Jaramillo P, and Yeh S. 'A Review of Learning Rates for Electricity Supply Technologies' (2015) 86 *Energy Policy* 198–218.

52. Abernathy WJ and Kenneth W. 'Limits of the Learning Curve' (1974) Sept-Oct *Harvard Business Review* 109–19.

53. Morgan MG. 'Use (and Abuse) of Expert Elicitation in Support of Decision Making for Public Policy' (2014) 111 *Proceedings of the National Academy of Sciences* 7176–84.

54. Meyer MA and Booker JM. *Eliciting and Analyzing Expert Judgment, A Practical Guide* (American Statistical Association and the Society for Industrial and Applied Mathematics, Washington DC: 2001).

55. Few S, Schmidt O, Offer GJ, Brandon N, *et al.* 'Prospective Improvements in Cost and Cycle Life of Off-Grid Lithium-Ion Battery Packs: An Analysis Informed by Expert Elicitations' (2018) 114 *Energy Policy* 578–90.

56. Nykvist B and Nilsson M. 'Rapidly Falling Costs of Battery Packs for Electric Vehicles' (2015) 5 *Nature Climate Change* 329–32.

57. Bertuccioli L, Chan A, Hart D, Lehner F, *et al. Development of Water Electrolysis in the European Union* (London: E4tech/Element Energy, 2014).

58. Wiser R, Jenni K, Seel J, Baker E, *et al.* 'Expert Elicitation Survey on Future Wind Energy Costs' (2016) 1 *Nature Energy* 16135.

59. Frith J. *2017 Lithium-Ion Battery Price Survey* (London: Bloomberg New Energy Finance, 2017).

60. Frith J. *2018 Lithium-Ion Battery Price Survey* (London: Bloomberg New Energy Finance, 2018).

61. Mainwood P. *A Modest Proposal to the International Energy Authority* (Quora, 2017). Available at: <https://tinyurl.com/mainwood-quora> accessed 5 November 2022.

62. U.S. Department of Energy. SunShot 2030 (2019). Available at: <https://www.energy.gov/eere/solar/sunshot-2030> accessed 5 November 2022.

5 Lifetime cost
Performing cost assessments

KEY INSIGHT	WHAT IT MEANS
There are two key lifetime cost metrics: levelized cost of storage (LCOS) for applications that value the provision of energy and annuitized capacity cost (ACC) for applications that value the provision of power. LCOS divides all costs incurred over the technology's lifetime by discharged energy. ACC divides these costs by power capacity and lifetime of the technology.	Lifetime cost is the correct measure for assessing the economics of a storage project with a specific technology and application. The first step in determining lifetime cost is to choose LCOS or ACC as the metric, based on what the application pays for.
For most electricity storage projects, the most important drivers of lifetime cost are the technology's investment cost, the application's annual cycle frequency, its discharge-duration requirement, and the applied discount rate.	Besides technology investment cost, the choice of application(s) and the project's financing conditions will strongly influence lifetime cost.
The lowest LCOS across major storage technologies is achieved for applications that require 4–10 hours discharge per cycle and continuous operation. The lowest ACC is achieved for applications that require less than 1 hour discharge each cycle and less than 300 cycles per year.	Lifetime cost is minimized by optimizing capital efficiency. This means optimizing energy-specific and power-specific investment cost for the application's discharge duration and then distributing the resulting total investment cost over as many discharge cycles as possible (LCOS) or as many lifetime years as possible (ACC). The different optima mean that there is no one storage technology that offers lowest cost for both types of service.
Matching the cost-efficiency of lithium ion in applications with less than 8 hours discharge and below 500 cycles per year is becoming increasingly difficult for the other major electricity storage technologies.	Alternative technologies may struggle to gain market share and achieve cost reductions in these applications due to the existing dominance of lithium ion and the continued wide-scale deployment of the technology.

(*Continued*)

Monetizing Energy Storage. Oliver Schmidt & Iain Staffell, Oxford University Press. © Oliver Schmidt & Iain Staffell (2023).
DOI: 10.1093/oso/9780192888174.003.0005

KEY INSIGHT | WHAT IT MEANS

In 2030, six electricity storage technologies may dominate seven distinct application categories based on current assumptions for cost and performance parameters and their expected improvement towards 2030:

	Category	Duration	Cycles per year	Technology
1	Short-to-medium discharge	1–8 hours	< 500	Lithium ion
2	Medium-to-long discharge	8–20 hours	< 30	Compressed air
3	Long discharge	> 20 hours	< 30	Hydrogen
4	High throughput—medium discharge	> 4 hours	> 500	Pumped hydro
5	High throughput—short discharge	1–4 hours	> 500	Vanadium flow
6	Power provision—few cycles	< 1 hour	< 1,000	Lithium ion
7	Power provision—many cycles	< 1 hour	> 1,000	Flywheels

Alternative technologies may displace the dominant technologies or emerge along the category borders. However, this analysis can act as a first guide for project developers, investors, or policy-makers in focusing their technology selection when planning storage projects.

5.1 Role of lifetime cost

There is consensus to use levelized cost of energy (LCOE) as a lifetime cost metric to compare energy generation technologies, such as solar, wind, or coal plants. However, there is no universally applied metric for calculating the cost of energy storage technologies. As a result, manufacturers have a hard time explaining cost advantages over their competitors, investors struggle to make educated decisions for financing, and end-users are unsure about which technology to choose.

Energy storage technologies can be used in a range of applications (e.g. frequency response, energy arbitrage, power reliability). These different applications have different operational requirements (e.g. duration of energy supply, number of activations per year) and each storage technology is differently suited to these applications based on their individual cost and performance parameters.[1,2] This further complicates technology selection.

For clearly defined application requirements, storage technologies can be compared using lifetime cost. This accounts for all technical and economic parameters affecting the cost of delivering stored electricity. There are two forms of lifetime cost which matter:[2–4]

a. **Levelized cost of storage (LCOS)** quantifies the discounted cost per unit of discharged electricity (e.g. USD/MWh) for a specific storage technology and application. It divides the total cost of an electricity storage technology across its lifetime by its cumulative delivered electricity.[3,5] By doing so, the metric describes the minimum revenue required for each unit of discharged energy for the storage project to achieve a net present value of zero. The metric is used for applications that value the provision of electric energy (e.g. MWh).

b. **Capacity cost** quantifies the discounted cost per unit of power capacity provided for a certain timeframe. If represented per year this gives the annuitized capacity cost (ACC).

KEY INSIGHT

There are two key lifetime cost metrics: levelized cost of storage (LCOS) for applications that value the provision of energy and annuitized capacity cost (ACC) for applications that value the provision of power. LCOS divides all costs incurred over the technology's lifetime by discharged energy. ACC divides these costs by power capacity and lifetime of the technology.

The concept of levelized cost or capacity cost for electricity storage technologies is analogous to LCOE or capacity cost for electricity generation technologies. Therefore, the lifetime cost for storage and generation technologies can be directly compared, keeping in mind that storage technologies are always time-limited in their provision of electrical energy or power. Figure 5.1 shows how lifetime cost of storage systems can be directly compared to LCOE of energy generation technologies, and how these have fallen rapidly in recent years.

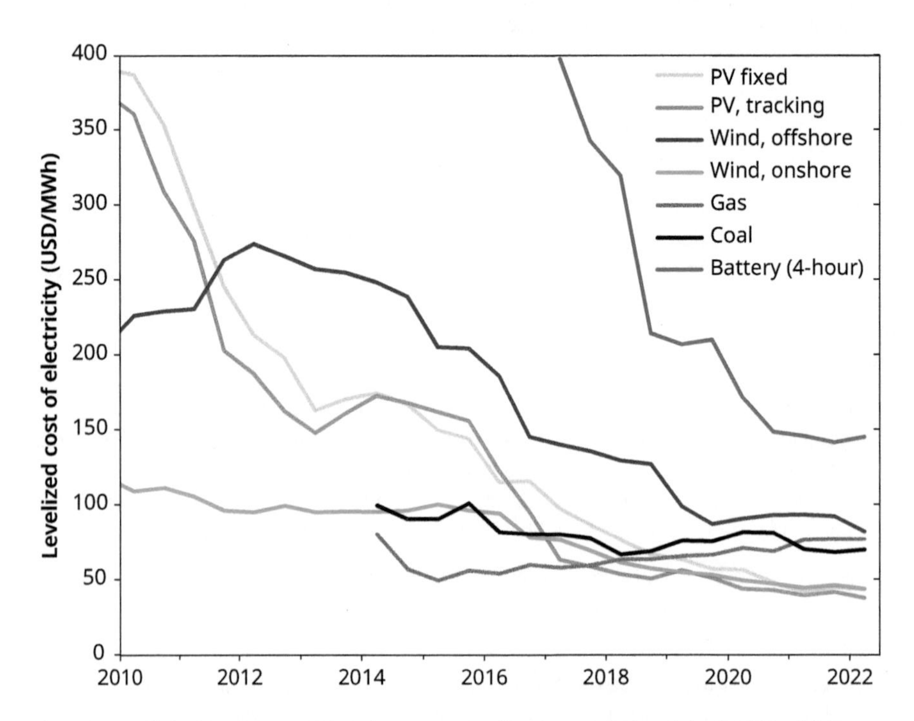

Figure 5.1 Global country-weighted average levelized cost of electricity (LCOE) of different renewable generation technologies compared to the 'storage LCOE' or LCOS of a battery system with 4-hour discharge duration, including charging cost. Values given in real 2021 USD. Data from BloombergNEF.[6]

5.2 Determining lifetime cost

The first step in identifying most cost-effective energy storage technologies is to choose the application these technologies are to operate in. If multiple applications are to be served, then a 'primary' application should be chosen. This is because the application sets key parameters that affect the lifetime cost of a technology, specifically: nominal power capacity, discharge duration, annual charge–discharge cycles, electricity price, and response time.

Required **power capacity** and **discharge duration** affect the size and energy-to-power ratio of the system. These have a significant impact on all cost components, such as investment cost. The number of **annual cycles** is a key driver of the lifetime energy discharged from the system, the denominator in the levelized cost of storage equation. **Electricity price** affects charging cost and **response time** determines the technology's eligibility for certain applications or affects investment cost through special requirements for power electronics.

ATTENTION It is inappropriate to compare lifetime cost of energy storage technologies across different applications. Each application has specific requirements that each technology can be optimized for differently based on their individual cost and performance parameters. So, the cheapest technology for one application is unlikely to be the cheapest for another. An analogy for LCOE would be: a diesel engine may offer lowest cost in backup applications where it runs less than 100 hours per year, but that does not mean it would be cheaper than a nuclear reactor for providing baseload generation all year round.

ATTENTION It is also inappropriate to compare lifetime cost of energy storage technologies serving the same application for very different projects. Project characteristics like size, location, and investor type influence investment cost (e.g. transport and installation cost are site-specific), charging cost (e.g. electricity prices are market-specific), and the discount rate (e.g. financing cost are country- and investor-specific).

Figure 5.2 provides a conceptual overview of the requirements for 13 archetypal electricity storage applications. It should be noted, however, that the nomenclature (e.g. frequency response vs primary response) and requirements (i.e. discharge duration, annual cycles) of storage applications vary across energy markets.

Figure 5.3 shows the lifetime cost for a vanadium redox-flow battery system replacing a 'peaker' power plant. In this example, investment and charging cost are the biggest contributors to lifetime cost. This is common for energy storage systems. Depending on specific energy and power cost and the required discharge duration, different technologies can optimize investment cost for certain applications. Investment cost can be expected to fall in future, but it is still likely to remain the biggest cost contributor. Charging cost strongly depends on electricity prices and the round-trip efficiency of the system. While it can be assumed that electricity prices are equal for all technologies, those with a high round-trip efficiency can minimize this cost component. Operation and maintenance (O&M), replacement, and end-of-life cost have a minor impact on total lifetime cost. Table 5.1 lists all input parameters.

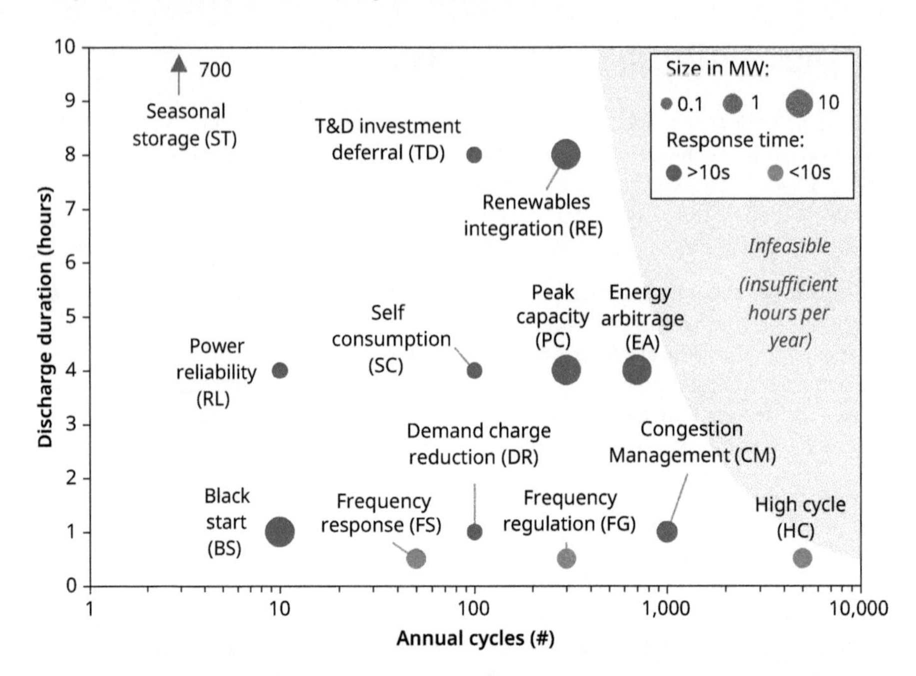

Figure 5.2 Thirteen electricity storage applications with illustrative requirements. The annual cycle and discharge duration requirements are chosen from within the common range of each application such that the entire spectrum for these parameter combinations is represented. Annual cycles refer to full equivalent charge–discharge cycles. Size in MW refers to the nominal power capacity of a typical electricity storage system serving the respective application. The quantitative values for discharge duration and annual cycles are derived in Chapter 8.

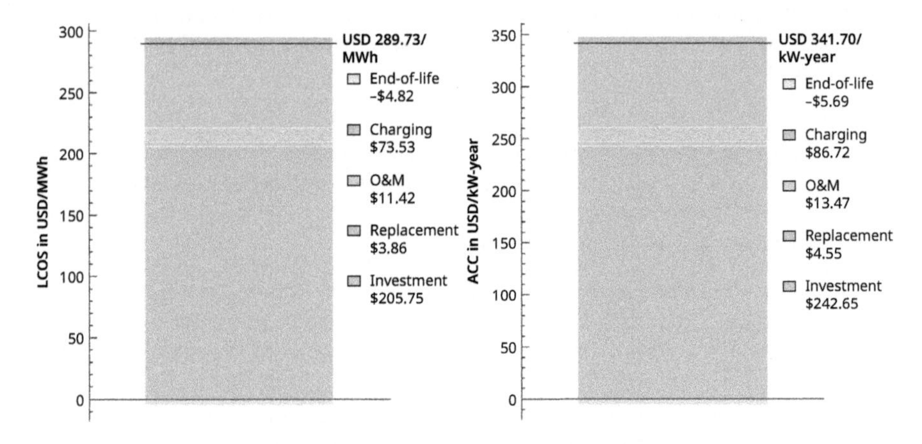

Figure 5.3 (a) Levelized cost of storage and (b) annuitized capacity cost for a vanadium redox-flow battery system providing peak capacity. Chart from <www.EnergyStorage.ninja>.

Table 5.1 Cost and performance input parameters for a vanadium redox-flow battery system providing peak capacity.

Parameter	Unit	Value
Technology cost		
Investment cost—power	USD/kW$_{cap}$	700
Investment cost—energy	USD/kWh$_{cap}$	450
Operation cost—power	USD/kW$_{cap}$-year	10
Operation cost—energy	USD/MWh$_{el}$	2
Replacement cost—power	USD/kW$_{cap}$	90
Replacement cost—energy	USD/kWh$_{cap}$	0
End-of-life cost—power	USD/kW$_{cap}$	20
End-of-life cost—energy	USD/kWh$_{cap}$	−100
Discount rate	–	8%
Construction time	years	1
Technology performance		
Replacement interval	cycles	3,500
Cycle lifetime	cycles	20,000
Round-trip efficiency	–	68%
Depth-of-discharge	–	100%
Self-discharge	–	0%
Temporal degradation	p.a.	0.15%
End-of-life capacity threshold	–	95%
Operational lifetime (*calculated*)	years	22.6
Application requirements		
Rated power capacity	MW	10
Rated energy capacity	MWh	40
Annual cycles	cycles	300
Electricity purchase price	USD/MWh$_{el}$	50
Electricity price escalator	p.a.	0%

Lifetime costs are calculated by first determining each individual cost component. These cost components are then divided by either the discounted lifetime energy discharged or discounted power capacity of the system to obtain individual lifetime cost components. These are then summed up to yield total LCOS or ACC.

Investment cost is the sum of the specific investment cost for power components multiplied with the system's power capacity and specific investment cost for energy components multiplied with the energy capacity (discharge duration × power capacity). In this example, the specific investment cost for power components mostly reflect the flow battery's cell stack and power electronics. The specific investment cost for energy components reflect the costs of the tank and the vanadium electrolyte. Costs that cannot be clearly identified as power or energy components (e.g. engineering, installation) can be allocated equally across both cost inputs.

Replacement cost follows the same logic using power-specific or energy-specific replacement cost. In this example, five replacements of the cell stack (power-specific cost) take place before the end of the system lifetime. When calculating total replacement cost, the present-day cost of the individual replacement occurrences must be discounted to reflect their future values.

O&M cost is determined on an annual basis by multiplying the fixed component with the power capacity and the variable component with the energy discharged per year. O&M cost based on annual energy discharged needs to account for the degradation of energy capacity. In this example, O&M cost accounts for scheduled inspections and minor maintenance of the cell stacks and power electronics (fixed cost) as well as pump inspections and cleaning of electrolyte spills (variable cost). To obtain total O&M costs, the annual values need to be discounted to reflect their future value.

Charging cost is also determined on an annual basis by multiplying nominal energy capacity with the depth-of-discharge, annual cycles, electricity price and electricity price escalator, and dividing by the round-trip efficiency (because energy capacity is defined as rated amount of energy that can be discharged). To obtain the lifetime charging cost, the annual values need to be discounted to reflect the future time-value of money. In addition, energy capacity degradation and the resulting reduction in the amount of energy charged must be accounted for.

End-of-life cost is determined from the end-of-life cost (or value) of the power and energy components of the storage system multiplied with the respective power capacity and (degraded) energy capacity at the technology's end-of-life. The cost needs to be discounted to reflect its future value. In this example, there is an end-of-life cost to recycle the cell stacks and power components, which is outweighed by the higher end-of-life value for the vanadium electrolyte. The total, discounted end-of-life value is relatively small, in part due to the long lifetime of the system.

Energy discharged accounts for the energy capacity of the storage system, depth-of-discharge, annual cycles, and self-discharge for annual values. The total amount is obtained by accounting for degradation of energy capacity throughout the technology's lifetime and discounting the annual values as representation for the future value of money that can be earned by selling the energy.

Levelized cost of storage (LCOS) is obtained by dividing the total, discounted individual cost components (or cashflows) by the total, discounted energy discharged. In this example, the total LCOS is 290 USD/MWh. This system would be profitable if it can discharge for > 290 USD/MWh in at least 300 periods of peak energy demand per year, assuming that it charges for 50 USD/MWh in off-peak periods. For comparison, gas peaking plants with 10–15% utilization rate are reported to have levelized cost of energy (LCOE) of 120–200 USD/MWh.[7-9] This means as of 2020 vanadium flow batteries were not yet competitive with the incumbent technology in this application. This, however, does not account for increasing gas peaker LCOE due to rising gas and carbon emission costs and reducing LCOS due to cheaper electricity purchase prices (e.g. solar PV). Thus, storage solutions for peak capacity may already be competitive today when considering future fuel and emission cost changes.

Annuitized capacity cost (ACC) is determined by dividing total, discounted costs by the discounted system's power capacity. In this example, total ACC amounts to 343 USD/kW-year, which means the system would need to earn more than 343 USD each year for each kW of power capacity to be profitable. This is not intuitive in the context of peak power provision, but this metric is useful for ancillary service applications or capacity contracts that are often reimbursed on a USD/kW basis.

FREQUENTLY ASKED QUESTION

The media is talking about levelized cost of storage of below 20 USD/MWh. Can this be true?

A lifetime cost significantly below 100 USD/MWh for a complete electricity storage system is unlikely with current cost and performance inputs. Chances are high that these low quotes refer to a renewable power plant with co-located energy storage. In that case, the capacity of the storage system will be significantly smaller than the total generation capacity of the renewable generator. As a result, the total cost of the storage system is scaled over the total output of the renewables plant. This does not reflect the lifetime cost of the storage system, but rather the increase of LCOE of the renewables plant by adding storage capacity. However, actual lifetime cost of the storage system can be reverse engineered by:[10]

1. Multiplying the quoted cost with the ratio of energy delivered from the renewables plant over the energy discharged from the storage system (e.g. per year)
2. Adding the electricity purchase price of the storage system (i.e. the LCOE of the renewables plant)

5.3 Assessing uncertainty

There is a high degree of uncertainty in determining lifetime cost due to the large number of input parameters that must be considered. For example, a project developer may have multiple offers from technology providers, or quotations for investment cost may be indexed to raw material costs. Similarly, the average electricity purchase price could vary widely based on the charging schedule and overall developments in the market. Such uncertainties are

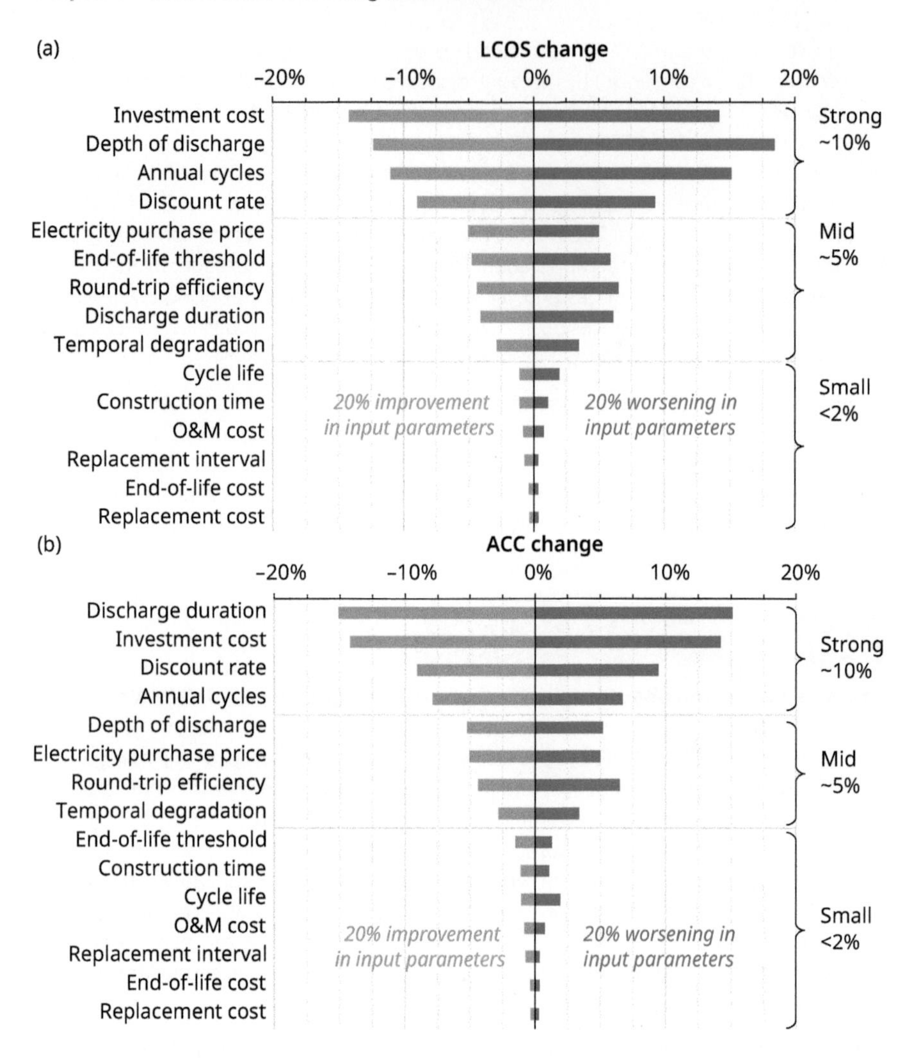

Figure 5.4 Sensitivity analysis of lifetime cost displayed as percentage change to (a) LCOS and (b) ACC resulting from a 20% improvement (green) or 20% worsening (red) of the input parameters displayed in Table 5.1. Parameters are sorted by the percentage impact for 20% improvement.

often overlooked when determining lifetime cost, yet they can have significant influence on outcomes.

Figure 5.4 shows the sensitivity of lifetime cost to input parameters. It quantifies the percentage impact on LCOS and ACC when changing each input parameter by ±20%, using the example of a vanadium redox-flow battery system providing peak capacity. Investment cost, annual charge–discharge cycles and discount rate have the strongest impact on lifetime cost. A 20% change to these parameters changes both LCOS and ACC by ~10%. The reasoning is simple for investment cost, as it is the major cost contributor to lifetime cost in this example (see Figure 5.3). A change in annual cycles significantly impacts the overall lifetime of the system. Due to discounting and annual degradation, this then affects the lifetime energy discharged or power provided. If an electricity price escalator applies, it also affects charging cost. Similarly, the discount rate has a strong impact on lifetime cost as it directly affects the value of the lifetime energy discharged or power provided.

Discharge duration has the strongest impact on ACC but only a moderate impact on LCOS. For LCOS, the increased investment cost for more energy capacity (resulting from longer discharge duration) is balanced by the greater amount of energy discharged over the lifetime. In contrast, for ACC the additional energy capacity only increases investment cost with no impact on the power capacity provided. This explains why electricity storage systems for frequency response or frequency regulation applications are limited to 0.5 or 1 hour discharge duration.[11,12]

Similarly, depth of discharge has the second strongest impact on LCOS, but only a moderate impact on ACC. Since it acts as limitation for the installed energy capacity, it significantly affects energy discharged over lifetime for LCOS.

A 20% change in electricity purchase price and round-trip efficiency affects lifetime cost moderately, by ~5%. Both parameters influence charging cost, which is the second strongest cost contributor in this example (see Figure 5.3). Other parameters which influence technology lifetime, like temporal degradation or the energy capacity threshold at which the technology reaches its end-of-life, have a similar impact.

Since flow batteries have a long lifetime of 20,000 cycles, a 20% change only has a small impact of less than 5% on lifetime cost. O&M cost, replacement cost, and end-of-life cost also fall into this category as their cost contributions in this example are limited (see Figure 5.3).

Both the choice of technology and application will change the impact that each parameter has. For example, if an application has very high annual charge–discharge cycles then LCOS and ACC will be more sensitive to cycle life than to discount rate, as overall lifetime of the technology is relatively short. If the round-trip efficiency of a technology is low, the impact of a change in that efficiency or the electricity purchase price will be very strong. However, across the most common electricity storage technologies and applications introduced in Chapter 3, the impact categorization of parameters into strong, moderate, or small impacts will be similar to the example in Figure 5.4.

KEY INSIGHT

For most electricity storage projects, the most important drivers of lifetime cost are the technology's investment cost, the application's annual cycle frequency, its discharge-duration requirement, and the applied discount rate.

Since nearly all input parameters are subject to uncertainty, it is important to look at their combined impact on lifetime cost, for example if worst-case projections combine.

Monte Carlo analysis is an appropriate tool to deal with such scenarios. Lifetime cost is calculated multiple times, each time taking randomly chosen input parameters from a defined uncertainty range and distribution curve. Table 5.2 shows an example of uncertainty ranges

Table 5.2 Input parameters with central value, range, respective uncertainty, and relative increase of that uncertainty over time. These are modelling assumptions for vanadium flow batteries.

Parameter	Unit	Value	Range	Uncertainty	Relative increase
Technology cost					
Investment cost—power	USD/kW_{cap}	700	630–770	10%	+10% p.a.
Investment cost—energy	USD/kWh_{cap}	450	360–540	20%	+10% p.a.
O&M cost—fixed	USD/kW-year	10	9–11	10%	fixed
O&M cost—variable	USD/MWh_{el}	2	1.8–2.2	10%	fixed
Replacement cost—power	USD/kW_{cap}	90	80–96	10%	fixed
Replacement cost—energy	USD/kWh_{cap}	0	–	–	fixed
End-of-life cost—power	USD/kW_{cap}	20	0–40	100%	fixed
End-of-life cost—energy	USD/kWh_{cap}	-100	-200–0	100%	fixed
Discount rate	–	8%	4%–12%	50%	fixed
Construction time	years	1	–	–	fixed
Technology performance					
Replacement interval	cycles	3,500	2,800–4,200	20%	fixed
Cycle lifetime	cycles	20,000	18,000–22,000	10%	fixed
Round-trip efficiency	–	68%	65%–71%	5%	fixed
Depth-of-discharge	–	100%	–	–	fixed
Self-discharge	–	0%	–	–	fixed
Temporal degradation	p.a.	0.15%	0.13–0.17%	10%	fixed

(Continued)

Table 5.2 (*Continued*)

Parameter	Unit	Value	Range	Uncertainty	Relative increase
End-of-life capacity threshold	–	95%	–	–	fixed
Application requirements					
Rated power capacity	MW	10	–	–	fixed
Rated energy capacity	MWh	40	–	–	fixed
Annual cycles	cycles	300	240–360	20%	fixed
Electricity purchase price	USD/MWh_{el}	50	40–60	20%	fixed
Electricity price escalator	p.a.	0%	–	–	fixed

for each input parameter. In our example, it is assumed that these parameters are normally distributed. Also, uncertainty around a certain value may increase in future. Therefore, a relative increase in uncertainty is also assumed for selected parameters.

Figure 5.5 presents the results of a Monte Carlo analysis where levelized cost of storage were calculated 500 times with input parameters taken from the respective ranges in a random fashion. The dotted lines indicate the 90th and 10th percentile, with the former covering the lowest 90% and the latter the lowest 10% of the results. Thus, LCOS for the vanadium redox-flow battery system providing peak capacity could range from just 230 to 360 USD/MWh, excluding the most extreme 20% of results. This additional insight can significantly increase confidence in the lifetime cost results, which is required when using these results in further economic analyses to take an investment or policy decision.

FREQUENTLY ASKED QUESTION

The uncertainty on LCOS values is very high—how is it possible to make decisions with this information?

The uncertainty on LCOS is relatively high in the analyses presented here because input assumptions are based on a range of third-party studies in order to present results for nine technologies in 13 applications. For a specific application, individual data sheets and quotes may be obtained from technology manufacturers. In that case, we recommend using <www.EnergyStorage.ninja> to conduct a more precise LCOS and ACC assessment. The tools on the website are based on the same methodologies as presented in this book.

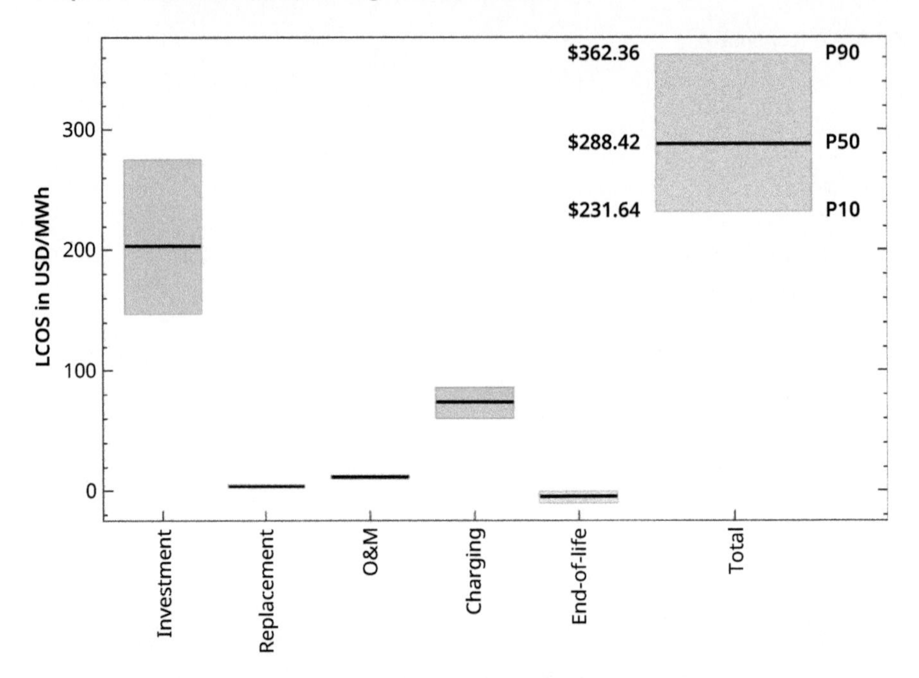

Figure 5.5 Results of a Monte Carlo analysis where LCOS for a vanadium redox-flow battery system providing peak capacity was calculated 500 times. For each calculation, input values were drawn randomly from the respective ranges in Table 5.3 following normal distributions. Chart from <www.EnergyStorage.ninja>. P10, P50, and P90 refer to the 10th, 50th, and 90th percentile.

5.4 Projecting future lifetime cost

The energy storage industry is highly dynamic with new technologies being developed and existing ones being improved continuously. Hence, there is significant potential for future investment cost reductions and performance improvements to positively affect lifetime cost.

Table 5.3 shows possible assumptions on future cost reductions and performance improvements for the example of vanadium redox-flow battery systems as well as the increase in uncertainty around the input assumptions. Figure 5.6 depicts the respective reduction of lifetime cost to 2040. For the exemplary vanadium flow-battery system, this would reduce from ~300 USD/MWh in 2020 to ~200 USD/MWh just after 2025 and ~100 USD/MWh by 2040, making it competitive with existing gas peaker plants.[7-9] The range around the central line indicates the 10th and 90th percentile of the Monte Carlo analysis for each time point based on the parameter uncertainty and the increase in uncertainty over time in Table 5.2.

Table 5.3 Relative changes per annum for the key cost and performance input parameter categories for energy storage technologies as well as relative change per annum of the uncertainty around the assumptions (see Table 5.2). The given values are exemplary for a vanadium redox-flow battery system.

Technology			*Application*		
Parameter	**Rel. change of value**	**Rel. increase uncertainty**	**Parameter**	**Rel. change of value**	**Rel. increase uncertainty**
Investment cost	–5% p.a.	+10% p.a.	Size	fixed	fixed
Operation cost	–2% p.a.	fixed	Discharge dur.	fixed	fixed
Replacement cost	–5% p.a.	fixed	Annual cycles	fixed	fixed
End-of-life cost	fixed	fixed	Electricity price	–5% p.a.	fixed
Round-trip eff.	+2% p.a.	fixed			
Self-discharge	–1% p.a.	fixed			
Lifetime	+5% p.a.	fixed			

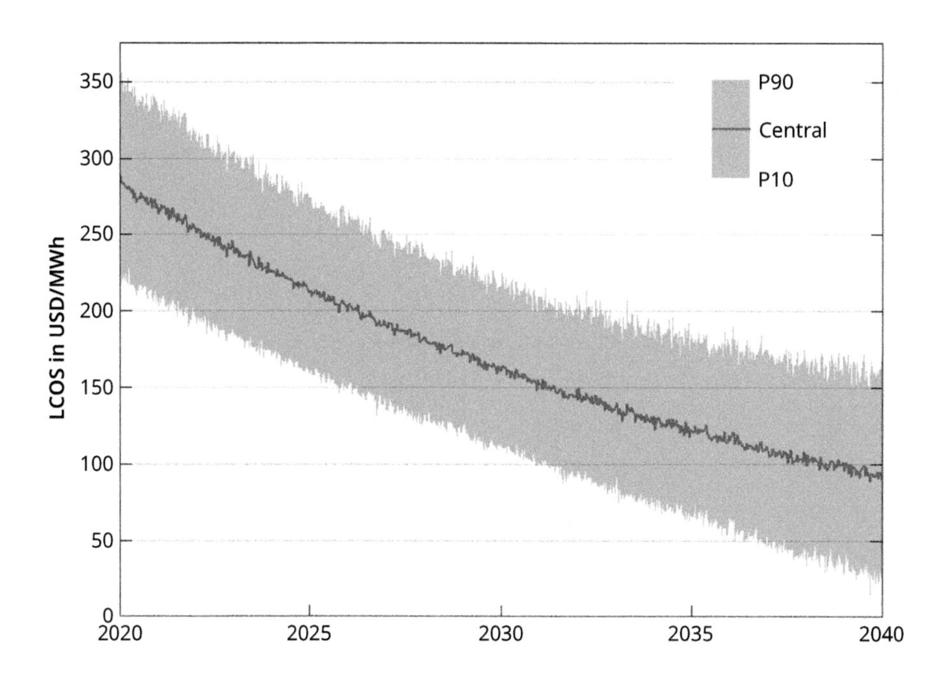

Figure 5.6 Reduction of lifetime cost (levelized cost of storage in USD/MWh) from 2020 to 2040 based on the 2020 input parameters in Table 5.1 and the relative change to these parameters given in Table 5.3. The 10th and 90th percentiles are based on the uncertainty and uncertainty increase indicated in Table 5.3. Chart from <www.EnergyStorage.ninja>.

5.5 Comparing future lifetime cost

Comparing future lifetime cost projections for multiple technologies allows probabilities to be calculated for each technology being the most cost-efficient in an investigated application.

Here, a Monte Carlo analysis can be conducted for multiple energy storage technologies. Comparing the outcomes gives the probability of one technology being cheaper than another. Figure 5.7 provides a schematic depiction of the approach. More details can be found in Chapter 8.

Figure 5.8 compares the LCOS of the four most competitive storage technologies in the application *peak capacity* (*discharge duration: 4 hours, annual cycles: 300*). It also shows the probabilities for each technology to be most cost-efficient calculated at one-year intervals from 2015 to 2040. The benchmark in this application would be a gas peaking plant with a 10–15% utilization rate, which is reported at levelized cost of energy (LCOE) of 120–200 USD/MWh.[7–9]

Pumped hydro had the lowest LCOS in 2015 at just below 200 USD/MWh median (range: 150–225 USD/MWh), followed by compressed air at 250 USD/MWh median (range: 200–300 USD/MWh). However, the strong anticipated investment cost reductions for battery technologies mean that by 2030 vanadium redox flow and lithium ion are likely to be the most cost-efficient solutions for this application.

The median LCOS of the most cost-efficient technology reduces from just below 200 USD/MWh (the current upper LCOE bound of gas peaker plants) in 2015 to 175 and 150 USD/MWh in 2030 and 2040 respectively. This is in line with findings of other studies and means that from 2030 energy storage solutions may be the most cost-effective solution to provide peak capacity services, in particular when accounting for the uncertainty in future natural gas prices.[8] When charging for less than 50 USD/MWh (e.g. solar PV in sunny locations) and

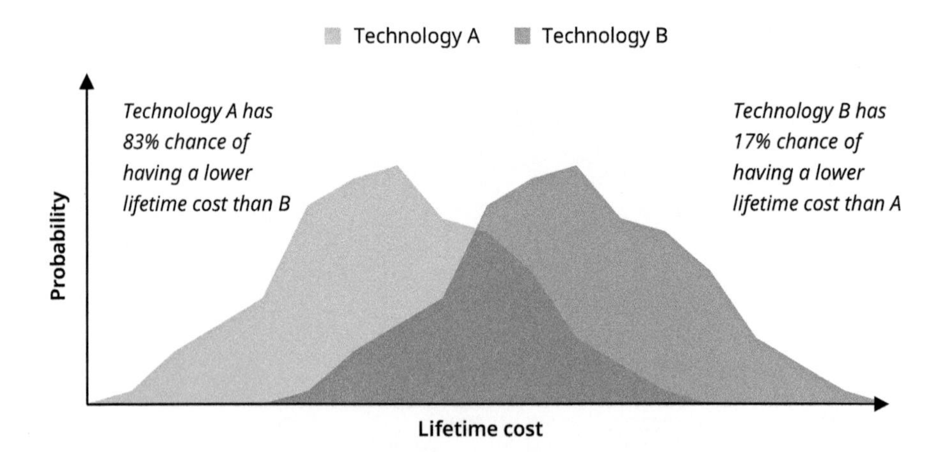

Figure 5.7 Schematic depiction of the methodology to determine the probability with which a technology (Technology A) will exhibit lower lifetime cost than an alternative technology (Technology B) based on the results of a Monte Carlo analysis of both. This method can be extended to incorporate multiple technologies.

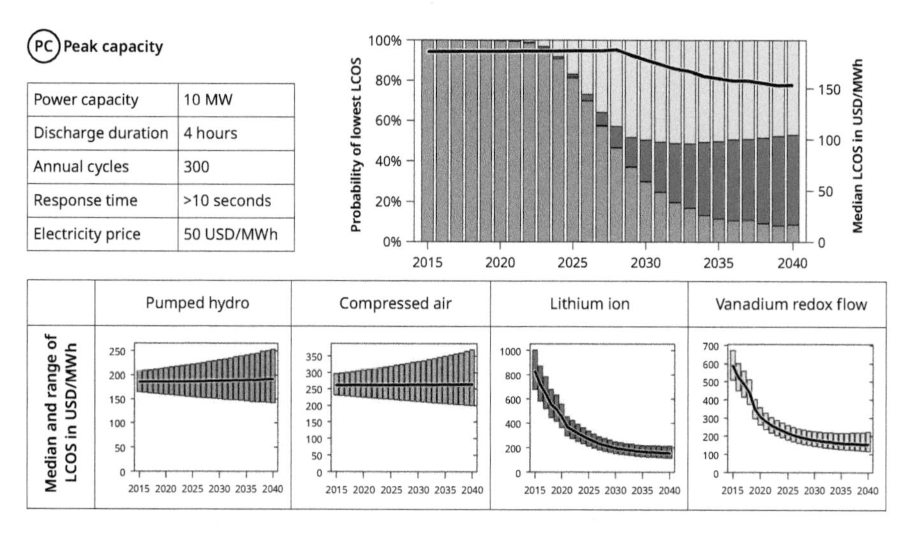

Figure 5.8 Lifetime cost projections for providing peak capacity. Top left: Application requirements. Bottom: Explicit LCOS projections for the four most competitive technologies, including uncertainty ranges based on Monte Carlo simulation of LCOS calculation. Top right: The probability of these technologies having the lowest LCOS, with the median LCOS of the technology with highest probability to be most cost-efficient (black line). Note that LCOS projections are based on future investment cost reductions only (see Table 8.2) and disregard potential performance improvements.

providing additional grid services, battery solutions may become the most cost-efficient solution much earlier.[13]

The same analysis can be conducted for the nine electricity storage technologies introduced in Chapter 3 and the 13 applications shown in Figure 5.2. Figure 5.9 summarizes all technologies and applications, showing each technology's probability of having lowest lifetime cost, and the median LCOS or ACC of the most cost-efficient technology out to 2040.

The overview reveals that the incumbent technologies which dominated electricity storage applications in the past will lose their competitiveness, for example pumped hydro for peak capacity, compressed air for seasonal storage, or lead acid for power reliability.

Instead, by 2030 lithium-ion batteries will be the most cost competitive option in seven out of the 13 applications. Note that these are all the applications with < 4 hours discharge and < 300 annual cycles. For specific applications with requirements outside of these ranges, other storage technologies will come to dominate:

- High throughput, short discharge: Vanadium redox-flow batteries for congestion management (1,000 cycles × 1 hour)
- High throughput, long discharge: Pumped hydro for renewables integration (300 cycles × 8 hours)
- Very long discharge: Hydrogen systems for seasonal storage (10 cycles × 700 hours)
- Very high cycling frequency: Flywheels for high cycle (5,000 cycles × 0.5 hours)

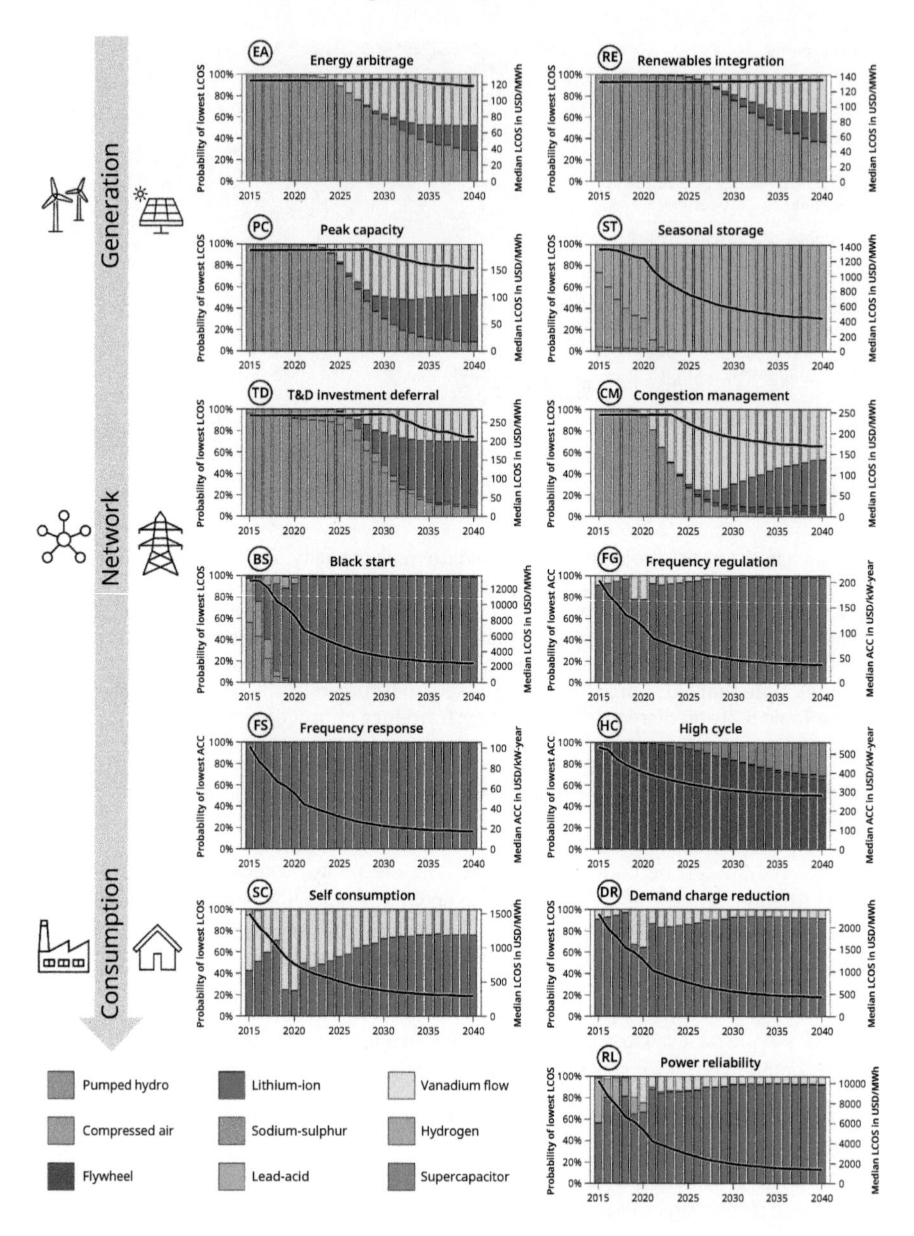

Figure 5.9 Probability of lowest lifetime cost for nine electricity storage technologies in 13 applications from 2015 to 2040. Probabilities reflect the frequency with which each technology has lowest cost accounting for the uncertainty ranges identified with Monte Carlo simulations. Within each panel, the left axis displays probability and right axis displays median lifetime cost of the technology with highest probability for lowest cost. Costs are usually displayed as levelized cost of storage (LCOS), but annuitized capacity cost (ACC) is used for services which are reimbursed for power capacity. Note that there are different scales between panels. Circled letters in panel titles correspond to applications in Figure 5.2. Bespoke probability charts for newly defined applications and/or distinct selections of technologies can be created at <www.EnergyStorage.ninja>.

Please note that lifetime cost for frequency regulation, frequency response, and high cycle is displayed as annuitized capacity cost. These services are usually reimbursed for the power they provide instead of energy. Also, for network services that require < 10 seconds response time and for services that are usually provided at the customer site, pumped hydro and compressed air are excluded from the analysis.

5.6 Lifetime cost drivers

Further insights can be derived when moving away from the concept of clearly defined applications with discrete discharge and cycle requirements. This allows a more overarching view to be taken on technology competitiveness and lifetime cost variability.

Figure 5.10 and Figure 5.11 show the technologies with lowest LCOS and their explicit LCOS for all possible combinations of discharge duration and cycling frequency. The positions of the previously discussed applications are indicated by circled letters in the spectrum.

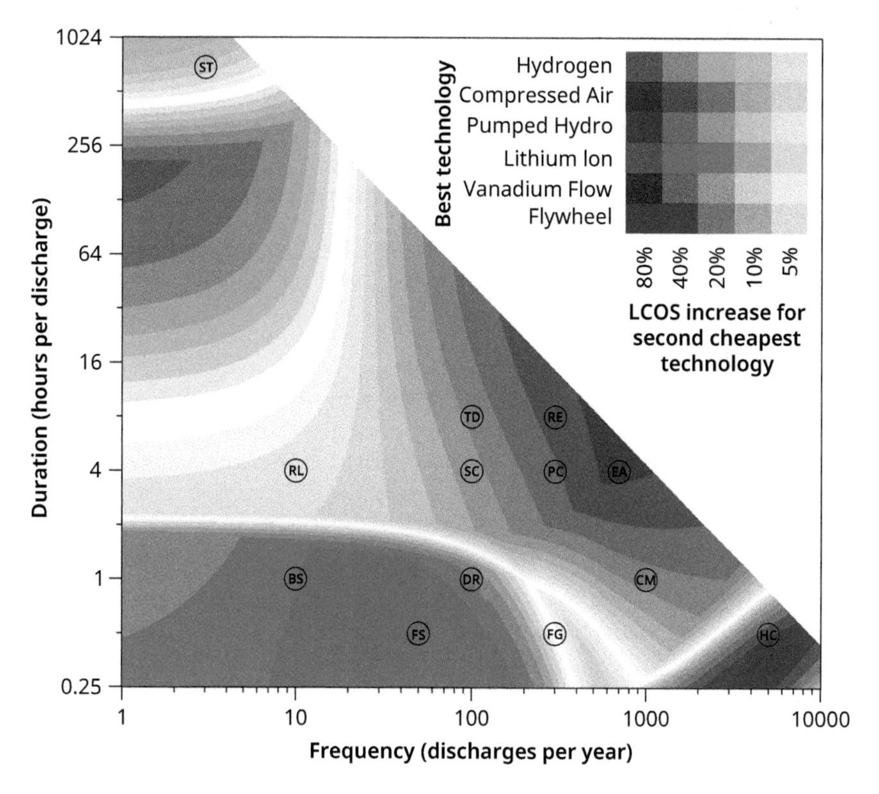

Figure 5.10 Competitive landscape showing energy storage technologies with highest probability of having lowest LCOS relative to discharge duration and annual cycle requirement in 2020. Circled letters represent the requirements of the 13 archetypal applications introduced in Figure 5.2. Colour indicates the technology with the lowest LCOS. Shading indicates how much higher the LCOS of the second most cost-efficient technology is; meaning lighter areas are contested between at least two technologies, while darker areas indicate a strong cost advantage of the dominant technology. Both axes are on logarithmic scale: x-axis with base 10 and y-axis with base 2.

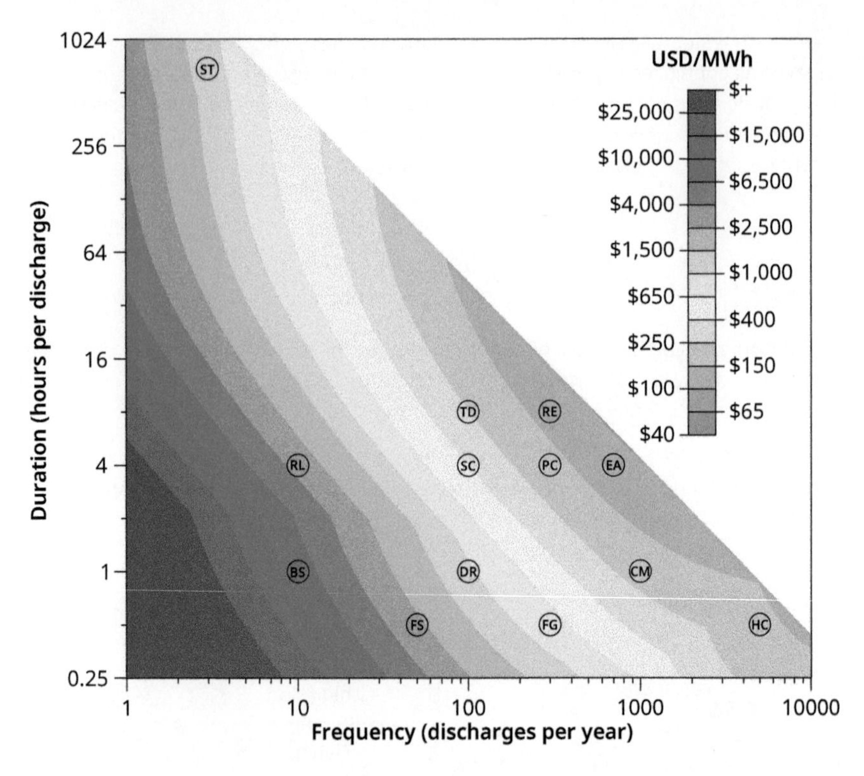

Figure 5.11 LCOS of the most cost-efficient technologies relative to discharge duration and annual cycle requirement in 2020. Circled letters represent the requirements of the 13 applications introduced in Figure 5.2.

Figure 5.10 shows that pumped hydro and compressed air are most cost-efficient for applications with more than 2 hours discharge duration due to relatively low energy-specific investment cost. Above ~300 hours discharge, hydrogen with even lower energy-specific cost takes the lead. Lithium ion is most cost-efficient in applications with below 2 hours discharge and below 300 cycles per year. The longer cycle life of vanadium redox flow makes it more cost-efficient between 300 and 1,000 annual cycles. Above that, flywheels take the lead due to even higher cycle life.

Naturally, cycle life and energy-specific investment cost are not the only determining factors for technology competitiveness. However, the above analysis shows that both are a good indicator for energy storage technology competitiveness in different application regimes.

Figure 5.11 shows that LCOS falls with higher cycles (i.e. discharge frequency). This is intuitive since more energy is discharged for the same energy/power capacity installed (i.e. investment capital deployed). The strong impact is a result of the high share of investment cost in the LCOS, which gets diluted with higher energy throughput. In addition, LCOS falls with increasing discharge duration (i.e. energy-to-power ratio). This increases the energy discharged with each cycle. However, investment cost does not increase proportionately as only the energy-specific cost component are affected. This effect diminishes at higher discharge durations where energy-specific cost already makes up the majority of the total investment cost.

Overall, LCOS reduces with increasing utilization (i.e. discharge hours per year). This is driven by hours per discharge (energy capacity) and discharges per year (cycle frequency). However, at maximum utilization, energy storage applications with lower duration requirements enable lower cost solutions. Given current technologies, it is cheaper to purchase a 1-hour system and discharge it 4,380 times per year than to pay for a 4,380 hour system and discharge it once per year. The lowest LCOS in 2020 is achieved by pumped hydro at maximum utilization between 4 and 10 hours discharge duration (seen from comparing Figure 5.10 and Figure 5.11).

KEY INSIGHT

The lowest LCOS across major storage technologies is achieved for applications that require 4–10 hours discharge per cycle and continuous operation.

These insights are summarized graphically in Figure 5.12.

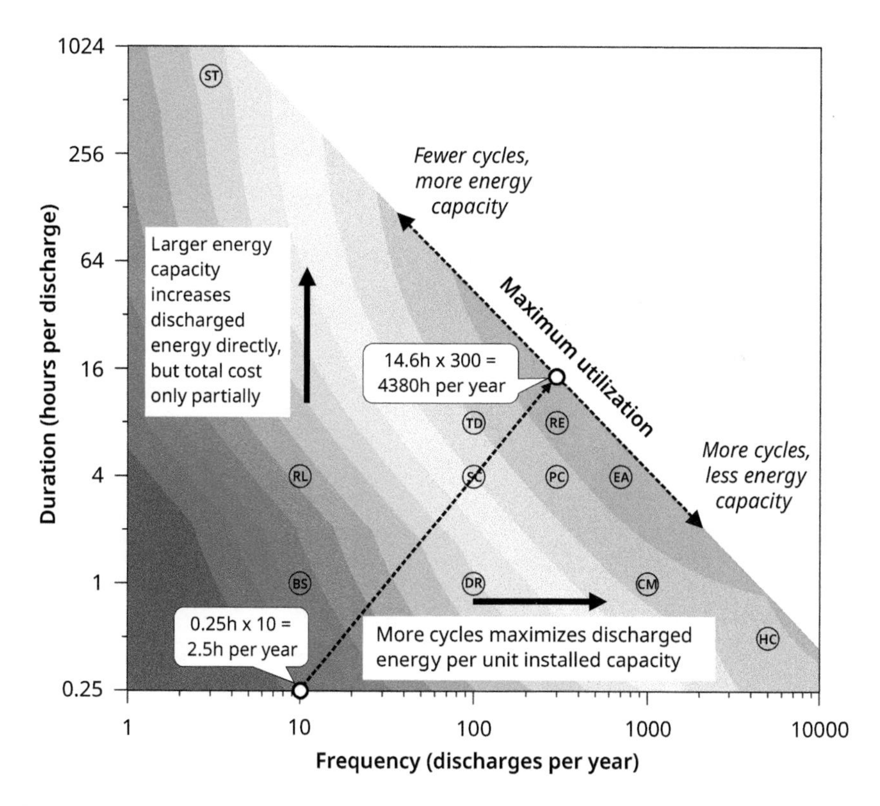

Figure 5.12 Graphic representation of LCOS drivers. Increase in duration (energy capacity) reduces LCOS. Increase in frequency (annual cycles) also reduces LCOS. The combination of both gives the lowest LCOS due to optimization of investment cost (i.e. high share of energy-specific cost) and high number of annual cycles to recoup the investment. Speech bubbles indicate hours of discharge per year (i.e. hours per discharge multiplied with discharges per year).

This view is focused on applications that reimburse energy. Figure 5.13 shows the lowest cost technologies for power provision and the corresponding minimum annuitized capacity cost (ACC) in USD/kW-year.

The dominance of lithium ion for short-duration applications is more pronounced in ACC, and even expands to applications up to 4 hours discharge and 1,000 annual cycles. This confirms the high uptake seen already in ancillary service applications around the world. In contrast, vanadium redox-flow batteries are not competitive anymore. This is because depth-of-discharge is not relevant for the monetization of this service as it only reimburses power capacity provided and not energy discharged. It is therefore not accounted for in this metric. This is to the advantage of lithium ion, which has a lower depth-of-discharge than its 'direct' competitors (i.e. pumped hydro, compressed air, vanadium redox flow).

Low ACC values are achieved in applications with short-duration discharges and few annual cycles. Because storage technologies in power applications get reimbursed for available power capacity, rather than energy discharged, any additional cycle reduces lifetime without leading to additional revenues. Black start and frequency response are cases in point. Also, any additional energy capacity increases investment cost without directly enabling additional revenues. Indirectly, more energy capacity allows power to be provided for longer, which is an advantage in some services (e.g. de-rating in capacity markets).[14] This is not accounted for in ACC and should be considered as a boundary condition if applicable. However, the majority of services that reimburse for power are in the ancillary services market and require discharge durations below 1 hour.

Figure 5.14 again provides a graphical presentation of the drivers for ACC.

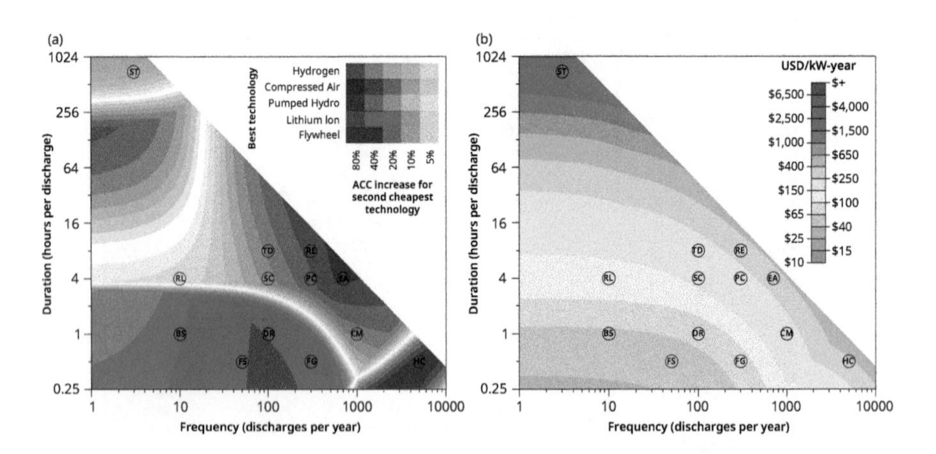

Figure 5.13 (a) Competitive landscape showing energy storage technologies with highest probability of having lowest ACC relative to discharge duration and annual cycle requirement. (b) ACC of most cost-efficient technologies. In both panels, circled letters represent the requirements of the 13 archetypal applications introduced in Figure 5.2. Colour indicates the technology with the lowest ACC. Shading indicates how much higher the ACC of the second most cost-efficient technology is; meaning lighter areas are contested between at least two technologies, while darker areas indicate a strong cost advantage of the dominant technology.

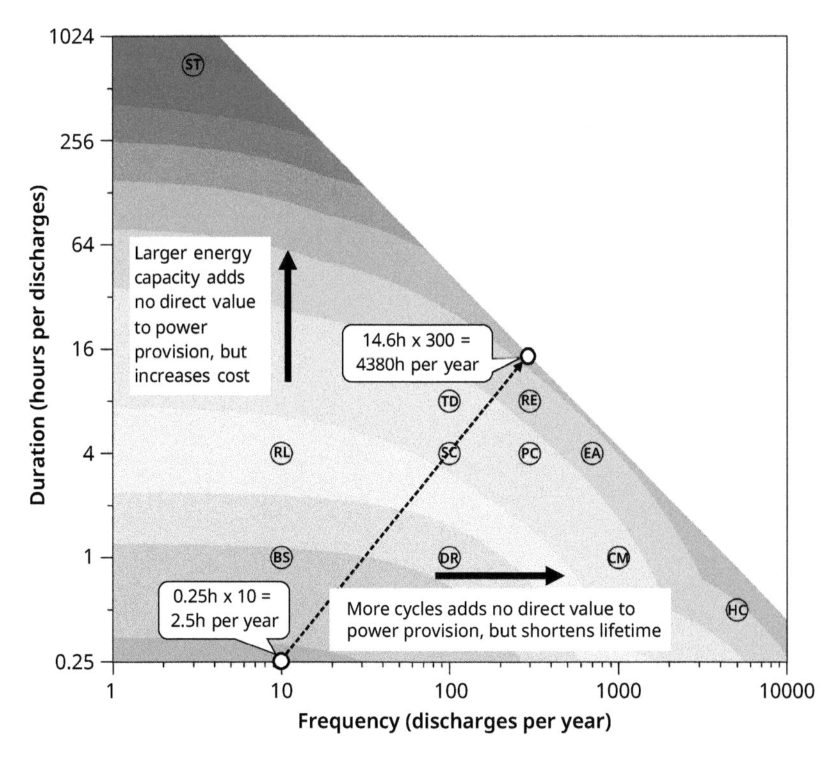

Figure 5.14 Graphic representation of ACC drivers. Longer discharge duration (energy capacity) or higher discharge frequency (annual cycles) add no value to power provision. Therefore, lowest ACC is achieved when minimizing both. Black start capability (BS) is an example of a service where this can be achieved.

The different locations of cost optima for LCOS and ACC already indicate that it will be challenging to provide both types of services cost-efficiently with the same storage system, in other words value stacking. This is discussed further in Chapter 6.

KEY INSIGHT

The lowest ACC is achieved for applications that require less than 1 hour discharge each cycle and less than 300 cycles per year.

5.7 The competitive landscape

Figure 5.15 and Figure 5.16 project the technology competitiveness 'landscape' up to 2040 for LCOS and ACC respectively. The left-hand panels include all storage technologies, while the right-hand panels exclude pumped hydro and compressed air. They are excluded because building these technologies may not be an option for selected projects due to:

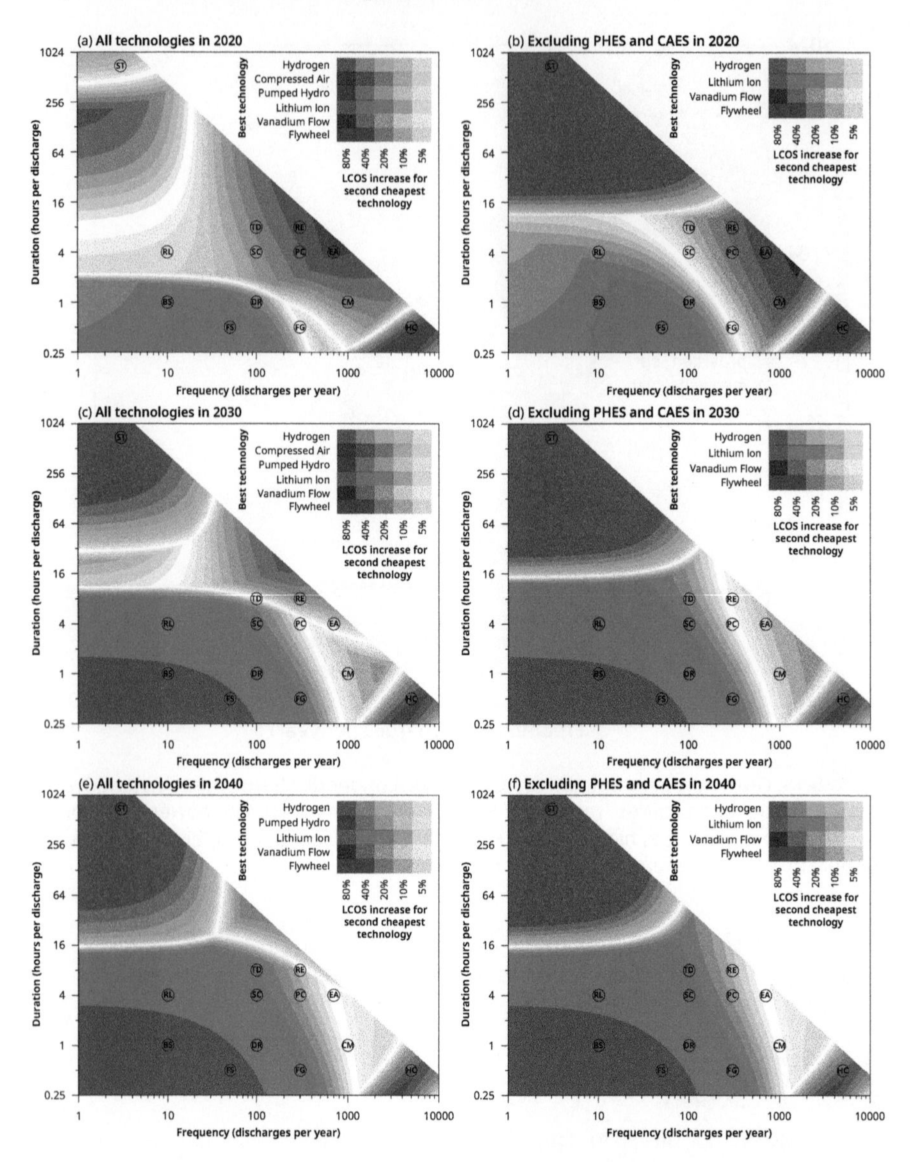

Figure 5.15 Competitive landscapes showing the storage technologies with lowest LCOS relative to discharge duration and annual cycle requirements for all modelled technologies (panels a, c, e) and excluding pumped hydro and compressed air (panels b, d, f). Circled letters represent the requirements of the 13 archetypal applications introduced in Chapter 3. Colours represent technologies with lowest LCOS. Shading indicates how much higher the LCOS of the second most cost-efficient technology is; meaning lighter areas are contested between at least two technologies, while darker areas indicate a strong cost advantage of the dominant technology. White spaces mean the LCOS of at least two technologies differ by less than 5%. Modelled with an electricity price of 50 USD/MWh and a discount rate of 8%.

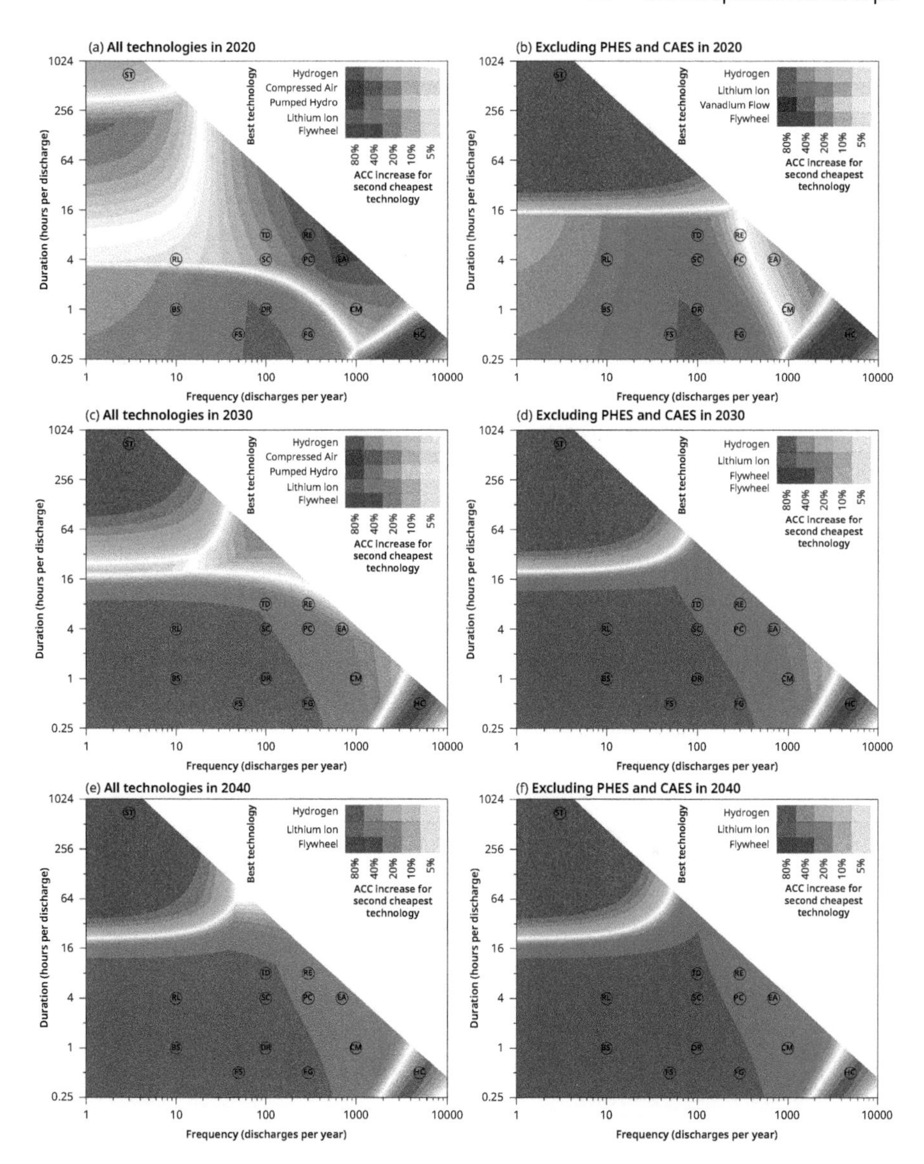

Figure 5.16 Competitive landscapes showing the storage technologies with lowest ACC relative to discharge duration and annual cycle requirements for all modelled technologies (panels a, c, e) and excluding pumped hydro and compressed air (panels b, d, f). Circled letters represent the requirements of the 13 archetypal applications introduced in Chapter 3. Colours represent technologies with lowest ACC. Shading indicates how much higher the ACC of the second most cost-efficient technology is; meaning lighter areas are contested between at least two technologies, while darker areas indicate a strong cost advantage of the prevalent technology. White spaces mean the ACC of at least two technologies differ by less than 5%. Modelled with an electricity price of 50 USD/MWh and a discount rate of 8%.

- limited response times: these technologies may not qualify for services that require ramp-up from idle state to nominal power output in less than 10 seconds
- limited geographic flexibility: it may be impossible to build these technologies at the requested location due to a lack of height differential and/or storage reservoirs

The competitive landscape for 2020 was described already in section 5.6. With continued investment cost reduction, lithium ion is projected to outcompete pumped hydro and compressed air below 8 hours discharge to become the most cost-efficient technology for most of the 13 archetypal applications by 2030. At the same time, hydrogen storage becomes more cost-efficient than compressed air for long-discharge applications. Vanadium redox-flow batteries dominate in high-throughput applications with 300–3,000 annual cycles and up to 4 hours discharge.

The initial increase and subsequent decrease in cost efficiency of vanadium redox flow between 2020 and 2040 reveals a distinct cost-reduction dynamic compared to lithium ion. As a less mature technology, flow batteries can realize significant cost reductions in the near-term assuming similar deployment levels to lithium ion (i.e. doublings in cumulative deployed capacity are achieved faster).[15] However, the experience rate for lithium ion is higher, translating to stronger cost reductions in the long-term. Lithium ion is therefore likely to further increase its competitiveness over vanadium redox-flow batteries and 'regain territory' towards 2040.

Excluding pumped hydro and compressed air reveals that hydrogen storage would have already been most cost-efficient in 2020 for discharge durations beyond 12 hours. The remaining application space that would have been covered by pumped hydro is then dominated by lithium ion and vanadium redox flow.

For ACC, the dominance of lithium ion appears more pronounced. By 2040, only three technologies cover the full application space, with hydrogen most cost-effective in applications above 16–64 hours, flywheels above 1000–3000 cycles, and lithium ion taking all the rest.

KEY INSIGHT

Matching the cost-efficiency of lithium ion in applications with less than 8 hours discharge and below 500 cycles per year is becoming increasingly difficult for the other major electricity storage technologies.

A more detailed look at the results for 2030 reveals dominant storage technologies along seven application categories (Figure 5.17). These categories may be helpful when ranges instead of explicit discharge and frequency requirements are considered. They can act as a first guide for project developers, investors, or policy-makers in focusing their technology selection when planning storage projects or procuring storage capacity.

The dominance of established technologies does not mean that only these technologies have a 'right to play' in a given application category. New technologies may displace them or emerge along the category borders where current technologies compete, in other words where they are not optimally suited. For example, mechanical storage alternatives like gravity storage may compete with pumped hydro and vanadium redox flow directly or in between the two 'high throughput' categories.

KEY INSIGHT

In 2030, six electricity storage technologies may dominate seven distinct application categories based on current assumptions for cost and performance parameters and their expected improvement towards 2030:

	Category	Duration	Annual cycles	Technology
1	Short-to-medium discharge	1–8 hours	< 500	Lithium ion
2	Medium-to-long discharge	8–20 hours	< 30	Compressed air
3	Long discharge	> 20 hours	< 30	Hydrogen
4	High throughput—medium discharge	> 4 hours	> 500	Pumped hydro
5	High throughput—short discharge	1–4 hours	> 500	Vanadium flow
6	Power provision—few cycles	< 1 hour	< 1,000	Lithium ion
7	Power provision—many cycles	< 1 hour	> 1,000	Flywheels

The future projection of LCOS for the most cost-efficient technology at all discharge and frequency combinations is displayed in Figure 5.18. The lowest LCOS is achieved at maximum utilization of the storage systems between discharge durations of 1–64 hours and discharge frequencies of 100–5,000 cycles per year. The LCOS range of 100–150 USD/MWh corresponds to the LCOS from new pumped hydro facilities.[16]

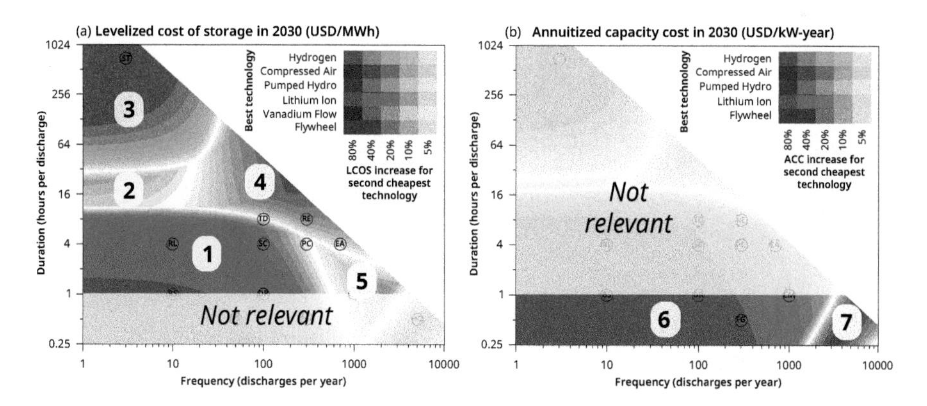

Figure 5.17 Dominant technologies and respective application categories. (a) Most cost-effective electricity storage technologies in terms of levelized cost of storage. This metric is relevant for the area above 1 hour discharge as services with this requirement are usually reimbursed for energy provided. (b) Most cost-effective electricity storage technologies in terms of annuitized capacity cost. This metric is relevant for the area below 1 hour discharge as services with this requirement are usually reimbursed for power provision.

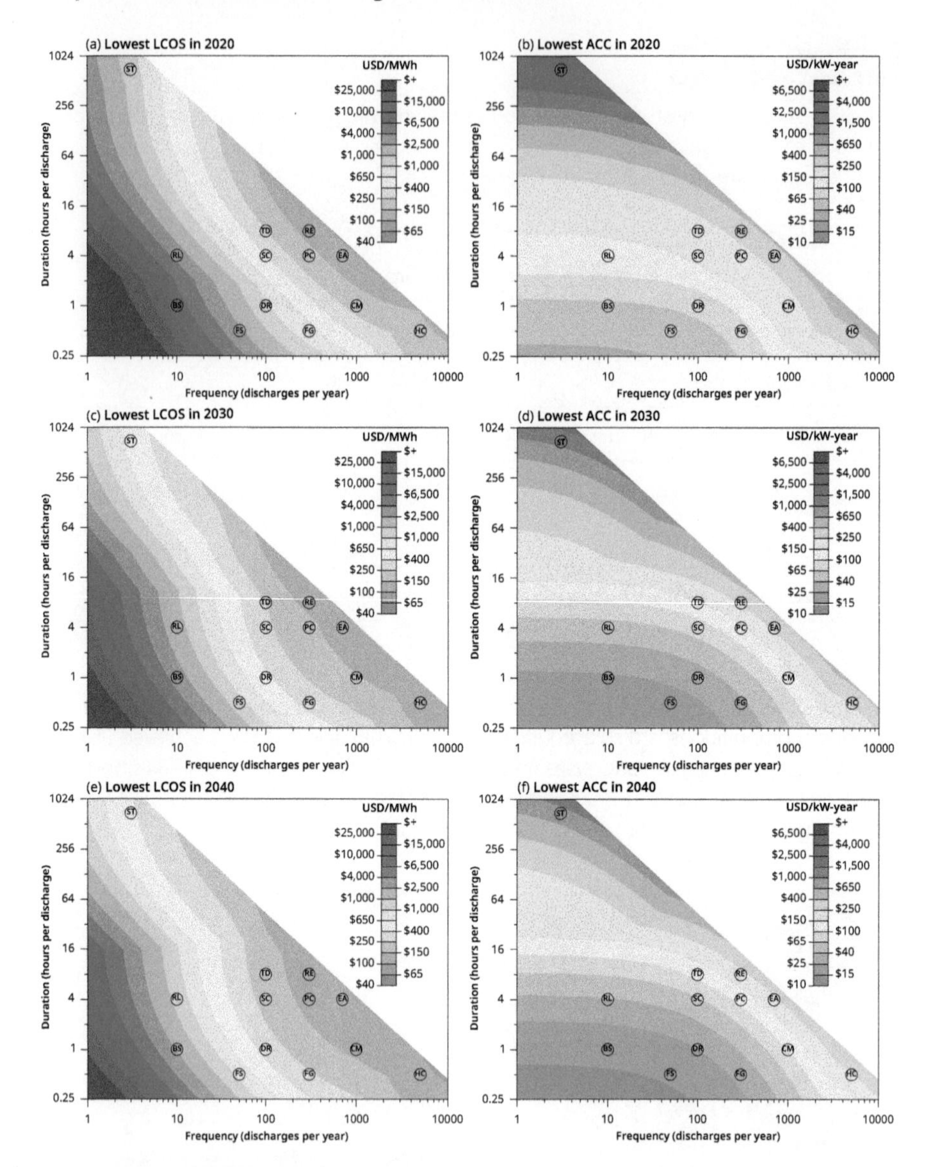

Figure 5.18 LCOS (panels a, c, e) and ACC (panels b, d, f) of the most cost-efficient technologies relative to discharge duration and annual cycle requirements for all modelled technologies. Circled letters represent the requirements of the 13 archetypal applications introduced in Chapter 3. Colours represent LCOS or ACC range. Modelled with an electricity price of 50 USD/MWh and a discount rate of 8%.

The future projection of LCOS shows a proportional reduction across the entire discharge and frequency spectrum, despite the changing technologies that achieve the lowest cost (see Figure 5.15). As a result, LCOS of 100–150 USD/MWh will be achieved in five of the 13 modelled archetypal applications by 2040.

The analogous plot for ACC is also shown in Figure 5.18. The lowest ACC is achieved for applications with short discharge duration and few annual cycles. For example, black start could be serviced for 65 USD/kW-year in 2020 and below 25 USD/kW-year by 2040.

5.8 Scenario analyses

Charging cost is usually the second highest cost contributor to LCOS after investment cost. Hence, it is interesting to explore the impact on the competitive landscape of technologies and absolute LCOS at different electricity purchase prices. Figure 5.19 shows the competitive landscape and LCOS in 2030 with a price of 100 USD/MWh, reflecting the high electricity prices seen in 2021–22, and with a price of 0 USD/MWh to account for situations where electricity is available at no cost, such as from excess renewable generation that would otherwise be curtailed.

Higher electricity purchase price increases the relative importance of round-trip efficiency. Technologies with relatively low round-trip efficiencies (hydrogen, compressed air, and

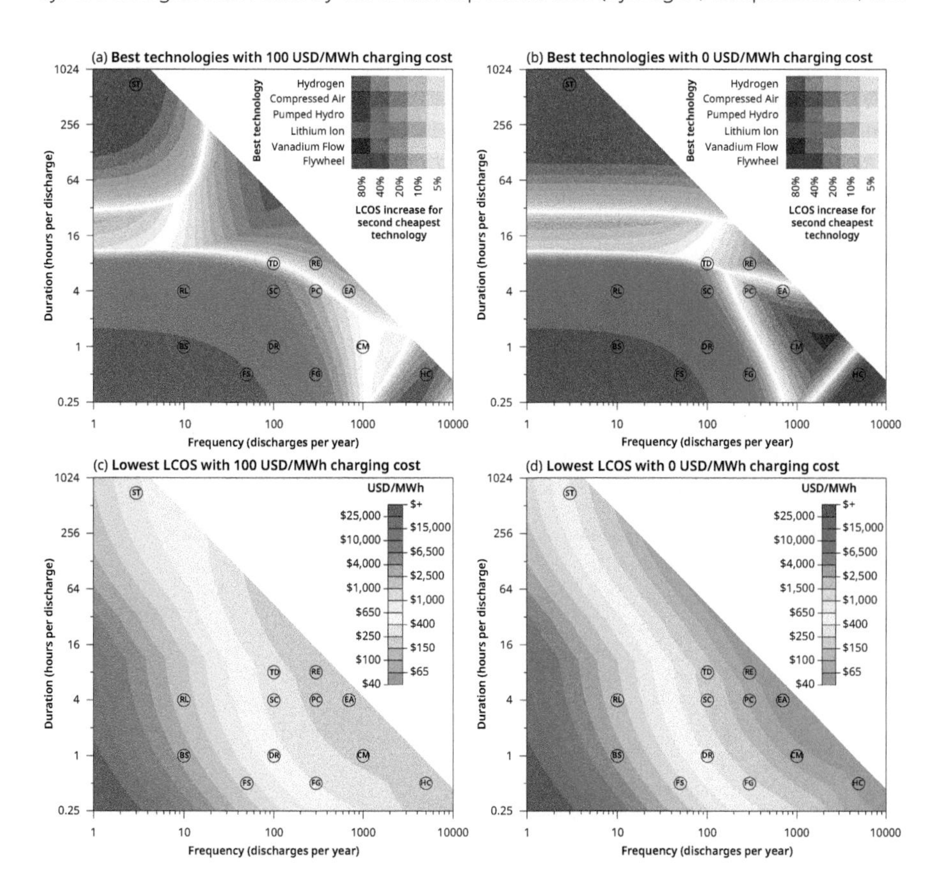

Figure 5.19 Sensitivity of the competitive landscape and absolute LCOS to electricity price in 2030. Panels (a) and (c): Competitive landscape. Panels (b) and (d): Absolute LCOS.

vanadium flow) see reduced competitiveness with high power prices and expand their competitiveness when there is no cost for charging. The competitiveness of lithium ion is largely unaffected, except for peak capacity (PC) and congestion management (CM) services which it trades with vanadium flow. The overall LCOS for high throughput applications in 2030 rises above 150 USD/MWh with high power prices, and drops to below 65 USD/MWh with zero-cost electricity for the most cost-efficient technologies. This also shows that the impact of charging cost on LCOS increases with total throughput (i.e. the product of duration and frequency).

Electricity storage projects will have different financing costs based on the maturity of the technology, the business case, the investor type, and the markets they sell into. Figure 5.20 shows the importance of the discount rate on technology competitiveness and LCOS in 2030. It shows the discount rate increased from 8% to 12% to reflect projects in developing countries or using novel technologies,[17,18] and reduced to 4% to represent a social cost of capital in government-backed projects.[7]

Lithium ion increases its competitiveness with high discount rates, becoming the lowest cost provider for 8 of the 13 services. This is because its comparatively low investment cost becomes more important than long lifetime. Lithium ion is therefore a stronger option in situations with higher risk, or with less availability of low-cost finance. Conversely, pumped hydro, vanadium redox flow, flywheels, and supercapacitors improve their competitiveness with low discount rates, in particular towards lithium ion. For example, for peak capacity, it is cheaper to build pumped hydro plants rather than lithium ion at a discount rate of 4% in 2030. This is because revenues occurring far into the future, corresponding to a high technology lifetime, have a larger impact on LCOS at lower discount rates. All these technologies have longer lifetimes than lithium ion, which is valued more highly at a discount rate of only 4%.

Another source of uncertainty is future performance improvement for the investigated technologies. These could lead to lower LCOS than in Figure 5.18. LCOS is most sensitive to investment cost and energy discharged. Hence, besides different investment cost, round-trip efficiency, depth-of-discharge, and lifetime, which is determined by cycle life and temporal degradation, will have a significant impact on LCOS. For example, if lead acid manages to improve in all these dimensions towards 2030, it could replace lithium ion in applications between 50 and 200 cycles (see Figure 5.21).

Similarly, sodium-sulphur could outcompete lithium-ion and vanadium redox-flow systems in applications requiring up to 6 hours discharge and 500–1,000 cycles if its performance in round-trip efficiency, cycle life, depth-of-discharge, and temporal degradation improves by 2030 (see Figure 5.22).

Note that these scenarios consider the impact of performance improvements for one technology in isolation. It is more likely that all technologies will experience some degree of performance improvement, including lithium ion, which may further improve its cost advantage.[19,20]

Also, investment cost represents the largest lifetime cost component for nearly all technologies. Thus, any additional reduction that goes beyond the experience curve-based projections used in this book would have the most significant impact on lifetime cost reduction.

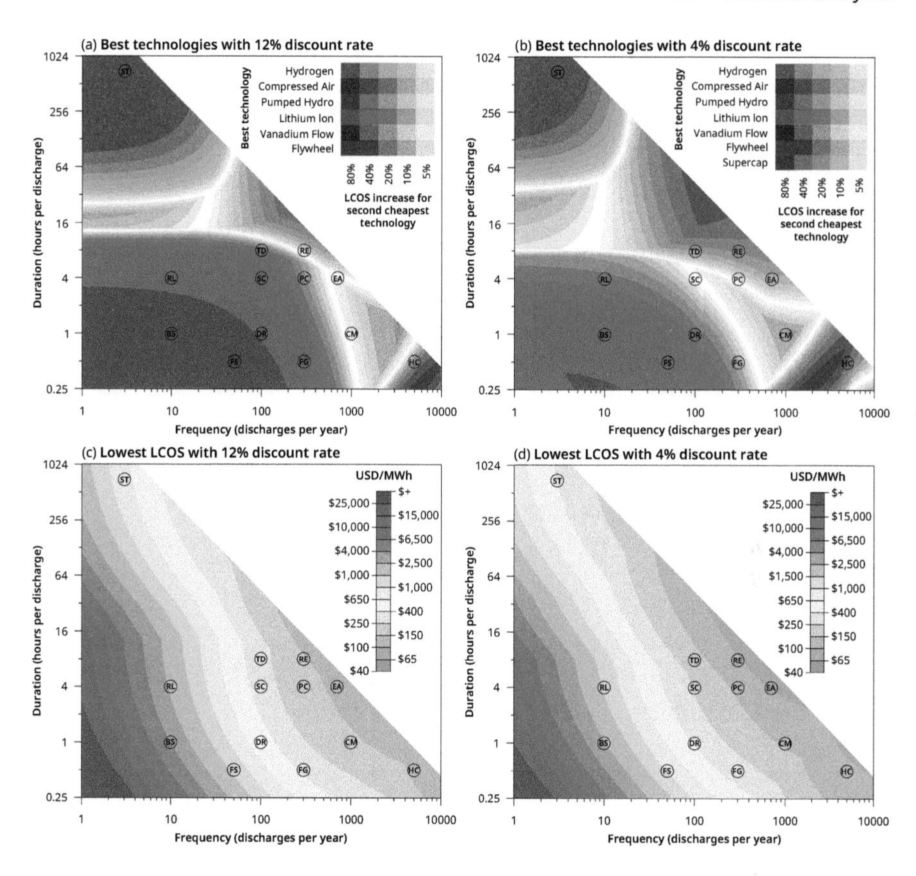

Figure 5.20 Sensitivity of the competitive landscape and absolute LCOS to discount rate in 2030. Panels (a) and (c): Competitive landscape. Panels (b) and (d): Absolute LCOS.

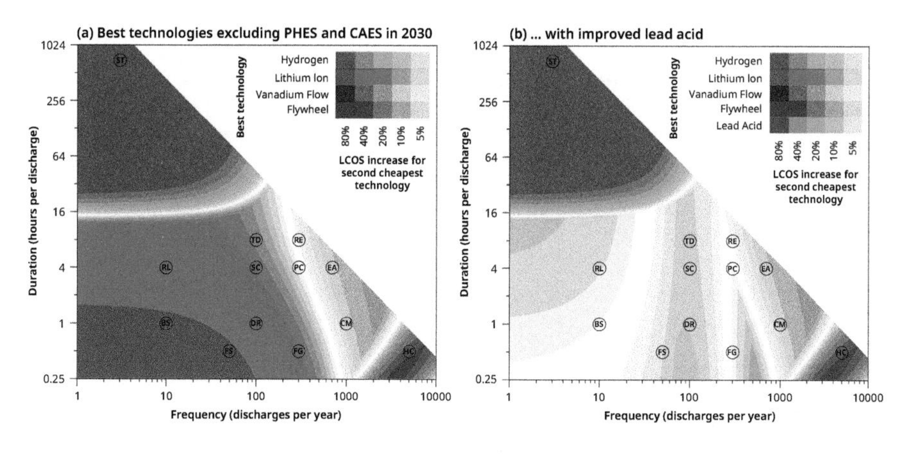

Figure 5.21 Sensitivity of the competitive landscape to various performance parameters for lead acid in 2030. (a) Lead acid at 72% round-trip efficiency, 80% depth-of-discharge, 900 cycle lifetime and 1% temporal energy capacity degradation per year. (b) Lead acid at 86% round-trip efficiency (like lithium ion), 100% depth-of-discharge, 3,500 cycle life (like lithium ion), and 0% temporal energy capacity degradation per year.

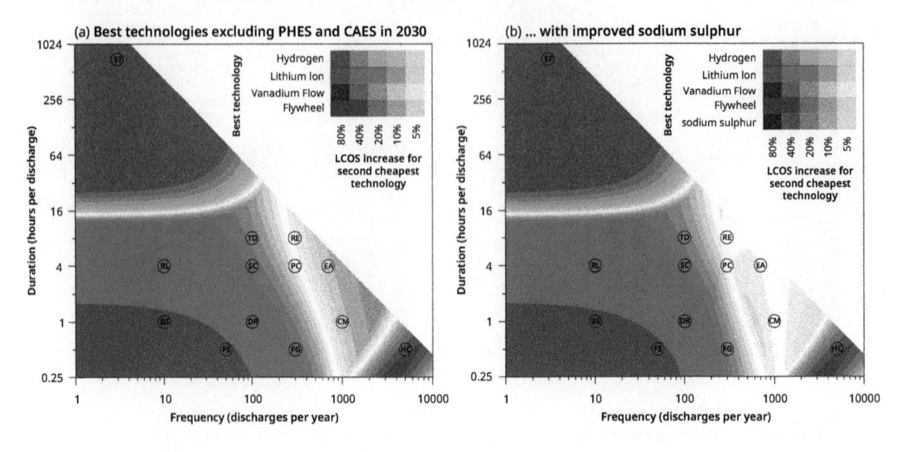

Figure 5.22 Sensitivity of competitive landscape to various performance parameters for sodium sulphur in 2030. (a) Sodium sulphur at 75% round-trip efficiency, 80% depth-of-discharge, 4,000 cycle life, and 1% temporal energy capacity degradation per year. (b) Sodium sulphur at 86% round-trip efficiency (same as lithium ion), 100% depth-of-discharge, 12,000 cycle life, and 0% temporal energy capacity degradation per year.[19,20]

5.9 Technology egalitarianism

If total cost were no object, how would the specific cost and the performance characteristics of each storage technology determine where they naturally sit on the competitive landscape? This section conducts a thought experiment to answer this question. In Figure 5.23 the investment costs of the nine focus technologies are balanced such that each technology plays an equally important role across the spectrum of discharge durations and cycle frequencies. It shows where technologies would sit based on their specific cost and performance characteristics if each technology improves their total cost such that they 'stay in the race' and play an equal role: in other words, technology egalitarianism.

Under these conditions, hydrogen storage would be used for seasonal storage with discharge durations of more than 1.5 weeks. Compressed air and pumped hydro would provide medium-duration storage with between 8 hours and 1.5 weeks of discharge duration. In terms of cycle frequency, compressed air would discharge less than 30 times per year and pumped hydro up to twice per day. Sodium-sulphur, lithium-ion, and flow batteries, and flywheels would provide for the bulk of present-day energy storage applications with up to 8 hours discharge and between 10 and 3,000 charge-discharge cycles per year. Lead-acid batteries and supercapacitors would cover extreme applications like black start with less than 10 cycles per year and high cycle with more than 1,000 per year.

This analysis shows the 'natural' application for each one of the nine electricity storage technologies based on their ratio between energy-specific and power-specific cost, and their performance characteristics. It also confirms the classification of the nine technologies into four technology groups highlighted in Chapter 3. Figure 5.24 expands this classification by qualitatively defining the discharge duration and cycle frequency, where these technology groups outperform the competition. Regardless of the details on cost and dominance of

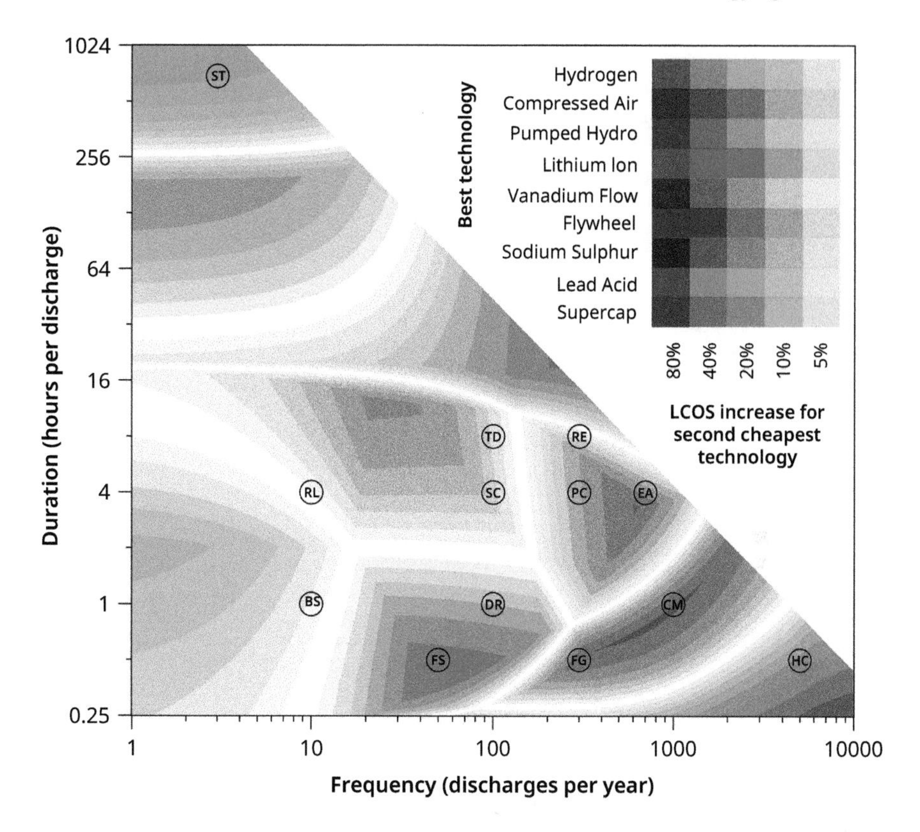

Figure 5.23 Competitive landscape showing the storage technologies with lowest LCOS relative to discharge duration and annual cycle requirements if each technology's investment cost was balanced, such that each technology occupies an equal area on the chart. Unlike previous charts in this chapter, the combination of discharge duration and annual cycle requirement for each technology is determined purely by the ratio of cost, its technical characteristics, and the constraint that it should occupy an area equal to all other technologies.

specific storage technologies, businesses and policy-makers can use this classification to identify a group of technologies for further investment analysis or policy support.

FREQUENTLY ASKED QUESTION

You are modelling lifetime cost for nine electricity storage technologies. Does that mean other technologies will never be competitive?

No. Lifetime cost are modelled for the nine most widely deployed stationary electricity storage technologies (as of 2020). That does not mean that new technologies (e.g. liquid-metal batteries, novel gravity solutions, thermal electricity storage) stand no chance. If you are aware of a novel storage technology and know its cost and performance parameters, feel free to explore its lifetime cost and how it compares on the 'competitive landscape' at <www.EnergyStorage.ninja>.

(Continued)

One thing to consider, though, is the time it takes for a new technology from R&D stage to market, which has historically been longer than 10 years. Lithium-ion batteries have already achieved a huge advantage from economies of scale and will continue to do so driven by the electrification of transport. New technologies must offer lower cost and/or higher performance (i.e. be breakthrough products) to stand a realistic chance of success.

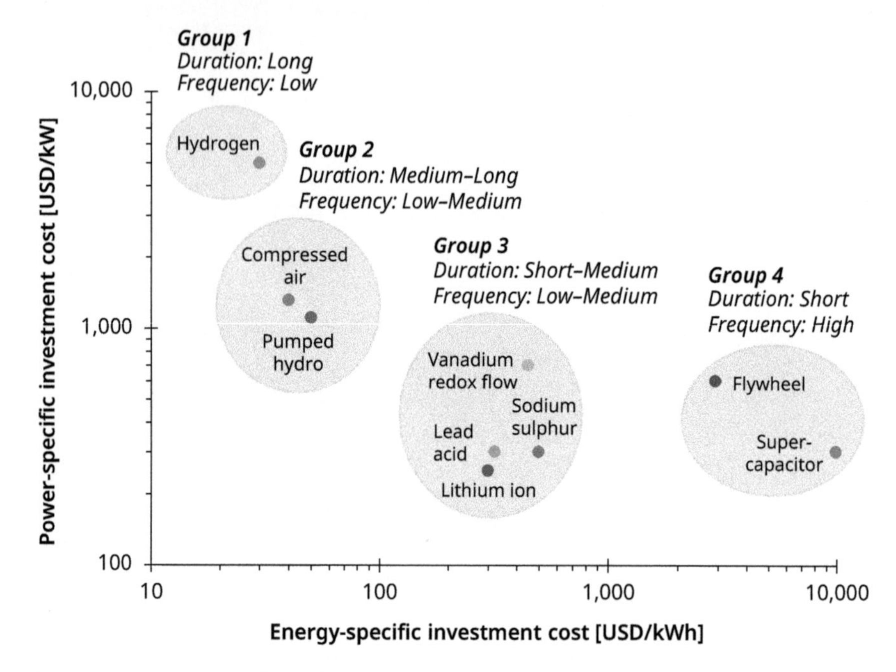

Figure 5.24 Power-specific and energy-specific investment cost for the nine most widely deployed stationary electricity storage technologies. Technologies can be classified into four groups that indicate their positioning in Figure 5.23.

5.10 Discussion

5.10.1 Dominance of lithium ion

Lithium ion is projected to dominate in applications that require less than 500 cycles per year and less than 8 hours discharge. This is the result of good performance parameters, such as a relatively high round-trip efficiency and solid cycle life matched with very strong investment cost reductions. As highlighted in Chapter 4, the investment cost reductions are the result of a high experience rate coupled with very high historical and projected deployment levels for components (cells, packs) for the electric vehicle industry. This, in turn, reduces cost for stationary systems, which facilitates initial deployment in stationary applications. As a result, the technology has entered a reinforcement loop where additional deployment leads to further cost reductions, which lead to further deployment. Figure 5.25 shows that the cost reduction trajectory for lithium-ion battery packs to date matches the one for crystalline silicon solar cells.

It follows that the development of alternative electricity storage technologies that directly compete with lithium ion might become futile due to the challenge in matching the cost and performance advancement lithium ion has achieved to date and is expected to achieve in the future. This would mirror the continuing dominance of first-generation (crystalline silicon) solar cells despite significant investments in alternative solar cell technologies which were initially expected to be cheaper, but failed to achieve the same economies of scale.[21] Just like crystalline silicon solar cells, 'lithium ion' is collective for a range of technologies,[22,23] offering the possibility of chemistry or design improvements that ensure the projected cost reduction for the technology group.

It appears more probable that technologies with distinct advantages in low energy-specific investment cost, high cycle life, or both can outcompete lithium ion in non-standard applications. These would be applications that require very long discharge (> 8 hours; e.g. weekly, seasonal storage), very high cycles (> 500; e.g. congestion management), or very high overall energy throughput (near continuous utilization; e.g. energy arbitrage, renewables integration).

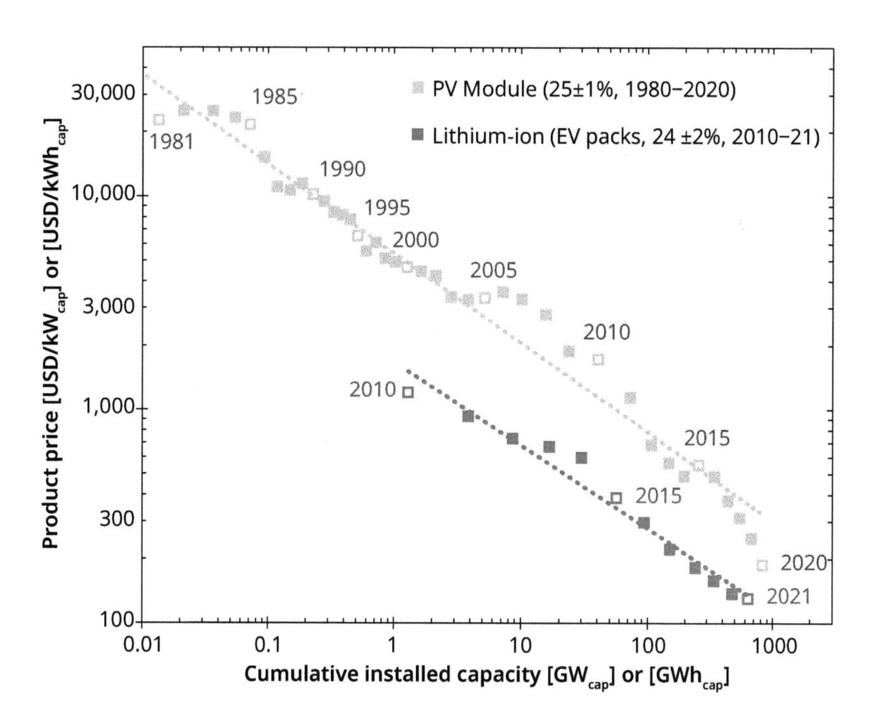

Figure 5.25 Comparison of experience curves for lithium-ion EV battery packs (~10 years) and crystalline silicon solar PV modules (~30 years). The data points for lithium-ion packs are given in units of GWh and USD/kWh, while the data points for solar PV modules are given in GW and USD/kW.

5.10.2 Impact of discount rate

The analyses in this book reveal the sensitivity of LCOS to the underlying discount rate, which even determines which technologies are most cost-effective. The discount rate is usually represented by the weighted average cost of capital (WACC) for the investor in most levelized cost studies.[16,24] The cost of capital is determined by the financing structure of the project (the ratio of debt vs. equity) and usually affected by country (e.g. regulatory and fiscal policy), technology (e.g. capital-intensity, maturity), project (e.g. local authority permission), and market risks (e.g. electricity price volatility). Depending on investor type and market environment, the financing structure and exposure to these risks varies. For example, a government-owned utility in countries with good bond ratings may obtain debt funding close to the social cost of capital at around 4%.[7] Investors with low risk of default in stable investment environments, such as electricity utilities in regulated markets may face a WACC of 8%. Investors facing substantial finance, technology, and market risks, such as private energy storage funds in liberalized markets, may face 12%.

So, who should invest in electricity storage technologies and what could policy do to reduce financing cost? The analysis in this book suggests that government-owned utilities with their lower cost of capital would build pumped hydro plants and private energy storage funds with higher cost of capital would build lithium-ion or vanadium redox-flow battery systems for peak capacity in 2030. As such, the cost and technologies driving electricity system transformation are not primarily determined by technology parameters but rather investment conditions. To limit this effect, governments could provide stable, transparent policy frameworks, debt guarantees, or revenue stabilization schemes. In addition, public financial institutions can provide finance at low cost.[25] These could be among the most effective contributions to the sustainable transformation of the electricity system, because the future should not be discounted too strongly when aiming for 'development that meets the needs of the present without compromising the ability of future generations to meet their own needs'.[26]

FREQUENTLY ASKED QUESTION

Are long-lived technologies disadvantaged when using lifetime cost metrics?
Quite the opposite. Lifetime costs account for the energy or power output of a technology over its entire life, which is an advantage for long-lived technologies, unlike comparing investment cost only. This advantage diminishes with higher discount rates since the additional output enabled by the longer lifetime may be valued much lower than output in the first years. That's why it is important to choose a discount rate that carefully reflects the return requirement of the investor and the risk associated with the technology.

5.10.3 Multiple applications

A possible route to improving the business case for electricity storage is by providing multiple services with one device and thereby stacking multiple revenue streams.[27,28] The methodology presented here can be used to assess LCOS for these 'revenue-stacking' use cases by determining the application requirements that allow to optimize revenues through the provision of multiple services with the same device:

- Nominal power capacity would be based on the service that requires most power (sequential stacking) or the sum of all services provided at the same time (parallel stacking)
- Discharge duration should reflect the duration required by the longest-discharging service
- Full equivalent cycles should reflect the sum across all services provided
- Average electricity price could be the sum of prices captured when charging for individual services weighted by the full equivalent cycles attributed to them

Chapter 6 provides a more detailed discussion on the different options for revenue stacking and their implications on storage system design and operation.

5.10.4 Limitations

It should be reiterated that all results presented in this chapter are subject to the investment cost projections made with experience curves (Chapter 4). These are based on historical price reduction trends and are thus uncertain by nature. Another limitation of this study is that the experience-based cost reductions are exogenous, assuming that all technologies take the entire future stationary storage market individually. It thereby explores the full lifetime cost reduction potential for each technology based on investment cost reductions. In reality, a mix of technologies will be deployed, limiting individual investment cost reductions along experience curves.[29]

Similarly, the results presented here are based on assumptions for distinct cost and performance parameters for each technology. These were compiled through a comprehensive literature review and conversations with industry experts. Nonetheless, actual parameters may differ from the ones assumed here, especially going forward. That is why the online version of the presented lifetime cost methodology at <www.EnergyStorage.ninja> allows LCOS and ACC to be modelled with customized input parameters. It produces most of the graphs presented in this chapter to enable easy comparison of technology cost and competitiveness.

FREQUENTLY ASKED QUESTION

Can lifetime cost be used for system planning?
No. Lifetime cost analysis is suitable for comparing individual technology options or assessing the investment attractiveness for a technology in a specific application. In system planning, the interplay between various technologies in a power system must be assessed in more detail. Hence, power system models are required to analyse total system cost. These models use investment and operating costs, plus performance parameters as inputs for all technologies, including energy storage systems (see Chapter 7).

5.10.5 Why it matters

The results in this chapter explore future lifetime cost potentials for the most widely deployed stationary storage technologies and establish a quantitative foundation for the discussion of storage competitiveness and its drivers. These insights can help guide research, policy, and investment activities to ensure a cost-efficient deployment of electricity storage technologies for a successful transition to a secure and affordable low-carbon energy system.

5.11 Worked examples

This section features two worked examples. In the first one, we will analyse the lifetime cost of vanadium redox-flow batteries providing peak capacity and thereby replicating the results and charts shown in sections 5.2 to 5.4. In the second one, we will put these lifetime cost in context by analysing which technology can provide a service like peak capacity most cost-effectively.

Worked example 1

1. Open <www.EnergyStorage.ninja> and go to the 'Lifetime cost' tab
2. Model the lifetime cost of vanadium flow for peak capacity and thereby reproduce Figure 5.3:
    a. Choose 'Vanadium flow' as technology and click '2020 values' to load the respective cost and performance parameters
    b. Choose 'Peak capacity' as application and click '2020 values' to load the respective application requirements
    c. Click 'Calculate' to perform the lifetime cost calculation and reproduce Figure 5.3

3. Go to the section 'Cost variation' and include the parameters from Table 5.2 (technology & application) and click 'Calculate' to reproduce Figure 5.5. Note that the values you will see on the website will differ slightly due to the stochastic nature of the Monte Carlo analysis, and these will change if you hit 'Calculate' a second time.

4. Go to the section 'Cost projection' and include the parameters from Table 5.3 (change in value & change in uncertainty) and click 'Calculate' to reproduce Figure 5.6.

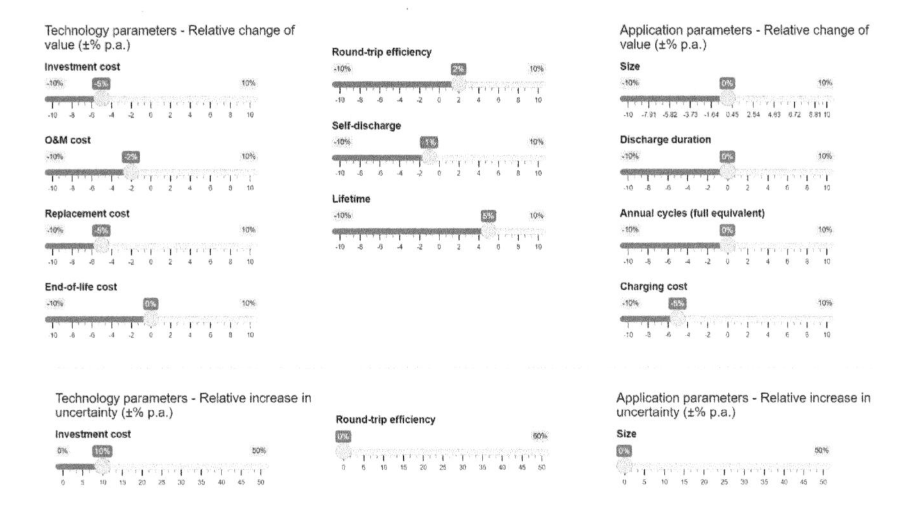

Worked example 2

Let's now check which technology can provide peak capacity services most cost effectively.

1. Please go to the tab 'Landscape'
2. Click '2020 values'

   You will now see that pumped hydro is dominating the area around 4 hours discharge duration and 300 discharges per year.

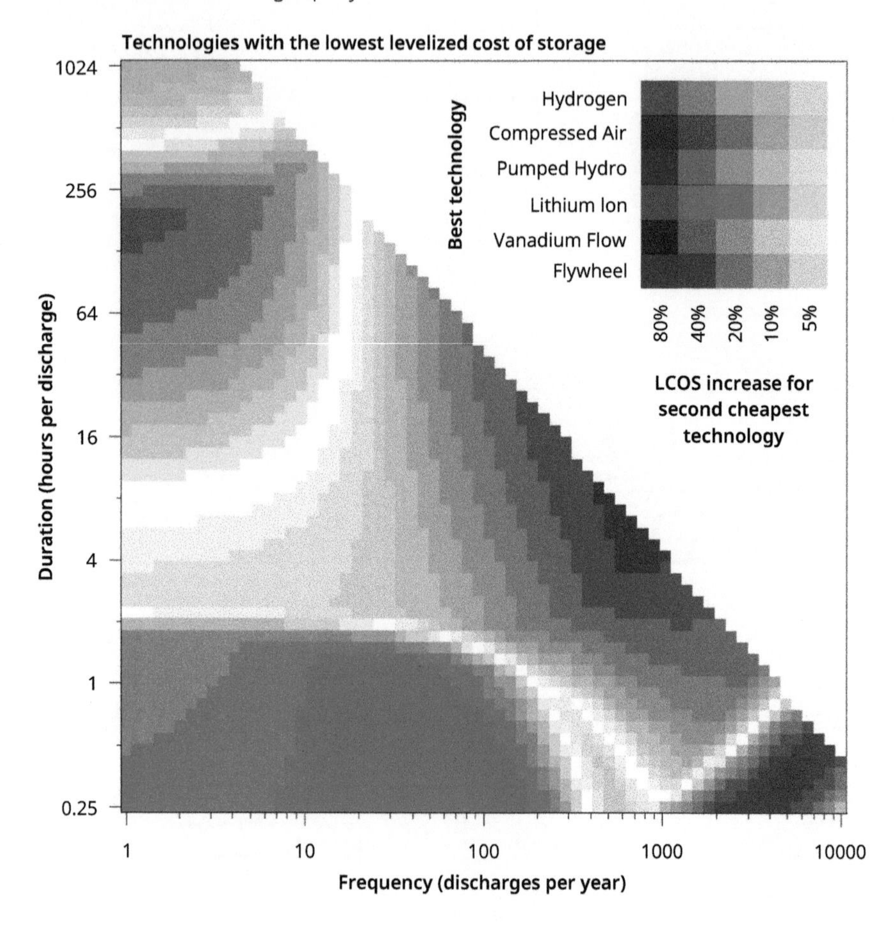

However, in our worked example it is infeasible to build pumped hydro or compressed air storage due to geographical limitations. Therefore, please

3. Set the specific cost of 'pumped hydro' to '9999' to make it infeasible

a. Choose 'pumped hydro' as technology and click 'Load values' to load the respective cost and performance parameters

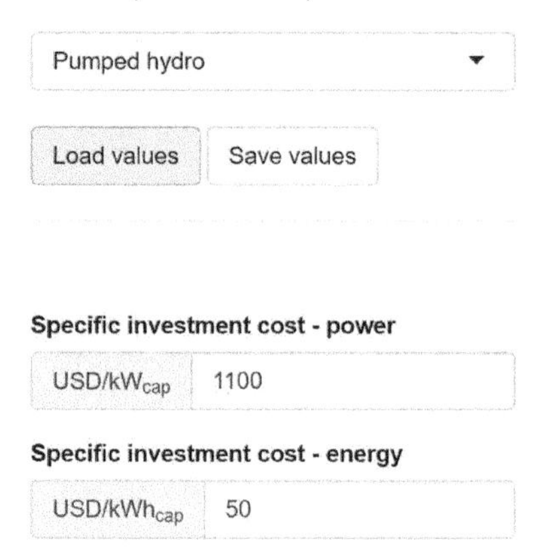

b. Change both specific cost values to '9999' and click 'Save values'

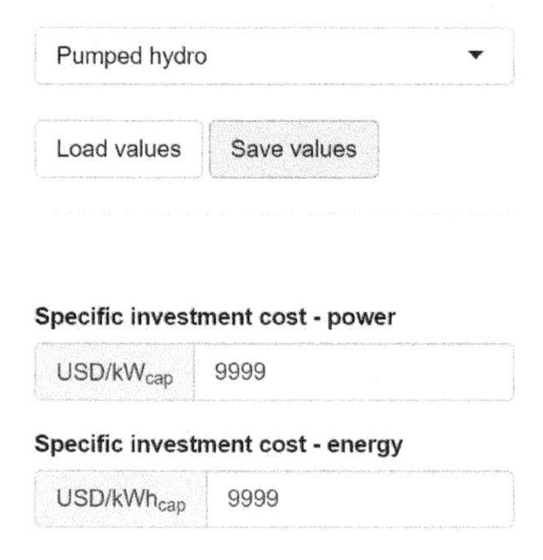

c. Repeat the steps a) and b) for 'compressed air'

d. Click 'Calculate' to perform an updated competitive landscape analysis

The competitive landscape will now show that vanadium flow with the lifetime cost determined in worked example 1 is the most competitive technology for applications with ~4 hours discharge duration and ~300 discharges per year, for example peak capacity provision.

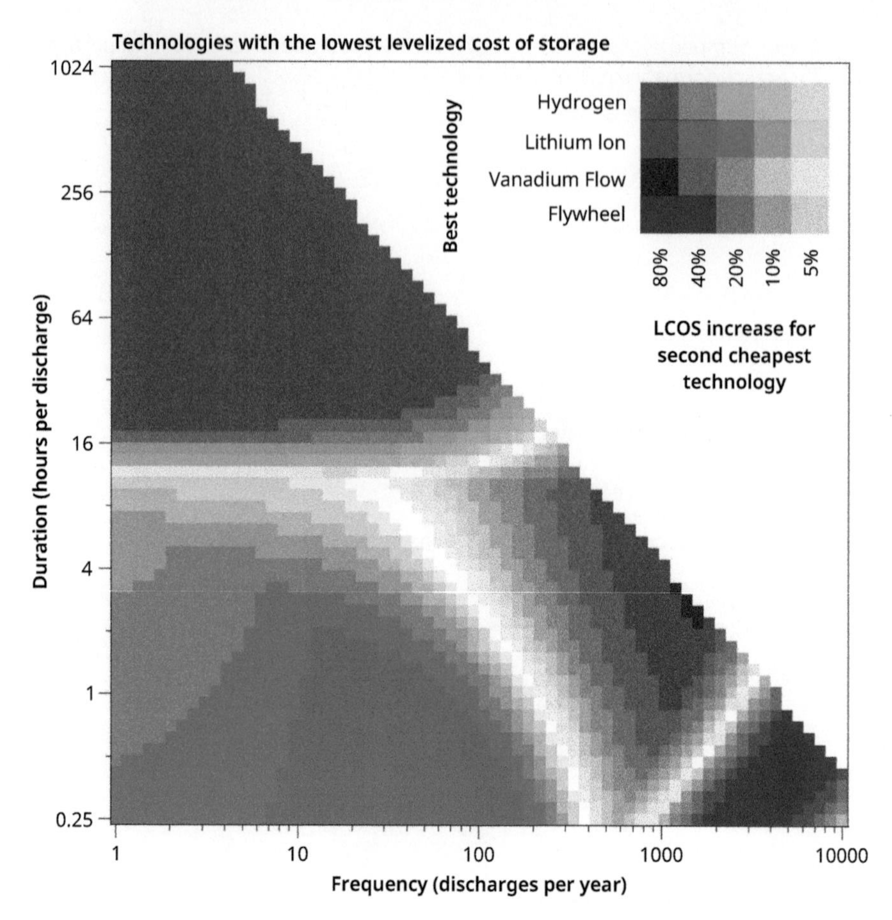

However, in 2021 a lithium-ion battery system was claimed to be 'the first standalone battery energy storage system specifically procured to replace a natural gas peaker plant in the U.S'.[13] So, how can that be? On the one side, cycle life improvement is at the heart of lithium-ion technology research. It has been found that under ideal cycling conditions, commonly deployed LFP lithium-ion batteries can be cycled more than 7,000 times.[30] In addition, stationary lithium-ion batteries are more widely deployed than vanadium redox-flow batteries (15 GW vs 0.1 GW in 2020, see Chapter 3). As a result, a lower technology risk may be associated with lithium-ion technology, which could translate to a lower discount rate, for example 6%. Please implement these changes:

a. Select 'lithium ion' and 'load values'

Lithium ion ▾

Load values | Save values

b. Change 'Cycle life' to '7000' and 'Discount rate' to '6'

Cycle life (@DoD)

cycles	7000

Temporal degradation

% p.a.	1

End-of-life threshold in % of capacity

%	80

Construction time

years	1

Discount rate

%	6

c. Click 'Save values' and then click 'Calculate' to perform the updated competitive landscape analysis

You will now see that the competitiveness of lithium ion has improved. Based on the updated cost and performance parameters (and the exclusion of pumped hydro and compressed air) it is the most competitive technology for applications with ~4 hours discharge and ~300 discharges per year, that is, peak capacity. Vanadium flow remains as most competitive technology for higher throughput applications (e.g. 4 hours, 1000 cycles).

You've done it!

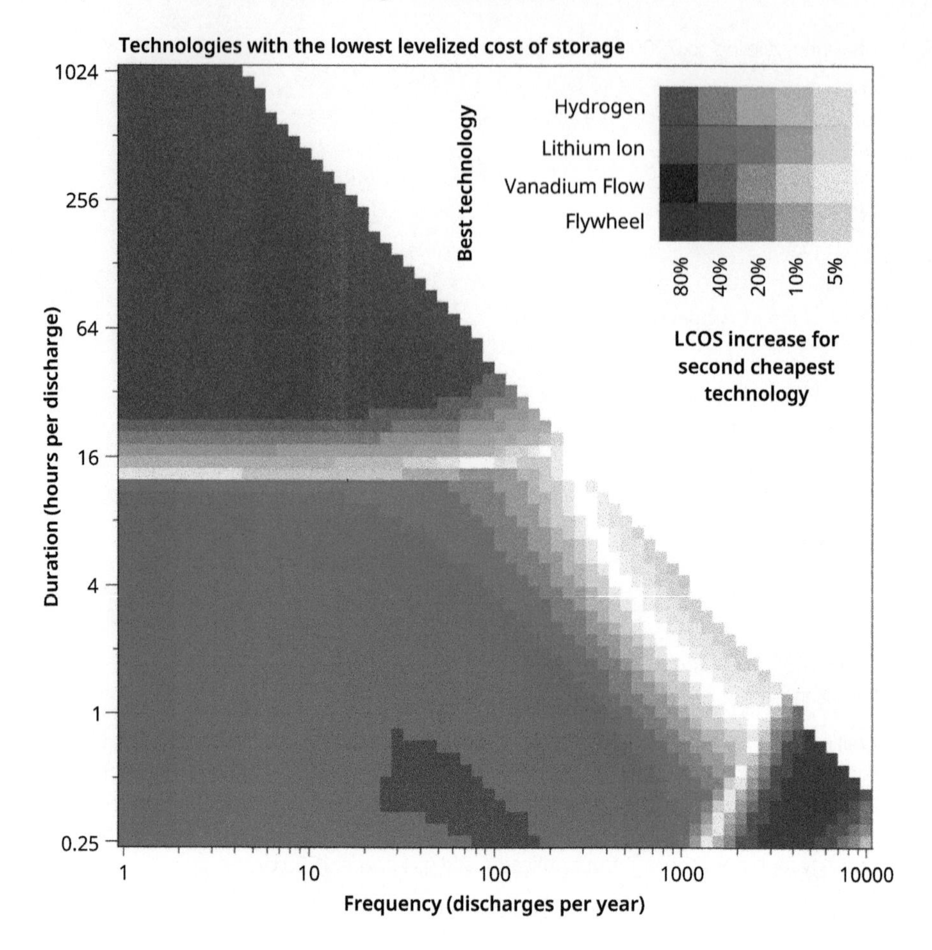

5.12 References

1. Braff WA, Mueller JM, and Trancik JE. 'Value of Storage Technologies for Wind and Solar Energy' (2016) 6 *Nature Climate Change* 964–69.
2. Mayr F and Beushausen H. 'Navigating the Maze of Energy Storage Costs' (2016) 5 *PV Magazine* 84–8.
3. Pawel I. 'The Cost of Storage—How to Calculate the Levelized Cost of Stored Energy (LCOE) and Applications to Renewable Energy Generation' (2014) 46 *Energy Procedia* 68–77.
4. Belderbos A, Delarue E, Kessels K, and D'haeseleer W. *The Levelized Cost of Storage Critically Analyzed and its Intricacies Clearly Explained* (Leuven: KU Leuven, 2016).
5. Jülch V. 'Comparison of Electricity Storage Options Using Levelized Cost of Storage (LCOS) Method' (2016) 183 *Applied Energy* 1594–606.
6. BloombergNEF. Cost of New Renewables Temporarily Rises as Inflation Starts to Bite (2022). Available at: <https://about.bnef.com/blog/cost-of-new-renewables-temporarily-rises-as-infla tion-starts-to-bite/> accessed 5 November 2022.

7. International Energy Agency. *Projected Costs of Generating Electricity* (Paris: International Energy Agency, 2020).
8. McCarthy R. From Proposal to Execution: 5 Challenges on the Path to Net Zero. *Wood Mackenzie* (2020). Available at: <https://www.woodmac.com/news/opinion/from-proposal-to-execution-five-challenges-on-the-path-to-net-zero/> accessed 5 November 2022.
9. Lazard. *Lazard's Levelized Cost of Storage Analysis—Version 13.0* (New York: Lazard Ltd, 2020).
10. Mayr F. Storage at 20 USD/MWh? Breaking Down the Low-Cost Solar-Plus-Storage PPAs in the USA. *Apricum.* Available at: <https://apricum-group.com/storage-at-20-usd-mwh-breaking-down-the-low-cost-solar-plus-storage-ppas-in-the-usa/> accessed 5 November 2022.
11. Figgener J, Stenzel P, Kairies KP, Linßen J, *et al.* 'The Development of Stationary Battery Storage Systems in Germany—Status 2020' (2021) 33 *Journal of Energy Storage* 101982.
12. Wood Mackenzie. *Global Energy Storage Outlook H1 2021* (Edinburgh: Wood Mackenzie, 2021).
13. AES. The AES Alamitos Battery Energy Storage System Made History. Here's Why It Matters. Available at: <https://www.aes.com/aes-alamitos-battery-energy-storage-system-made-history-heres-why-it-matters> accessed 5 November 2022.
14. Ofgem. *The Capacity Market (Amendment) Rules* (London: The Office of Gas and Electricity Markets, 2018).
15. Schmidt O, Hawkes A, Gambhir A, and Staffell I. 'The Future Cost of Electrical Energy Storage Based on Experience Curves' (2017) 2 *Nature Energy* 17110.
16. International Energy Agency. *Projected Costs of Generating Electricity* (Paris: 2015).
17. Ondraczek J, Komendantova N, and Patt A. 'WACC the Dog: The Effect of Financing Costs on the Levelized Cost of Solar PV Power' (2015) 75 *Renewable Energy* 888–98.
18. Hern R, Radov D, Carmel A, Spasovska M, and Guo J. NERA Economic Consulting. *Hurdle Rates update for Generation Technologies* (London: 2015).
19. Few S, Schmidt O, Offer GJ, Brandon N, *et al.* 'Prospective Improvements in Cost and Cycle Life of Off-Grid Lithium-Ion Battery Packs: An Analysis Informed by Expert Elicitations' (2018) 114 *Energy Policy* 578–90.
20. Schmidt O, Gambhir A, Staffell I, Hawkes A, *et al.* 'Future Cost and Performance of Water Electrolysis: An Expert Elicitation Study' (2017) 42(52) *International Journal of Hydrogen Energy* 30470–30492.
21. Swanson RM. 'A Vision for Crystalline Silicon Photovoltaics' (2006) 14 *Progress in Photovoltaics: Research and Applications* 443–53.
22. Schmuch R, Wagner R, Hörpel G, Placke T, *et al.* 'Performance and Cost of Materials for Lithium-Based Rechargeable Automotive Batteries' (2018) 3 *Nature Energy* 267–78.
23. Hesse H, Schimpe M, Kucevic D, and Jossen A. 'Lithium-Ion Battery Storage for the Grid—A Review of Stationary Battery Storage System Design Tailored for Applications in Modern Power Grids' (2017) 10 *Energies* 2107.
24. Lazard. *Lazard's Levelized Cost of Storage Analysis—Version 3.0* (New York: Lazard Ltd, 2017).
25. Bassi S, Boyd R, Buckle S, Fennell P, *et al. Bridging the Gap: Improving the Economic and Policy Framework for Carbon Capture and Storage in the European Union* (London: Centre for Climate Change Economics and Policy & Grantham Institute for Climate Change, 2015).
26. Brundtland GH. *A Report of the World Commission on Environment and Development: Our Common Future* (New York, NY: UN, 1987).
27. Stephan A, Battke B, Beuse MD, Clausdeinken JH, *et al.* 'Limiting the Public Cost of Stationary Battery Deployment by Combining Applications' (2016) 1 *Nature Energy* 16079.

28. Gardiner D, Schmidt O, Heptonstall P, Gross R, *et al*. 'Quantifying the Impact of Policy on the Investment Case for Residential Electricity Storage in the UK' (2020) 27 *Journal of Energy Storage* 101140.
29. Schmidt TS, Battke B, Grosspietsch D, and Hoffmann VH. 'Do Deployment Policies Pick Technologies by (Not) Picking Applications?—A Simulation of Investment Decisions in Technologies with Multiple Applications' (2016) 45 *Research Policy* 1965–83.
30. Preger Y, Barkholtz HM, Fresquez A, Campbell DL, *et al*. 'Degradation of Commercial Lithium-Ion Cells as a Function of Chemistry and Cycling Conditions' (2020) 167 *Journal of The Electrochemical Society* 120532.

6 Market value

Making money

KEY INSIGHT	WHAT IT MEANS
The economic value of electricity storage varies greatly within individual applications in the same market, across different applications within a market, in the same application across different markets, and over time.	There is no stand-out best market or application for deploying storage. This makes developing a generic, international business model for storage deployment impossible: bespoke plans are required.
In general, the revenue potential for energy storage in USD/kW-year increases with longer discharge duration and higher cycle frequency, both of which give greater utilization.	Utilization can provide an initial guidance on the revenue potential of specific services relative to each other based on their respective requirements.
There are two clusters of applications where the lowest cost electricity storage technologies are particularly profitable. The first cluster covers 'energy applications' with 100–1,000 annual cycles and 4–8 hours discharge duration. The second cluster covers 'power applications' with 30–200 annual cycles and less than 1 hour discharge duration.	The analysis on pockets of opportunity can guide investors towards application requirements where electricity storage technologies are likely to be profitable.
When providing arbitrage, the duration and efficiency of storage will influence how much value it can access. Longer duration and higher efficiency will enable higher utilization as more actions will be profitable.	Higher efficiency and longer discharge duration have a clear value in energy arbitrage. However, the additional revenue potential needs to be evaluated against additional investment cost.
The profitability of energy arbitrage varies by plus or minus 25% from year to year within most major electricity markets.	This presents a financial risk to investors unless it can be hedged, as unstable revenues will reduce access to low-cost borrowing.
The standard deviation of hourly power prices is a reasonable predictor for how profitable energy arbitrage will be.	This provides an initial screening test for how attractive different markets will be for providing arbitrage, or how volatile profits may be over time within a market.

(*Continued*)

Monetizing Energy Storage. Oliver Schmidt & Iain Staffell, Oxford University Press. © Oliver Schmidt & Iain Staffell (2023).
DOI: 10.1093/oso/9780192888174.003.0006

KEY INSIGHT	WHAT IT MEANS
Most of the variability in electricity prices happens over the diurnal cycle, so there are diminishing returns to increasing storage duration, and beyond 8 hours there is currently very little additional profit to be gained from arbitrage.	Generally, longer duration storage will see greater profits from arbitrage, but the added benefit of higher performance must be carefully weighed up against additional investment cost.
Every new storage system added to a market will lead to 'revenue cannibalization' for arbitrage, worsening the profitability of all existing storage due the smoothing of power prices.	A little storage will go a long way in markets. Project owners must be wary of the threat posed by new investments on their profitability. Investors may install less storage than would be 'societally optimal'.
Energy storage systems can maximize their value by 'stacking' the revenues of multiple applications they serve within a specified time frame.	Project developers, investors and operators should not limit their focus on one application only, but assess which other applications could be provided with the same storage system.

There are four principal archetypes of revenue stacking:

- **Parallel stacking:** Power capacity is separated into individual parts which serve different applications simultaneously.
- **Sequential stacking:** The same power capacity is provided to different applications in different time periods.
- **Sequential stacking in opposite directions:** The same power capacity is provided to different applications in different time periods. These applications cover both charging and discharging.
- **Overlapped stacking:** The same power capacity is provided to multiple applications at the same time.

KEY INSIGHT	WHAT IT MEANS
The standard profitability metrics are key outputs from the financial modelling of energy storage projects: net present value (NPV), internal rate of return (IRR), and payback period of the investment.	Modelling project finances for energy storage projects is no different to other investment projects. The focus should be on representing the cost and performance parameters of storage systems correctly over time.
The two parameters which most strongly affect the profitability of energy storage projects are the annual revenue and investment cost.	This highlights the importance of identifying application(s) with highest revenues and closely following technology cost developments.

6.1 Sources of value

Storage will not be built because of its technical superiority or its ability to improve the efficiency and environmental impacts of energy production. It will be built because it can generate profits, offering a competitive rate of return against other assets.

To recap Chapter 3, electricity storage creates economic value through four fundamental services:[1]

1. **Power Quality**: Keeping frequency and voltage within permissible limits
2. **Power Reliability**: Providing electricity in case of supply reduction or interruption
3. **Increased utilization**: Optimizing use of existing assets in the power system
4. **Arbitrage**: Exploiting temporal price differentials

Of these, it is the former services related to power quality and power reliability that have proven more lucrative so far, hence most new storage projects up to 2020 were for ancillary services and capacity.[2] However, price arbitrage (buying low and selling high) is considered to have the largest potential role for stationary storage,[3,4] for example with diurnal solar integration, which monetizes the need for reliability and arbitrage. Historically, pumped hydro was deployed for these two services to help utilize low-cost nuclear power generation at night.

All electricity markets are regulated, with tightly defined products controlling what can be monetized.[5] The services that can be provided differ across the world's electricity markets,[6] and the value available from providing these services differs strongly over time and between markets.[7] This value depends on many market characteristics such as the availability of flexible and inflexible generation sources, fuel prices, the penetration of variable renewables, and the weather.[8,9]

So, how much money can storage earn? How does this depend on the market and the requirements of each application? And, does the comparison of available value to the lifetime cost of storage reveal any 'pockets of opportunity'?

This chapter reviews the applications through which storage can access value in four major markets. These services are mapped onto the characteristics of storage duration and discharge frequency, showing the value that different storage technologies may access. This 'value landscape' for storage can then be compared to the lifetime cost landscape of storage from Chapter 5 to understand where profitability may be found, in terms of specific markets and application requirements. Many ancillary services markets are shallow and will be quickly saturated by storage, so the chapter then provides a deeper exploration of arbitrage, and the relative value accessible in different world regions, with different storage technologies, and different installed quantities.

FREQUENTLY ASKED QUESTION

What do you mean by the term 'shallow' ancillary services markets? Do you have an example?

Yes, let's take the German frequency response market as an example (German: *Primärregelleistung*). The market volume is ~600 MW in a system with ~80 GW peak demand.[10] Figure 6.1 shows that the cumulative battery storage capacity pre-qualified for this application increased from 0 MW in 2012 to ~450 MW in 2019.[11] Since batteries can offer this service at much lower cost than conventional bidders, the weighted price for weekly auctions has halved from ~3,000 EUR/MW-week before 2015 to ~1,500 EUR/MW-week in 2019. Weighted prices in the first quarter of 2020 stood at ~1,000 EUR/MW-week. Additional battery projects are likely to further reduce the weighted price down to the lifetime cost of the most expensive batteries in this application. So, a 'shallow' market for a specific application is one that can be easily saturated by a relatively small amount of energy storage. As a result, prices are reduced down to lifetime cost levels of energy storage in this application.

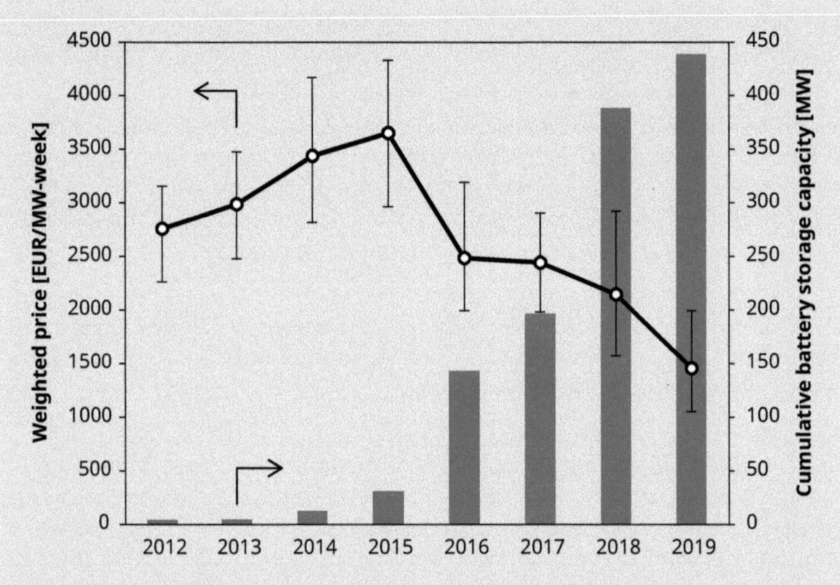

Figure 6.1 Development of mean weekly prices for frequency response in Germany (German: *Primärregelleistung*) and cumulative battery storage capacity pre-qualified for this application between 2012 and 2019.[11] Prices are weighted by contracted volume for each week in the respective year. Error bars represent 25th and 75th percentiles of weighted weekly prices in the respective year.

6.2 Value in international markets

There is no such thing as a 'typical' electricity market. Markets vary substantially around the world, with heterogeneity in their structure and ownership, level of vertical integration, regulations, and the services and products that are offered.[12,13] A thorough treatment of these aspects on how they relate to the business case for energy storage is the domain of bespoke consultancy work, and so this chapter seeks general insights that are common across the major electricity markets of the US, Europe, and Asia Pacific.

Figure 6.2 compares the economic value (i.e. revenue potential) of storage across electricity markets in the US, Great Britain, Germany, and Australia. These markets are chosen as they

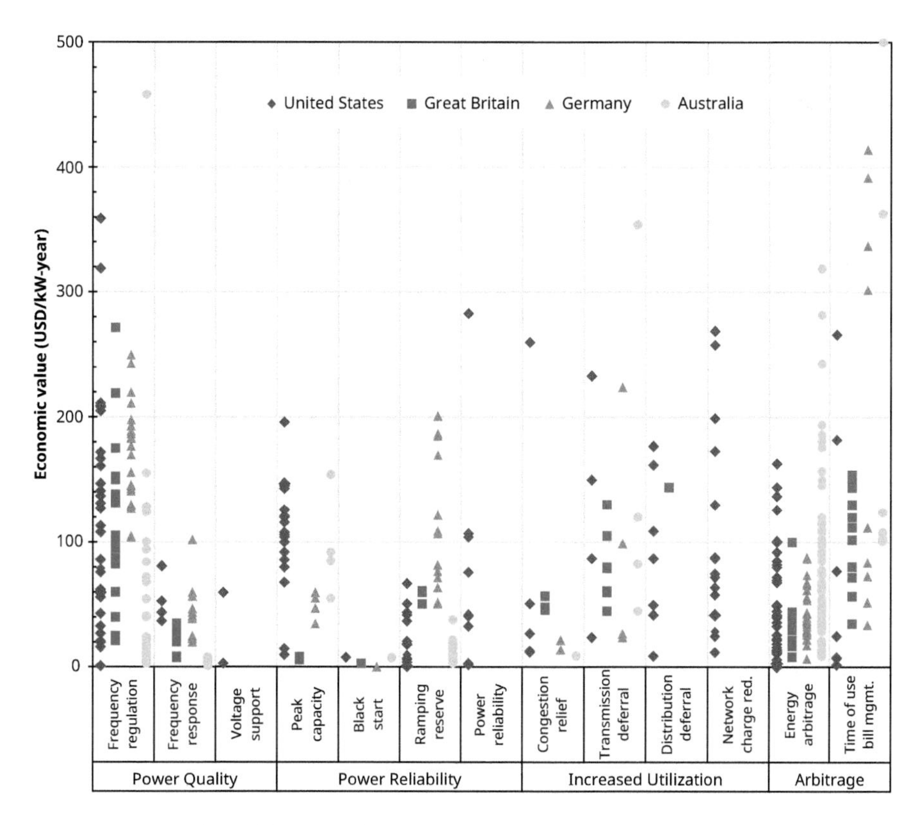

Figure 6.2 The value of different electricity storage applications across four major markets. Data are taken from 176 individual valuation studies and published market transactions as compiled by Balducci for the United States and Housden for Great Britain, Germany, and Australia.[14,15] The specifications of each application vary across the individual studies and are not necessarily aligned with the definitions given in this book. The scope of each study also varies in terms of the timeframe and market considered (as the United States and Australia have multiple electricity markets). All values are presented in USD/kW-year even for applications which are remunerated by energy discharged rather than power capacity, as this then incorporates device utilization. A value of 100 USD/kW-year can be interpreted as a 1 MW system receiving USD 100,000 annual revenue from service provision.

have been widely explored, and so this figure combines 176 values found for the major applications, drawn from recent studies and market analyses.

Three key features evident from this figure are:

- Variability within the same application and market (compare the points within a single column). For example, valuations of distribution network deferral within US markets range from 9 to 177 USD/kW-year.
- Variability within the same application across markets (compare the four columns of points within a given box). For example, ramping reserve averages just 19 USD/kW-year in the US versus 131 USD/kW-year in Germany.
- Variability across applications within the same market (compare columns of the same colour in different boxes). For example, black start is worth 3 USD/kW-year in Great Britain, whereas congestion relief is worth 50 USD/kW-year.

Table 6.1 summarizes the values provided in Figure 6.2, revealing some general trends despite the wide variability. The highest revenues, generally above 100 USD/kW-year, are available in all markets for providing frequency regulation and customer services (managing time-of-use charges). The latter is especially important in Germany and Australia, due to high residential electricity prices. These services are more valuable as they have fewer viable competitors than, for example, arbitrage. The value of arbitrage is broadly in the region of 40–80 USD/kW-year,

Table 6.1 Summary of electricity storage valuation studies and transactions in four major markets. For each market and service, the mean across all studies is given with the 25th and 75th percentiles in square brackets. All values in USD/kW-year.

	United States		Great Britain		Germany		Australia	
Frequency regulation	123	[59–171]	116	[77–151]	169	[143–189]	66	[17–86]
Frequency response	54	[42–60]	22	[11–33]	45	[25–57]	4	[3–6]
Voltage support	22	[3–31]						
Peak capacity	106	[86–126]	7.2	[7–8]	49	[44–57]	97	[78–108]
Black start	8		3.1		0.40		8	
Ramping reserve	20	[4–37]	57	[55–60]	114	[72–170]	15	[9–17]
Power reliability	77	[33–104]						
Congestion relief	73	[13–51]	50	[47–52]	18	[16–20]	9	

(Continued)

Table 6.1 (*Continued*)

	United States		Great Britain		Germany		Australia	
Transmission deferral	124	[71–171]	80	[61–92]	93	[26–130]	150	[73–178]
Distribution deferral	93	[48–122]	144					
Network charge red.	104	[54–141]						
Energy arbitrage	52	[15–81]	37	[22–44]	43	[29–56]	84	[38–102]
Time of use bill mgmt.	65	[7–77]	109	[78–144]	200	[73–337]	456	[107–655]

which is tied to the incremental cost of conventional generation technologies.[14] Transmission deferral is another service with similar and consistently high value across US, European, and Australian markets, due to the high economic cost and non-monetary barriers to building or upgrading transmission lines. Black start is also comparable between regions as the least valuable service, as it can be provided by many existing power stations, and the cost of providing it is minimal due to few cycles and short duration (see Chapter 5).

Some notable differences are also visible. The value of peak capacity is low in Great Britain relative to other markets (7 versus 49–106 USD/kW-year), due to power generation overcapacity in the market during the timeframe studied. Congestion relief is more valuable in the US and Great Britain than in Germany, as grid congestion is a greater problem due to limited grid interconnection. The US average of 71 USD/kW-year is skewed by one especially high value in ERCOT (Texas) which has particularly low interconnection to neighbouring markets. This explains the US mean being above its 75th percentile, and without this value the US mean falls to 26 USD/kW-year, in line with Germany. Australian markets also have limited interconnection in the south-western states, but it is difficult to comment on the value provided as it only comes from a single study.

KEY INSIGHT

The economic value of electricity storage varies greatly within individual applications in the same market, across different applications within a market, in the same application across different markets, and over time.

Revenue stacking describes the ability of energy storage systems to provide multiple services within a specified timeframe and to 'stack' the resulting revenue streams. There is a consensus that this approach can capitalize on the versatility of electricity storage and thereby increase its economic value.[16–19] See section 6.5 for further discussion.

6.3 Value and profitability landscape

6.3.1 Mapping value and specifications

Assessing the economic value of electricity storage requires transparency around the variation of this value along application requirements. Figure 6.3 matches the potential revenues for storage in different applications to the respective discharge duration and cycle frequency requirements in the US markets.[14,20]

While power reliability applications have up to six hours discharge duration but less than 100 full discharges per year, power quality applications are characterized by less than 1 hour discharge duration at various different discharge frequencies. Applications that deliver increased asset utilization have between 1 and 8 hours, and up to 500 discharges. For arbitrage, there are two types: discharge duration below 1 hour at up to 350 cycles and discharge duration up to 6 hours at below 250 annual cycles.

There seems to be a positive relationship between economic value and increasing discharge and frequency requirements (i.e. increasing number of running hours). Applications with up to 8 hours discharge duration or 10,000 cycles are valued at around 125 USD/kW-year respectively, and those with a moderate mix of discharge duration and cycle frequency are valued at around 105 USD/kW-year. The values fall with a reduction in discharge duration

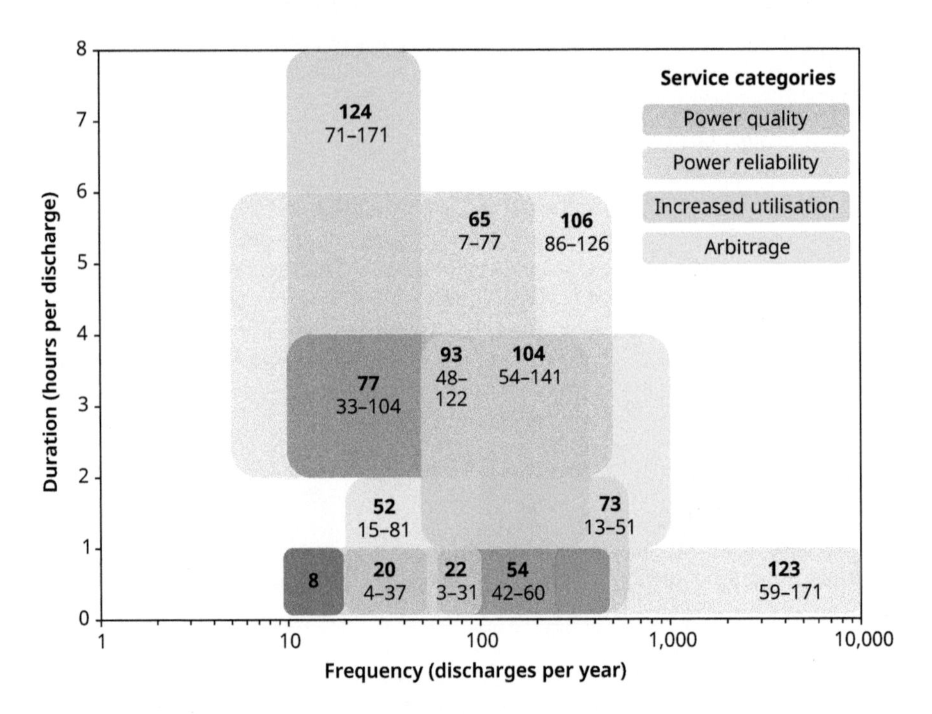

Figure 6.3 Mean economic value for different electricity storage applications in US markets, measured in USD/kW-year (from Figure 6.2 and Table 6.1) plotted along respective discharge duration and frequency application requirements.[14,20] The 25th and 75th percentile of economic values are shown in small font.

and annual cycle frequency, down to 8 USD/kW-year for *black start* at only 1 hour and 10 to 20 cycles. The residential arbitrage application time-of-use bill management represents an outlier with only 65 (7–77) USD/kW-year at up to 6 hours discharge and 250 annual cycles. However, a detailed review of the respective studies reveals that those valuing storage at the lower end of this range assume discharge durations below 4 hours,[14] more in line with the identified value-requirement relationship.

6.3.2 Landscape for storage revenue

Figure 6.4 expands this value analysis to all four major markets and to the entire spectrum of possible discharge duration and frequency requirements from 1 to 1,024 hours and 1 to 10,000 cycles. A Monte Carlo analysis accounts for the ranges in economic value per application and market (from Figure 6.2), and the ranges in discharge duration and cycling frequency for each application (from Figure 6.3).[15] Each discharge-frequency combination on the spectrum is assigned an economic value (in terms of USD per kW of power capacity per year) by interpolating the Monte Carlo results. See Chapter 8 for a more detailed discussion of the methods.

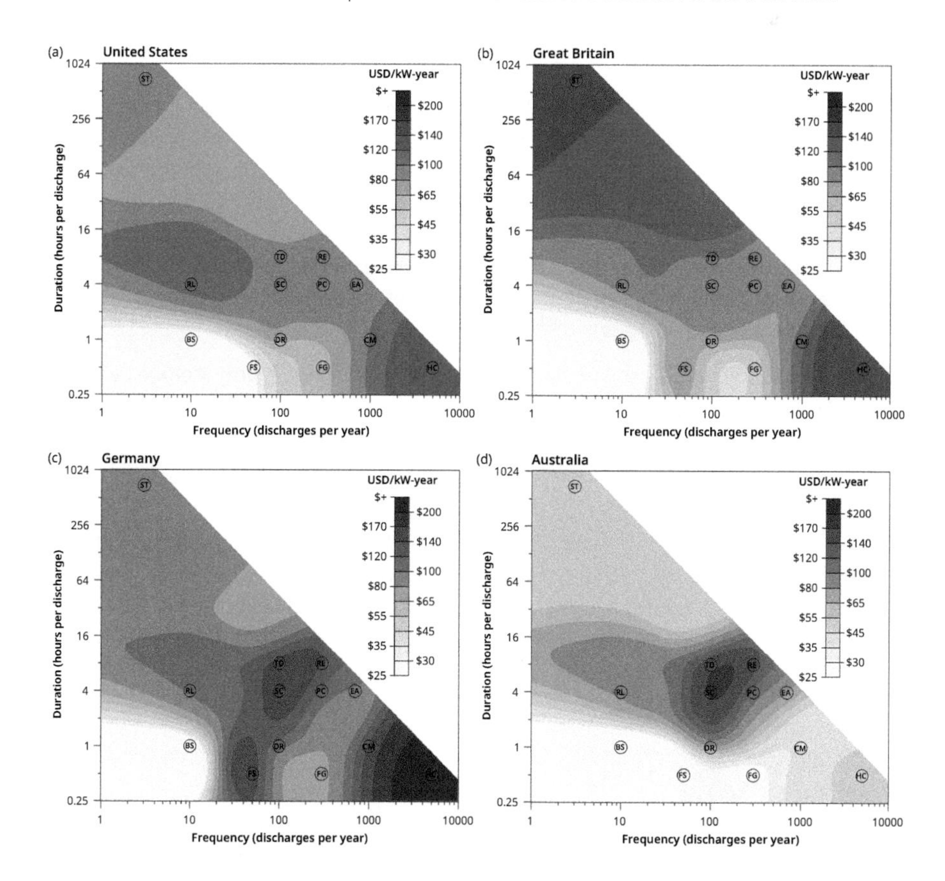

Figure 6.4 Economic market value for electricity storage power capacity across the full spectrum of discharge duration and discharge frequency requirements. Colours refer to economic market value in USD/kW-year. Circled letters represent the requirements of the 13 archetypal applications introduced in Chapter 3.

Panel (a) of Figure 6.4 shows that in the US, over 100 USD/kW-year can be earned for applications below 1 hour discharge duration and more than 1,000 full equivalent discharge cycles per year. This value falls with reducing frequency to below 30 USD/kW-year at 100 cycles per year. A similar pattern is seen in other markets.

Panels (a)–(d) of Figure 6.4 show that the revenue potential increases strongly with longer discharge duration when tracking applications black start (BS) to power reliability (RL) or frequency regulation (FG) to peak capacity (PC). The revenue potential also falls strongly with lower cycling frequency, both for applications with under 1 hour and those with 4 hour duration. This can be seen by tracking left from peak capacity (PC) to self-consumption (SC) to power reliability (RL) for Great Britain, Germany, and Australia.

There are three exceptions to these general trends, which indicate potential temporal 'sweet spots' for energy storage in the respective markets:

- In Great Britain and Germany frequency response (FS) at 50 cycles offers greater value than frequency regulation (FG) at 300 cycles. In Great Britain, successful bids for enhanced frequency response for 2017–21 were awarded 90 to 160 USD/kW-year, while in Germany frequency response in 2018 was valued at 100 to 150 USD/kW-year.[21,22] This shows that fault containment is valued higher than continuous regulation of grid frequency in these markets.
- Germany and Australia show revenue potentials of up to 200 USD/kW-year for applications with 4–16 hours discharge duration and 100–1000 cycles per year, that is, the cluster of services around renewables integration (RE), peak capacity (PC) and energy arbitrage (EA). It may indicate that power price volatility in those markets is particularly high and allows for high revenues (see section 6.4.2)
- The US and Australia show a particular increase in value for the cluster of applications around power reliability (RL), which may be a result of relatively weak power grids and more frequent power outages in these markets.

Figure 6.5 aims to generalize this analysis on revenue potential for energy storage by forming the average across the four studied markets. It confirms the positive relationship identified between revenue potential and increasing discharge and frequency requirements. Some exceptions to this trend also shine through: potential 'sweet spots' identified for selected markets and time frames.

While this analysis reveals high-level insights that support the general understanding of energy storage revenue potential, investment decisions should be based on detailed, market-specific data. This requires a bespoke assessment of detailed revenue data for a specific market, which would allow these landscape figures to be reproduced with greater coverage and validity.

KEY INSIGHT

In general, the revenue potential for energy storage in USD/kW-year increases with longer discharge duration and higher cycle frequency, both of which give greater utilization.

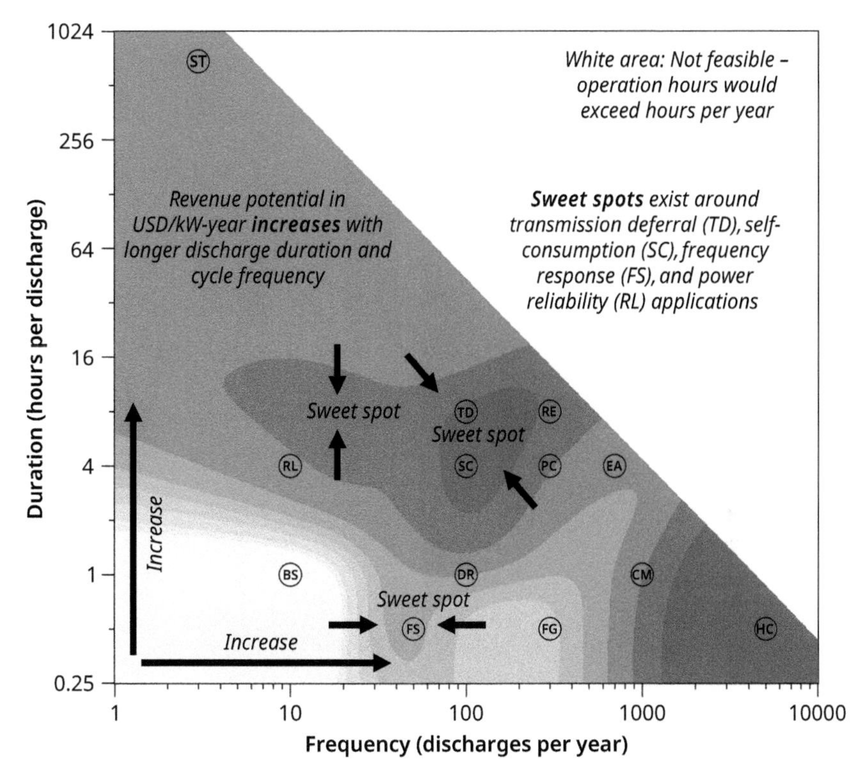

Figure 6.5 Graphic representation of trends in revenue potential for energy storage across the landscape of cycling frequency and discharge duration requirements.

For the purposes of comparing lifetime cost to revenue potential, Figure 6.6 shows the storage revenue landscape in terms of USD/MWh by converting from USD/kW-year using the specific annual utilization which is defined for every grid cell on the landscape plot (frequency multiplied by duration).

Figure 6.6 reveals the opposite relationship between revenue potential and application requirements as shown in Figure 6.5. Revenue in USD/MWh generally increases as discharge duration and cycling frequency decrease.

6.3.3 Landscape for storage profitability

The profitability of storage can be assessed by 'subtracting' the technology cost landscapes evaluated in Chapter 5 from the revenue landscapes derived in section 6.3.2.

Figure 6.7 shows the profitability of energy storage systems in applications that pay USD/MWh for delivered energy ('energy applications') based on subtracting the levelized cost of storage (see Figure 5.18) from the revenue potential (see Figure 6.6) for 2020. Alternatively, for applications that are reimbursed in USD/kW-year ('power applications'), annuitized capacity cost (see Figure 5.18) could be subtracted from the revenue potential in USD/kW-year

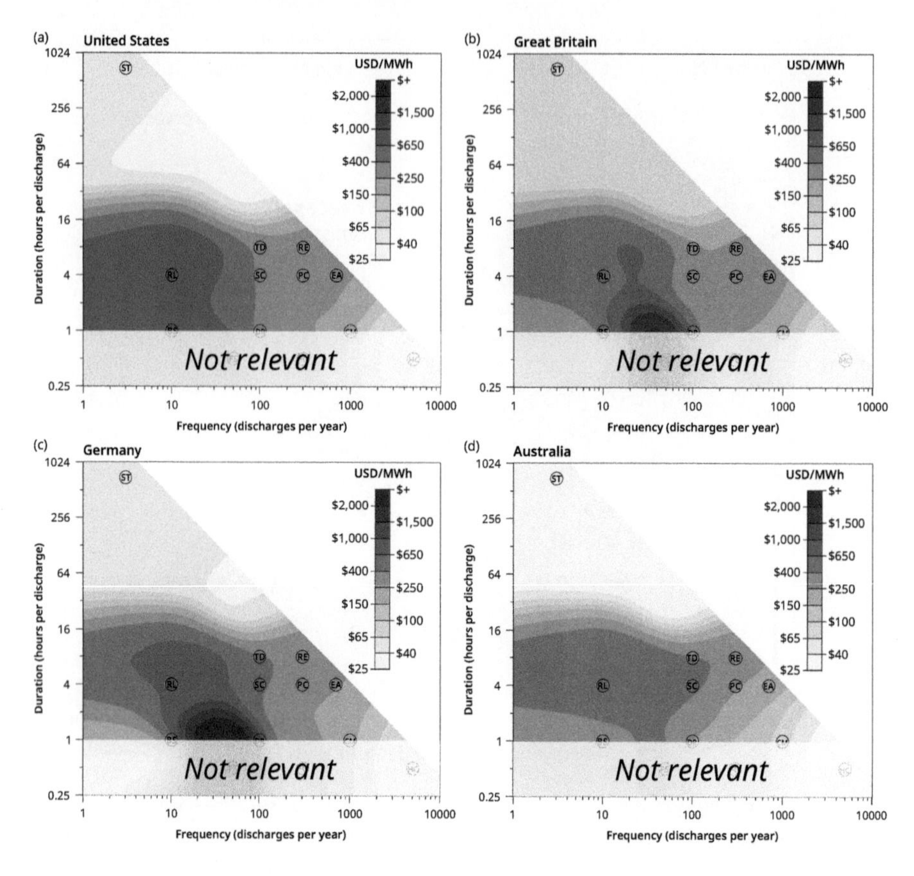

Figure 6.6 Economic market value for providing electricity storage energy discharge across the full spectrum of discharge duration and cycle frequency requirements. Colours refer to economic market value in USD/MWh. Circled letters represent the requirements of the 13 archetypal applications introduced in Chapter 3. Applications < 1 hour discharge are usually not reimbursed in USD/MWh and so are not relevant in this display.

(see Figure 6.4). It is important to differentiate between these two approaches, because the difference between ACC and LCOS is not merely the utilization in each application (as in revenue), but is also determined by the various individual technology cost and performance parameters.

Green areas in Figure 6.7 show 'pockets of opportunity', the operating space in which the most cost-effective storage technology in 2020 was profitable if it received the average revenue for the respective applications in each market. There is a broad pocket across all markets surrounding the cluster of renewable integration (RE), peak capacity (PC), and energy arbitrage (EA). This cluster centres around 100–1000 full discharge equivalent cycles per year with 4–8 hours of discharge duration. The technology with lowest LCOS in this cluster is pumped hydro (see Figure 5.15.a), which is widely deployed worldwide to provide these services.

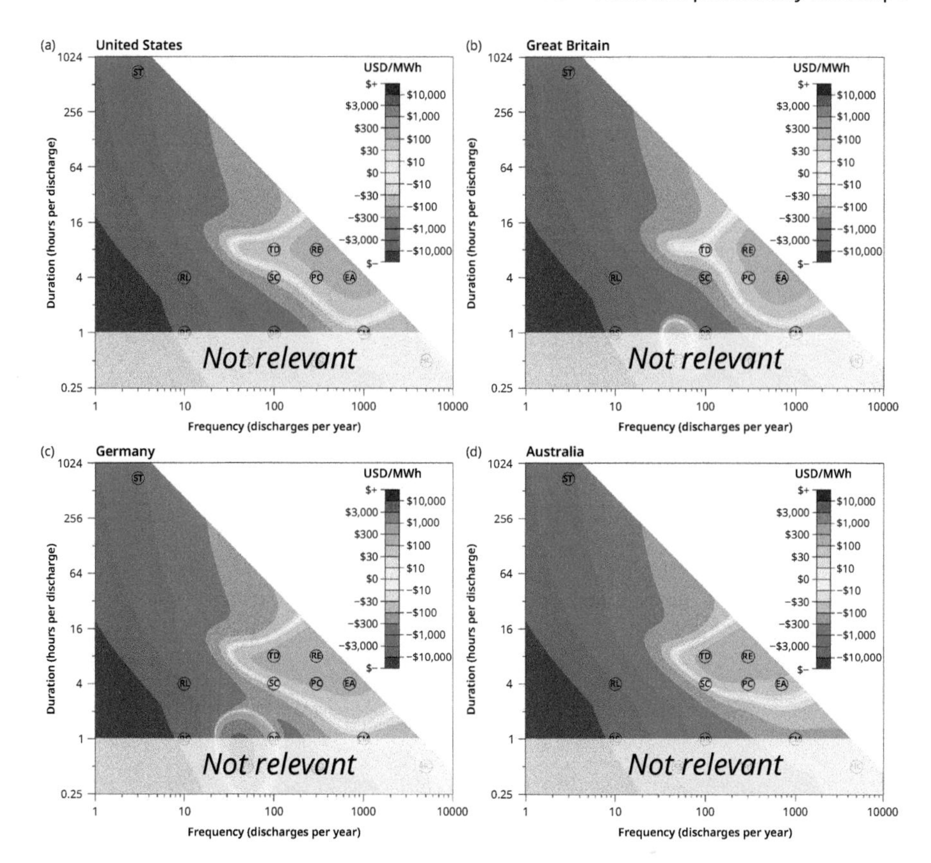

Figure 6.7 Profitability of providing electricity storage energy discharge (in USD/MWh) for applications with various discharge duration and annual cycle requirements across four markets in 2020. Colours indicate the mean revenue per unit of electrical energy discharge (see Figure 6.6) minus the mean levelized cost of storage (LCOS) of the most cost-efficient technology (see Figure 5.18). Green signifies areas where the cheapest storage technology is profitable, other colours signify it is loss-making. Circled letters represent the requirements of the 13 archetypal applications introduced in Chapter 3. Applications with less than 1 hour discharge duration are usually not reimbursed in USD/MWh and so are not relevant in this display.

New pumped hydro systems face many non-financial barriers (geographical suitability, ecological concerns, public opposition), and they are not suitable for certain applications which are gaining traction recently, such as peak capacity by combining solar + storage. The method described here can be used to create bespoke analyses that assess such cases by building a profitability landscape chart for a specific technology, for example comparing anticipated revenues with the LCOS for lithium-ion batteries, rather than the lowest LCOS across all technologies.

This analysis can also be conducted using projected *future* lifetime cost with the assumption that future revenue potentials do not change. Figure 6.8 shows the profitability of the cheapest storage technology based on 2040 cost projections and assuming flat 2020 revenue

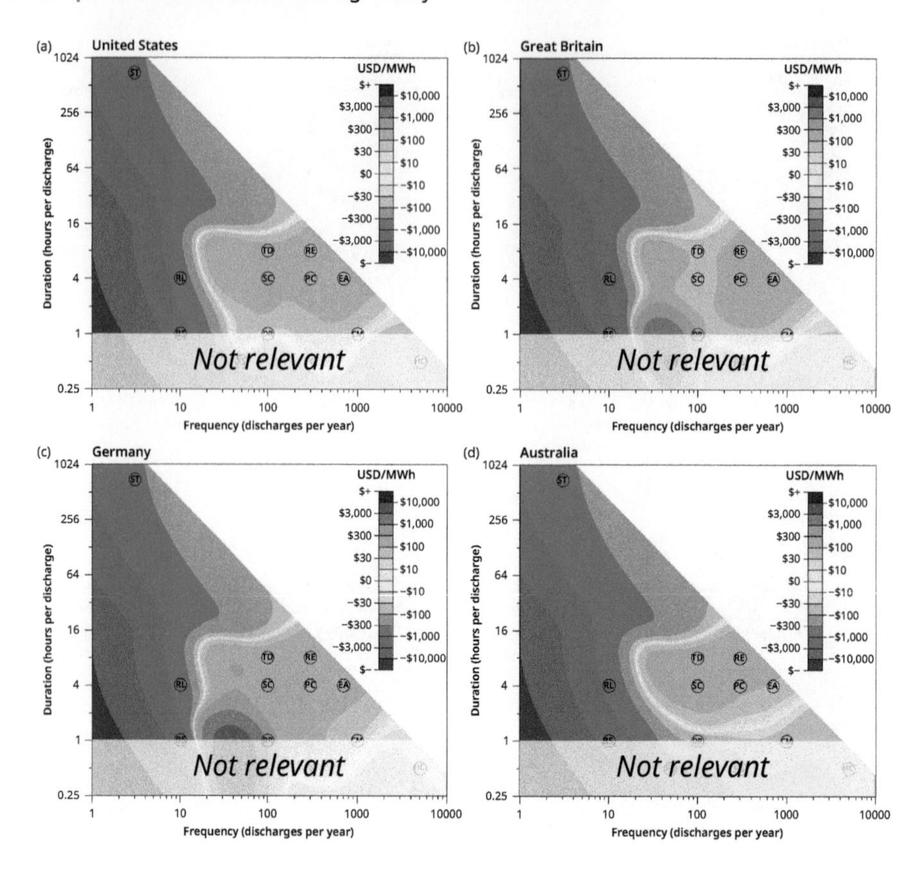

Figure 6.8 Profitability of providing electricity storage energy discharge (in USD/MWh) for applications with various discharge duration and annual cycle requirements across four markets, based on projected 2040 lifetime cost. Colours indicate the mean revenue per unit of electrical energy discharge (see Figure 6.6) minus the mean levelized cost of storage (LCOS) of the most cost-efficient technology (see Figure 5.18). Green signifies areas where the cheapest storage technology is profitable, other colours signify it is loss-making. Circled letters represent the requirements of the 13 archetypal applications introduced in Chapter 3. Applications with less than 1 hour discharge duration are usually not reimbursed in USD/MWh and so are not relevant in this display.

potential. The 'pocket of opportunity' around renewable integration (RE), peak capacity (PC), and energy arbitrage (EA) is enlarged to cover services with 100 discharges per year or less, namely transmission deferral (TD) and self-consumption (SC) across all four markets.

The assumption that revenues stay flat from 2020 through 2040 is highly unlikely though. The Frequently Asked Question in section 6.1 shows a clear example of how revenues diminish once a market is saturated with electricity storage technologies, which is just one factor of many that may affect future revenues.

When considering applications that are paid for power provision in USD/kW-year ('power applications', i.e. usually below 1 hour discharge duration), the cluster around frequency response (FS) was the only profitable 'pocket of opportunity' in 2020 ranging from ~30–200 charge–discharge cycles per year. This is in line with the significant deployment of lithium-ion batteries in frequency response applications, for example Enhanced Frequency Response and Dynamic Containment in Great Britain or *Primärregelleistung* in Germany.

KEY INSIGHT

There are two clusters of applications where the lowest cost electricity storage technologies are particularly profitable. The first cluster covers 'energy applications' with 100–1,000 annual cycles and 4–8 hours discharge duration. The second cluster covers 'power applications' with 30–200 annual cycles and less than 1 hour discharge duration.

6.4 Value of arbitrage

The markets for ancillary services are relatively shallow and will be easily saturated by small amounts of storage (see section 6.1). Arbitrage on the other hand, may not be the most lucrative market, but will be substantially larger and more durable. To give an example from Great Britain, the total cost of system balancing, congestion management, reserve, and response averaged around GBP 1 billion (bn) per year during the 2010s,[23] whereas the total wholesale market had an annual value of GBP 45 bn.[24] By 2022, three-fifths of utility-scale storage in the US was used for price arbitrage (up from one-fifth in 2019),[25] and in the longer term, analysts forecast thousands of GW of small-scale battery storage being installed worldwide,[26] primarily for this purpose to balance out the large uptake of renewable energy (see Chapter 7).

Arbitrage is also different from other services in that detailed historical electricity price data are readily available in many regions, or can be modelled for hypothetical or future power systems using a variety of free or commercial market models. This allows a long-term view on the economic attractiveness of energy arbitrage to be formed more easily than for other markets with opaque pricing.

6.4.1 Modelling arbitrage operation

A variety of optimization techniques can be used to model how storage would operate in a given market and the value it can obtain from its prices (see Chapter 8 for methods). Here we model the simplest case, where the operator has perfect foresight of future prices. This means there is one definite best course of action with no uncertainty due to the quality of price forecasts days or months ahead. This means it will give an upper bound on the expected profitability of storage, but this will consistently reflect the differences seen between markets, years, and technologies.

The optimal dispatch schedule will be to buy electricity during the cheapest periods and sell during the most expensive ones, subject to constraints on its operation. It cannot charge or discharge by more than its power capacity, which limits the amount of activity in a given period. Its state of charge can also not rise above its energy capacity or fall below zero, which imposes limits on the order in which actions occur (i.e. it cannot continue charging once its energy capacity is fully charged).

Figure 6.9 illustrates the optimal dispatch of a pumped hydro system during one week of German power prices.

Three features are evident:

1. The storage buys when power prices are low and sells when they are high, which is the cornerstone of arbitrage operation. During this example week, power for charging is purchased for 26 USD/MWh, and discharged power sold for 58 USD/MWh on average.
2. Not all the cheapest hours are exploited for charging (e.g. it sold power on Sunday daytime even though that was cheaper than Tuesday night when it was buying power). This is due to limitations on the storage duration, as it was already fully charged by Sunday morning and could accept no further energy. With longer duration, these lower-priced hours could be exploited, with all of Sunday used for charging, and all of Friday for discharging.

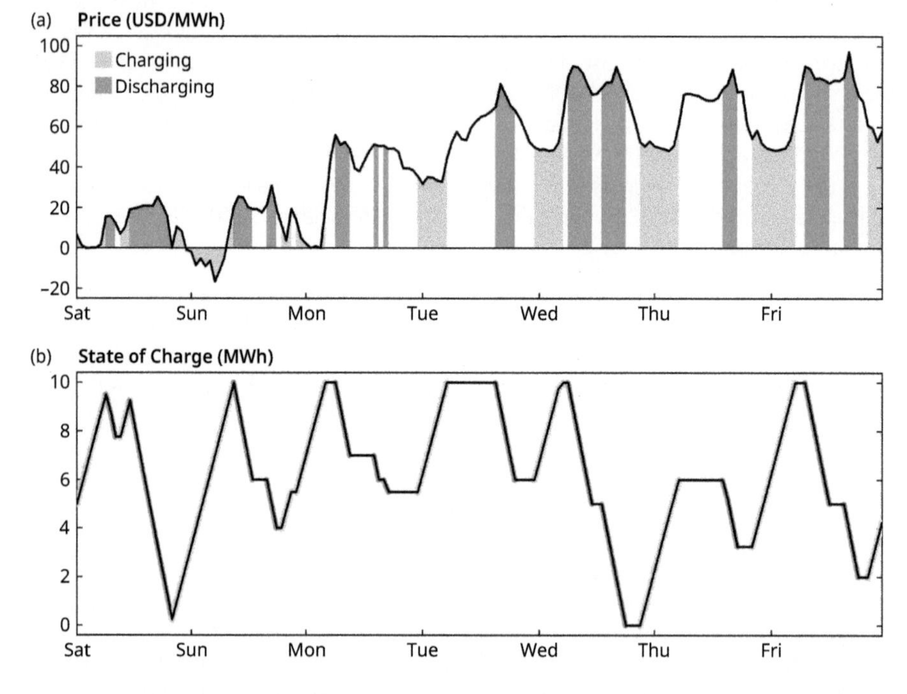

Figure 6.9 Example of the optimal dispatch schedule for a pumped hydro storage system (10 hours, 75% round-trip efficiency) during one week in the German electricity market (8–14 December 2018). Panels show (a) the hourly price in the market with colours signifying the operating schedule for the storage system, and (b) the state of charge of the system.

3. More power must be bought than is sold to cover the round-trip efficiency losses. It is therefore only worth charging if there are periods when it can sell that energy for 1.33 times higher price (1 ÷ 75%). With 100% efficient storage and no marginal costs of operation (e.g. due to operation and maintenance (O&M)), almost every hour would be filled with charge and discharge events, as even the smallest price difference would be enough to profit from. With lower efficiency, there would be more periods of no activity as the price differentials become too small to profit from.

Other market prices and technology configurations can be explored using the companion website at <www.EnergyStorage.ninja>. The worked example in section 6.8 provides a guided modelling exercise using this online tool.

KEY INSIGHT

When providing arbitrage, the duration and efficiency of storage will influence how much value it can access. Longer duration and higher efficiency will enable higher utilization as more actions will be profitable.

6.4.2 Value in current markets

When this process is applied to longer time series of power prices, it can be used to assess the annual revenues and profits, and the variation from year-to-year that could be expected. Figure 6.10 summarizes the value of providing arbitrage with a lithium-ion battery across 162 years of data from 36 electricity markets. These charts show the profit before fixed costs (i.e. the producer surplus), calculated as the revenue from electricity sold through discharging minus the cost of electricity bought for charging. Profit is shown both per unit of energy output and per unit of nominal power capacity, despite arbitrage being remunerated in energy terms. The per-power figure (Panel b) can be interpreted as the profit a device would earn per year factoring in its average utilization: a 100 MW system earning 50 USD/kW-year could anticipate USD 5 million annual profit.

Australian markets provide the greatest opportunity in this example, in part because high penetration of renewables and limited availability of other flexible technologies means power prices have very large spikes. Japan also offers relatively high profits with 2012–19 data as markets were still tight in the wake of the Fukushima nuclear disaster leading to price spikes during high demand periods, combined with a high penetration of solar PV giving low prices during summer daytimes. European and US markets offer typically half the revenues of the Asia-Pacific (APAC) markets shown here, as they have greater capacity margins, greater interconnection between markets, or greater availability of flexible technologies (including pumped hydro storage).

As with other services, the value of arbitrage sees large variation between world regions and also within them. This is most notable within the Australian markets. The Victorian and South Australian markets see consistently higher profits than other states' markets due to their high shares of wind power. Figure 6.11 shows the European markets in greater detail to explore this variability further.

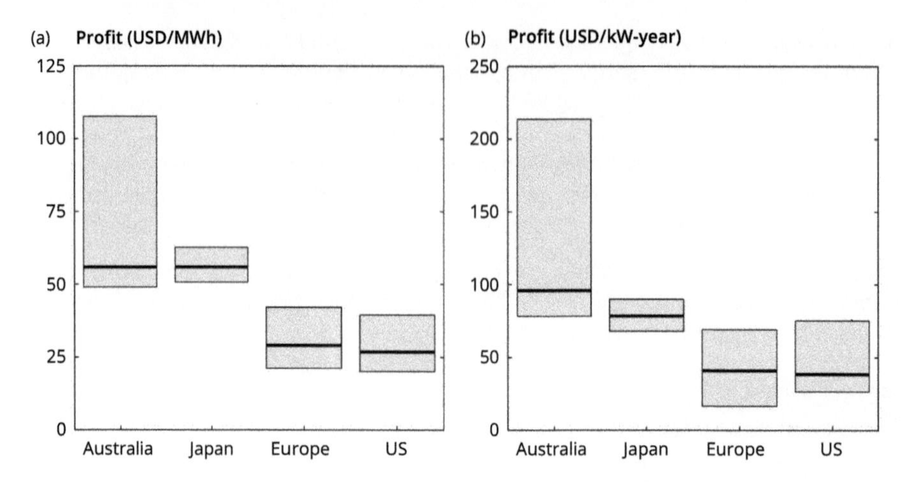

Figure 6.10 Profitability of electricity arbitrage from 2012–19 in various global electricity markets, considering a typical lithium-ion system (4 hours, 86% efficient). Panels show profit (a) per unit of electricity delivered, and (b) per unit of installed power capacity. Each bar covers the 10th to 90th percentile across the national/international markets within each country/region, and the thick line shows the median across markets and years. The value of arbitrage is calculated from the profit-maximizing dispatch against historical power prices, assuming perfect foresight and no uncertainty. Profit is calculated as revenue from discharging minus cost of charging (ignoring any fixed costs).

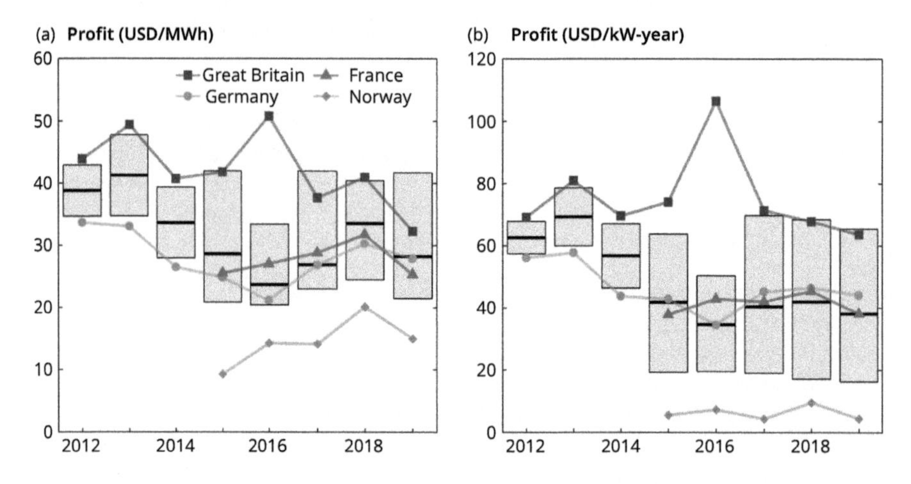

Figure 6.11 Profitability of electricity arbitrage from 2012–19 in various European electricity markets, considering a typical lithium-ion system (4 hours, 86% efficient). Panels show profit (a) per unit of electricity delivered, and (b) per unit of installed power capacity. Each bar covers the 10th to 90th percentile across the 25 national markets within Europe, with a thick line showing the median. Lines highlight the profitability within specific markets, which exemplify an island nation, and well-connected nations dominated by fossil fuels, nuclear, and hydro. The value of arbitrage is calculated from the profit-maximizing dispatch against historical power prices, assuming perfect foresight and no uncertainty. Profit is calculated as revenue from discharging minus cost of charging (ignoring any fixed costs).

During any given year there is a large range in profitability across markets due to their different circumstances. Lithium-ion storage could expect to earn 12 times more per year operating in Great Britain than it could in Norway. Great Britain is among the most lucrative markets in Europe as it is an island nation with relatively little interconnection to its neighbours and relatively little pumped hydro storage due to its geography, meaning power prices are volatile. At the other extreme, Norway offers consistently low profits for batteries as the electricity market is dominated by flexible hydro plants, yielding prices which change gradually over time with the shadow-value of the water they hold.

There is also a large variation in profitability from year to year within markets, as shown for example by the line for Great Britain in Figure 6.11. The median inter-annual volatility across all markets—defined as the standard deviation in annual profit across years, divided by the mean profit—is 24% (P10–P90 = 17–36%). For example, annual profit in Great Britain is 75 USD/kW ± 20%. Major markets such as Great Britain, Germany, and Japan lie close to this median at 20–24%, and most US markets see year-to-year volatility in the range of 19–26%. ERCOT (Texas) is an outlier, with a volatility of 69% (as profitability increased dramatically in 2018–19 compared to earlier in the decade), and Queensland in Australia at 62% (due to very high profitability in 2017).

KEY INSIGHT

The profitability of energy arbitrage varies by plus or minus 25% from year to year within most major electricity markets.

The profitability of storage is driven by variability in prices. The structure of these prices also matters (e.g. if high and low prices occur each day, or are separated by season), but in general the relative profitability of storage in different markets can be approximated by measures of the price volatility. Two such examples are the standard deviation of prices (across all hourly prices in the year), or the average daily price spread (the difference between maximum and minimum price within each day, averaged across all days in the year). Figure 6.12 shows a simple relationship between the logarithm of standard deviation in power prices and the profitability of arbitrage, measured across 162 years of market price data.

Figure 6.12 suggests there is a log-linear relationship between price variation and storage profitability, with each doubling in the standard deviation of hourly prices yielding an extra 35 USD/kW-year arbitrage profit. This relationship is best for describing the relationship across markets, but the data within each region are better described by linear relationships. For example, an additional 1 USD standard deviation in power prices yields an extra 2.10 USD/kW-year arbitrage profit within European markets, versus an extra 0.50–0.75 USD/kW-year within US and Australian markets. Which relationship is more appropriate depends on the underlying cause of the price variations. In the markets with the highest price variation (Australian states and Texas), the standard deviation is driven by a small number of hours with extremely high price spikes (e.g. 100× the annual average price). Unless storage has 100% round-trip efficiency and no variable operational costs, it cannot capture as much value from extreme but infrequent spikes as it would from a market where prices are less extreme, but vary over a wider range in most hours.

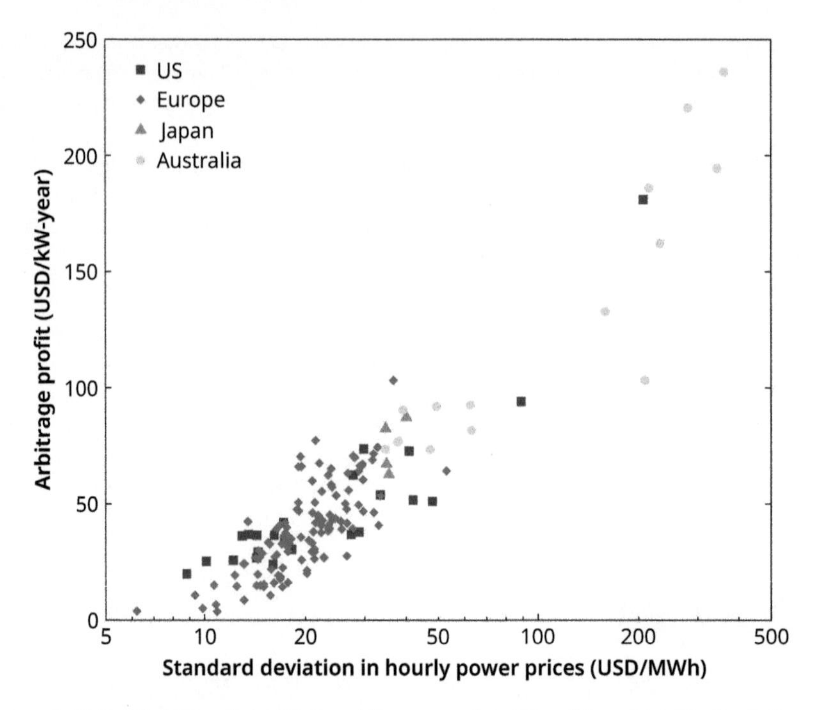

Figure 6.12 Profitability of electricity arbitrage as a function of the amount of variation in hourly market prices. Profitability is estimated for 4 hour, 86% efficient storage from 2012–19 in various electricity markets. Profit is calculated as revenue from discharging minus cost of charging (ignoring any fixed costs).

KEY INSIGHT

The standard deviation of hourly power prices is a reasonable predictor for how profitable energy arbitrage will be.

FREQUENTLY ASKED QUESTION

Are electricity price differentials too small for arbitrage to be a viable business model?

In many markets, yes this is true. However, Figure 6.10 and Figure 6.12 show that there is substantial variability between markets. Markets with large shares of variable renewables, which are short of conventional capacity, and/or have limited interconnection or other flexibility sources will typically have high price spikes and variability between peak and baseload prices. These will make for better investment cases for storage.

The validity of arbitrage as a business model can be confirmed by the rise in combined solar + storage investments, where the storage raises revenue by allowing the time of power sale to be shifted to later in the day to access higher prices.

The value available from arbitrage (or any other service) can be compared to the revenue requirements estimated in Chapter 5 for the relevant duration and cycling frequency. Figure 5.18 shows that the lowest lifetime cost for energy arbitrage is > 100 USD/MWh or > 250 USD/kW-year, for 4 hour duration and 500 cycles per year. Before 2020, this was exceeded only in South Australia (see Figure 6.2). More recently, higher electricity prices and volatility as a result of disrupted supply chains and geopolitical distortions lift arbitrage revenues above these thresholds in more markets. However, it is uncertain how long these price and volatility levels will persist.

It must be noted that storage will likely not only provide arbitrage, unless expressly limited to this by the prevailing regulations. Storage can provide other services in parallel, sequentially or overlapping such as frequency response, and thus access greater revenues through revenue stacking than from arbitrage alone.

6.4.3 Sensitivity to technology parameters

In most electricity markets the greatest price variation occurs over short timescales due to the day/night cycle of demand and solar PV output. Price variations over longer timescales (between days or months) are smaller, giving diminishing returns to increasing storage duration. The greater problem is that exploiting price differences over longer time periods gives fewer operating cycles per year, dramatically reducing the additional revenue that can be made. This presents a financial barrier to the idea that long-duration storage can 'fix' the issue of wind intermittency, which occurs over timescales of days or weeks.[27] There may not be sufficient revenue available to justify building such long-duration storage, unless adding extra duration is very cheap, power prices become more volatile between times of high and low wind output, or other revenue sources are available to longer-duration storage.

Figure 6.13 shows how profitability varies with storage duration, using the same historical market data as in Figure 6.10. Profit increases with storage duration as more periods are accessible due to reduced constraints on when charging and discharging can occur. In APAC markets, storage with 4 hours duration can earn double the profit of 1 hour duration, but over 40 hours duration is needed to earn triple the profit of 1 hour.

Both panels of Figure 6.13 show an elbow around 8 hours. Increasing storage duration beyond this yields very little increase in profit. This marginal profit from an extra hour of duration follows a log-log relationship, similar to the experience curves from Chapter 4. Figure 6.14 shows that for each doubling of storage duration, the extra profit from adding an hour of storage duration falls by around two-thirds (61% in Western and 66% in APAC markets). For example, moving from 4 to 5 hours adds USD 4.65/kW-year in Western markets, but moving from 8 to 9 hours adds USD 1.40/kW-year.

KEY INSIGHT

Most of the variability in electricity prices happens over the diurnal cycle, so there are diminishing returns to increasing storage duration, and beyond 8 hours there is currently very little additional profit to be gained from arbitrage.

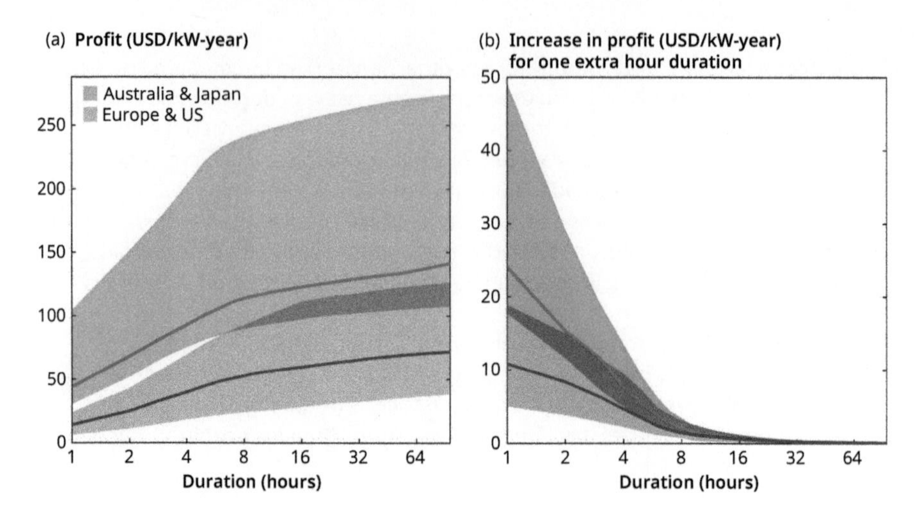

Figure 6.13 Profitability of electricity arbitrage from 2012–19 in various electricity markets, as a function of storage duration, assuming constant efficiency. Panels show (a) absolute and (b) marginal profit. Asia-Pacific and Western markets are shown separately due to their different price levels. Shaded areas cover the 10th to 90th percentile, with a thick line showing the median across markets and years. The value of arbitrage is calculated for storage with 86% round trip efficiency from the profit-maximizing dispatch against historical power prices, assuming perfect foresight and no uncertainty. Profit is calculated as revenue from discharging minus cost of charging (ignoring any fixed costs).

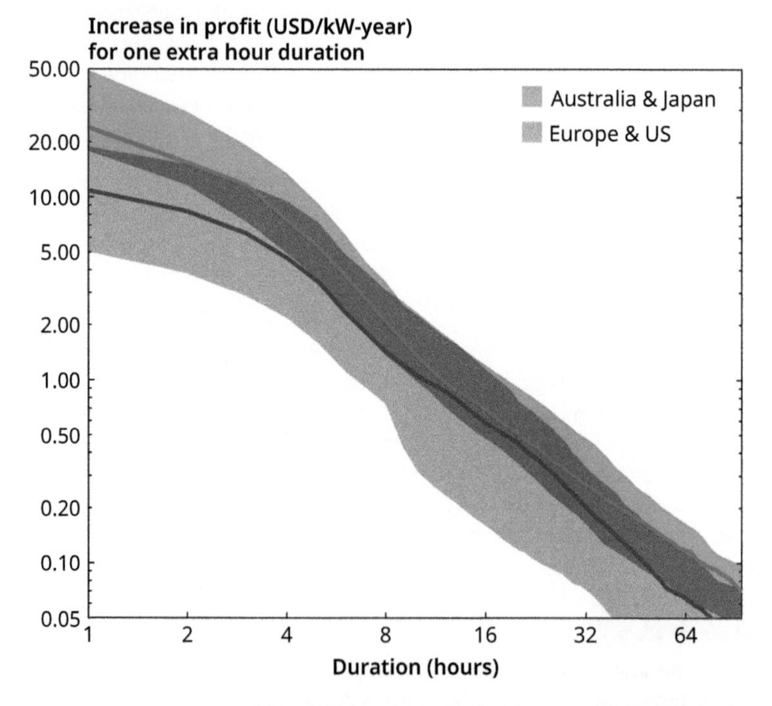

Figure 6.14 Marginal increase in profitability of electricity arbitrage from 2012–19 in various electricity markets, as a function of storage duration, assuming constant efficiency. This shows the same data as in Figure 6.13(b), but on log-log axes.

These figures show profit per kW of power capacity, even though arbitrage is remunerated by energy. This is because the profit per MWh of electricity discharged is not an intuitive metric in this context, as it does not account for utilization (i.e. how much power is bought and sold at this price differential). It is insensitive to storage duration; in APAC markets it averages USD 50/MWh with standard deviation of just ±1.30/MWh when ranging from 1 to 96 hours. In Western markets, it averages USD 30 ± 0.50/MWh.

Figure 6.15 shows how profit scales with storage efficiency using historical market data. More efficient storage is more profitable as less power must be bought for each unit sold (reducing costs), and smaller price differentials are needed to operate profitably (increasing utilization). In European and US markets, 100% efficient storage could earn twice as much as 75% efficient storage, which in turn earns twice as much as 55% efficient storage. The right panel of Figure 6.15 shows the marginal benefit of increasing round-trip efficiency. A rise of 1 percentage point increases profits by 0.5–2.2 USD/kW-year in APAC markets, or by 0.2–1.4 USD/kW-year in Western markets.

Again, profit per MWh discharged is not a helpful metric in this context. For example, 86% efficient storage with 4 hour duration would exploit price differentials of just USD 50/MWh in APAC markets, buying for USD 50/MWh and selling for double that. In contrast, 30% efficient storage would exploit price differences five times larger, buying electricity for 30/MWh on average and selling for over 300/MWh. That does not mean inefficient storage is more profitable though, as this is more than offset by a 92% reduction in device utilization, meaning that overall revenue is two-thirds lower.

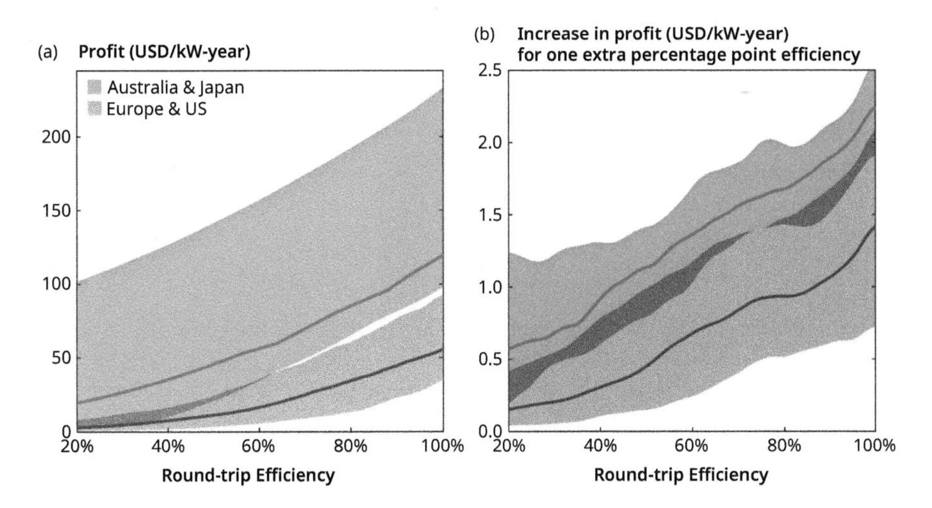

Figure 6.15 Profitability of electricity arbitrage from 2012–19 in various electricity markets, as a function of storage efficiency, assuming constant duration. Panels show (a) absolute and (b) marginal profit. Asia-Pacific and Western markets are shown as separate colours due to their different price levels. Shaded areas cover the 10th to 90th percentile, with a thick line showing the median across markets and years. Value of arbitrage is calculated for storage with 4-hour duration from the profit-maximizing dispatch against historical power prices, assuming perfect foresight and no uncertainty. Profit is calculated as revenue from discharging minus cost of charging (ignoring any fixed costs).

6.4.4 Impact of price cannibalization

The opportunity for arbitrage decreases as more storage systems (or other flexible technologies) enter a market. Adding storage will smooth out power prices, reducing the volatility that is needed for profit. Storage therefore suffers from the same effect as renewable generators, coined 'revenue cannibalization'.[28-30] Cannibalization is a normal market competition issue: when a new product is introduced, it reduces the sales volume or revenues of existing products on the market.

Renewables cannibalize their own revenues because of having near-zero marginal cost, they depress power prices when output is high. The average price captured by renewables is therefore lower than the average wholesale price, by an average of 9% across Europe and the US in 2019.[31] As more renewables are added to a market this effect strengthens, lowering the average price received not only by the new installation, but by all existing renewables too.

The same is true for energy storage, although revenue cannibalization is potentially a bigger issue than it is for renewable generation. Additional storage will both decrease peak power prices (and thus the revenue from discharging) by increasing supply, and increase off-peak prices (and thus the cost of charging) by increasing demand. So far, this has only been observed in electricity market modelling studies, as empirical studies of markets with large growth in storage capacity have not yet been conducted.[32,33] Figure 6.16 visualizes the impact that storage could have on electricity prices, which contrasts the simulated daily pattern of power prices in Great Britain with different penetrations of storage. The average power price is generally unchanged at 35 GBP/MWh, but the lowest overnight prices increase and the evening peak prices fall dramatically.

KEY INSIGHT

Every new storage system added to a market will lead to 'revenue cannibalization' for arbitrage, worsening the profitability of all existing storage due the smoothing of power prices.

Figure 6.17 shows the results of modelling larger uptake of lithium-ion storage in the British market, with the capacity of other generating technologies held constant. The profit of both new and existing storage decreases rapidly as more storage is added. Once 4 GW is installed (~10% of peak demand), profit is half the level that could be expected in a fresh market with no existing storage. Total market-wide profit decreases beyond 6 GW of storage installed, meaning each extra storage system reduces the profit of all existing systems by more than it generates itself. This signals over-competition in the market, and is brought about by both smaller price differentials and falling utilization of storage systems.

It has also be shown how the classic problem of monopoly ownership and lack of competition could affect the financial outcomes for energy storage.[32] A monopolistic owner could, for example, be an operator that owns a large share of the utility-scale or residential storage

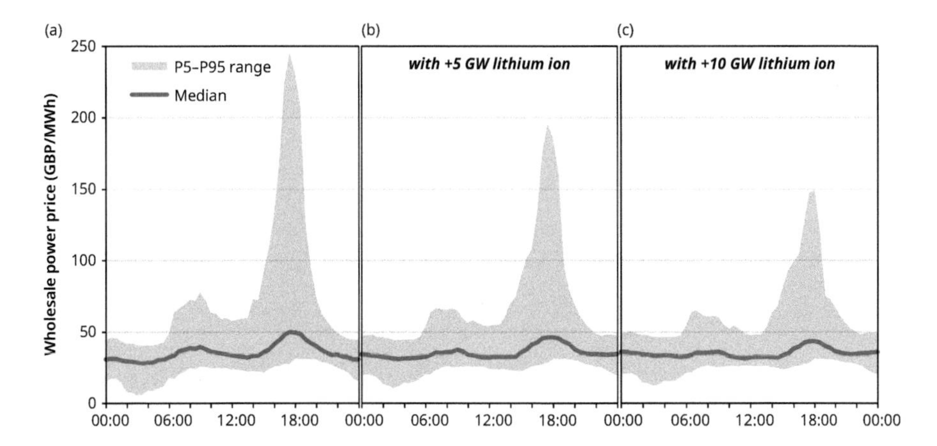

Figure 6.16 The simulated impact of storage on wholesale power prices in Great Britain. Each panel shows the diurnal profile of prices, averaged across all days in 2016. The dark line shows the median price in each hour, the shaded area shows the range from the 5th to 95th percentile (P5 to P95). Panel (a) shows prices modelled in the market as is (with no battery storage installed), while panels (b) and (c) show the prices modelled with additional storage providing arbitrage. Results derived using the MOSSI electricity market model.[34,35]

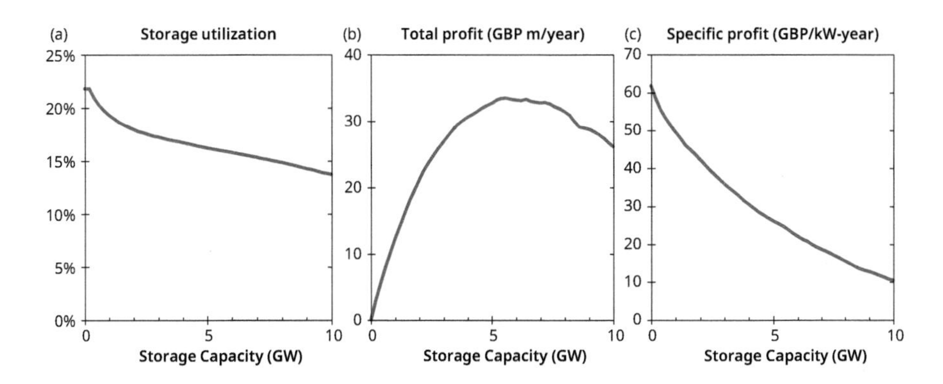

Figure 6.17 Simulations of the storage operation and profitability in the British electricity market. (a) As more storage is added to the market, utilization falls as the opportunities for profitable trade diminishes. (b) Total industry-wide profit increases, but at a decreasing rate until it plateaus and then begins to fall. (c) The profit earned per unit of storage falls rapidly with installed capacity. Data from Ward.[32]

in a market, or provides vehicle-to-grid (V2G) services to a large portfolio of electric vehicles. Such an operator could potentially coordinate the operation of a large number of individual storage devices to restrict utilization at times of peak demand to drive up its profits. With 8 GW of storage in the example above, total profits across the industry could be increased by 45% by restricting utilization by one-fifth. Such a situation could be avoided by regulators monitoring the level of concentration within their market, as they do for traditional power generation assets.

FREQUENTLY ASKED QUESTION

With large deployment of renewables in the coming decades, won't there be ample 'excess' electricity, meaning storage can always charge for zero cost?
It is true that as markets increase their share of wind and solar power, they see an increase in zero- or negative-priced electricity. This has been evident across much of Europe since the mid-2010s. However, 'free excess electricity' will only become commonplace in a market if there are not enough flexible solutions to make use of it.

If there was abundant free electricity, this would lower the cost of operating a storage system for arbitrage. Once this becomes a profitable service (given the capital and other costs which must be met), then as soon as developers build such systems, they will increase demand for electricity at the times when there is excess wind and solar, raising the price of power at those times. Some equilibrium will be reached in the long-term between the amount of renewables (which drive price volatility) and flexibility solutions such as storage (which reduce it).

6.5 Revenue stacking

Energy storage systems can maximize their value by providing multiple services within a specified time frame and 'stacking' the resulting revenue streams. This is called revenue stacking (or alternatively value stacking or benefit stacking) and has three major benefits that can help making energy storage projects profitable:[36,37]

- **Utilization**: Most applications do not require continuous availability or operation (e.g. frequency response) and some are even called upon rarely (e.g. black start). Serving multiple applications can therefore enable higher remunerated battery utilization.
- **Optimization**: Remuneration for most applications is volatile (e.g. auctions for frequency response, volatile wholesale prices for energy arbitrage). Serving multiple applications enables operators to optimize between them for highest revenues.
- **Resilience**: The specific applications that energy storage systems serve may be subject to regulatory changes (e.g. the 'Triad scheme' for demand charge reduction being withdrawn by Great Britain's regulator).[38] Avoiding dependence on a single revenue source can partially protect against the risk of these changes.

Figure 6.18 provides a conceptual depiction of revenue stacking. In this example, application 1 (app 1) is not sufficient to cover lifetime cost. Rather, the stacking of applications 1, 2, and 3 ensures that lifetime revenues exceed lifetime cost.

KEY INSIGHT

Energy storage systems can maximize their value by 'stacking' the revenues of multiple applications they serve within a specified time frame.

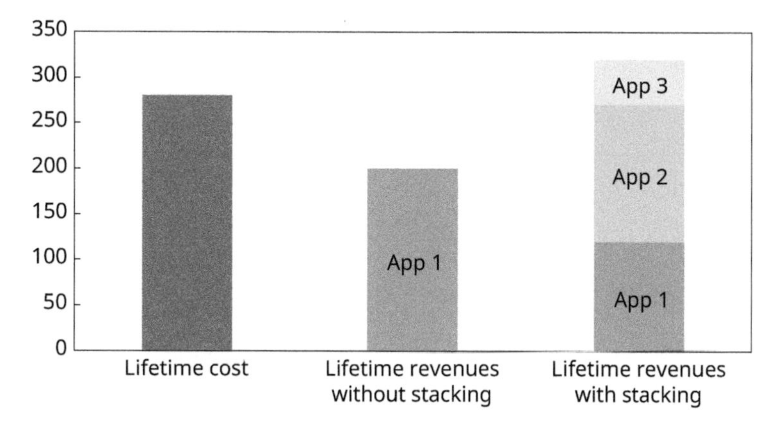

Figure 6.18 Conceptual depiction of revenue stacking. This schematic assumes that lifetime revenues of application 1 are reduced by also serving application 2 or 3; however, revenue stacking is also possible without reducing the lifetime revenues of the original application, depending on the revenue-maximizing operational schedule.

The implementation of revenue stacking in practice is more complex because energy storage systems can serve multiple applications in various ways. Figures 6.19 to 6.22 depict the four main archetypes of revenue stacking, including a description, real-world examples from Great Britain's power market, key considerations, and relevance. These archetypes can also be combined if technically feasible and allowed by regulators.

The archetype descriptions consider two key technical parameters:

- **Active power**: The provision of active power, that is, charging/discharging the storage system (stacked bars, primary y-axis)
- **State-of-charge**: The amount of energy stored in the system at any moment in time relative to full capacity (dashed line, secondary y-axis).

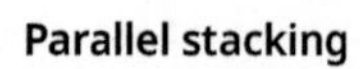

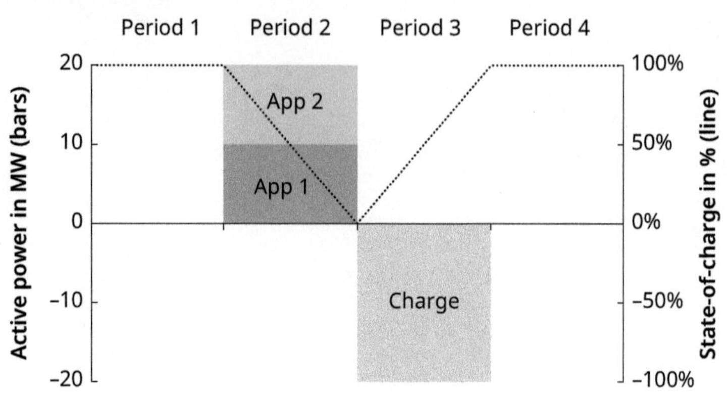

Figure 6.19 Revenue stacking through parallel provision of multiple applications. Exemplary schematic for a 20 MW electricity storage system.

Description:	Power capacity is separated into individual parts. These are provided in parallel to serve different applications. In the above schematic, the 20 MW electricity storage system provides 10 MW to application 1 and 2 each in period 2. As a result, state-of-charge (SoC) reduces from 100% to 0%. The system charges in period 3 to recover SoC from 0% to 100%.
Real-world examples:	Electricity storage operators in Great Britain can provide part of their available power capacity for frequency response services like Dynamic Regulation and bid their remaining capacity into the Balancing Mechanism.[39]
Considerations:	Operators should not exceed the power committed to the individual applications in order not to compromise any other application contracted for simultaneously.
Relevance:	Parallel stacking is among the most common types of revenue stacking. There are numerous examples for various applications in various geographies.

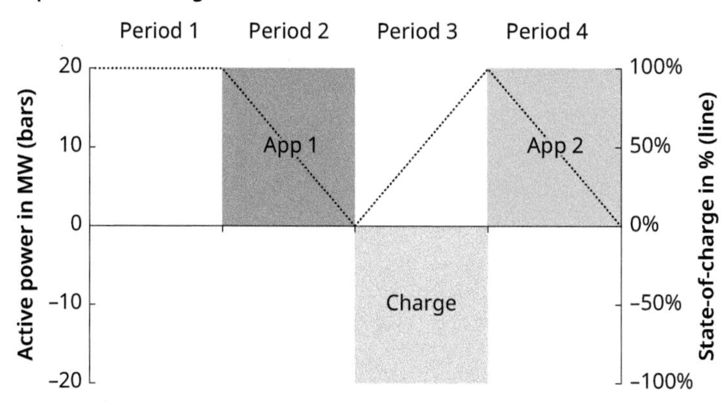

Figure 6.20 Revenue stacking through sequential provision of multiple applications. Exemplary schematic for a 20 MW electricity storage system.

Description:	Power capacity is provided to different applications in different time periods. In the above schematic, the 20 MW electricity storage system provides +20 MW to application 1 in period 2, then charges in period 3 to replenish SoC and then again provides +20 MW to application 2. Depending on the application, the storage system could be rewarded for providing +40 MW in application 2 as its operating point or *baseline* was at –20 MW in the previous period.
Real-world examples:	In 2016, Great Britain's National Grid procured 200 MW of a service called Enhanced Frequency Response. Selected operators did not bid for the entire year but excluded ~300 hours. In these periods they reduced network electricity consumption for large consumers instead, to reduce their demand charges as part of the Triad scheme.[21]
Considerations:	Operators must manage SoC appropriately to ensure that provision of one application does not affect provision of the other.
Relevance:	Sequential stacking is among the most common types of revenue stacking. There are numerous examples for various applications in various geographies.

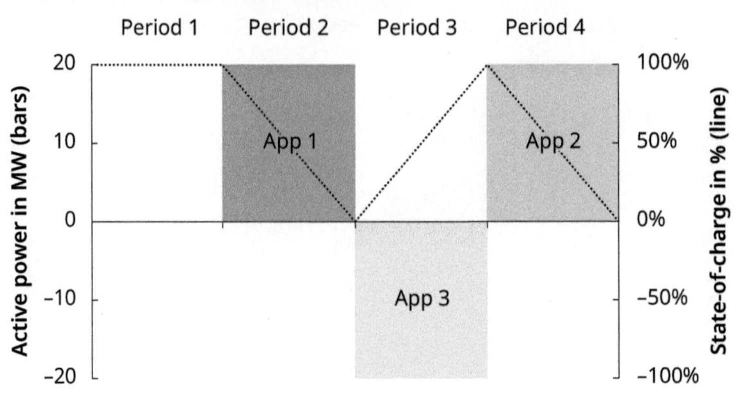

Figure 6.21 Revenue stacking through sequential provision of multiple applications with one application in opposite direction. Exemplary schematic for a 20 MW electricity storage system.

Description:	Power is provided to different applications in different time periods. At least one application is in the opposite direction, so that SoC levels can be managed while being remunerated. In the schematic, the 20 MW storage system provides +20 MW to application 1 in period 2, then –20 MW to application 3 in period 3, which also replenishes SoC. In period 4, +20 MW are provided to application 2 (again, depending on the application, the system could be rewarded for providing –40 MW and +40 MW to applications 3 and 2 respectively based on its *baseline* operating points in the previous periods).
Real-world examples:	In 2021, Great Britain's National Grid enabled the possibility to sequentially stack contracted frequency response services like Dynamic Containment with bids to the Balancing Mechanism (BM). The purpose of the BM bids is to manage SoC levels after providing frequency response.[40] The opposite approach is even more common: offer frequency services to charge the system and then discharge into the BM or intraday market.
Considerations:	It can be difficult to find two applications with opposite directions that fully enable managing SoC. Often, required power capacity and discharge duration vary strongly between applications. A starting point to identify opportunities are negatively correlated applications.
Relevance:	Opportunities are less common than parallel or sequential stacking.

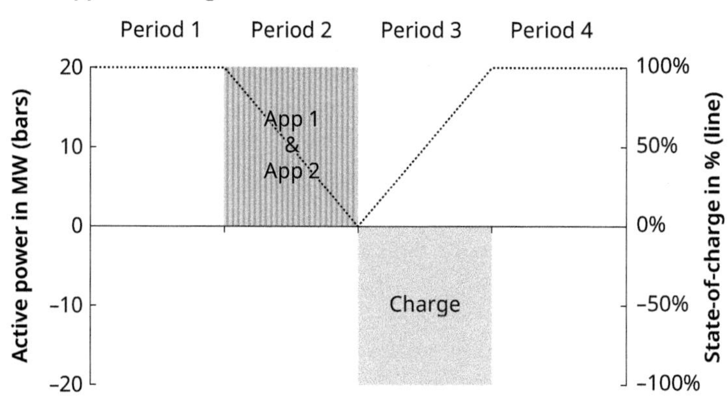

Figure 6.22 Revenue stacking through parallel provision of multiple applications with the same power capacity. Exemplary schematic for a 20 MW electricity storage system.

Description:	The same power capacity is provided to multiple applications at the same time. In the above schematic, the electricity storage system is serving its 20 MW power capacity to both applications in period 2. It then charges in period 3 to replenish its SoC.
Real-world examples:	Contracts in the Capacity Market and Firm Frequency Response (FFR) can remunerate the same MW of an electricity storage system in the same time period. During a Capacity Market stress event any power capacity provided to selected ancillary services like FFR are deducted from the power capacity required in the Capacity Market stress event without penalties.[41]
Considerations:	The ability to fulfil requirements of two applications simultaneously may be limited and there are high penalties if contractual obligations are broken. A good starting point to identify opportunities is to look for applications that are positively correlated.
Relevance:	Least relevant archetype, because opportunities are rare and regulations often inhibit providing for alternative applications when power capacity is already contracted.

KEY INSIGHT

There are four principal archetypes of revenue stacking:

- **Parallel stacking**: Power capacity is separated into individual parts which serve different applications simultaneously.
- **Sequential stacking:** The same power capacity is provided to different applications in different time periods.
- **Sequential stacking in opposite directions**: The same power capacity is provided to different applications in different time periods. These applications cover both charging and discharging.
- **Overlapped stacking**: The same power capacity is provided to multiple applications at the same time.

For simplicity, all four archetypes refer to the actual provision of active power. However, applications can differentiate in payments for the availability of power capacity only and the actual active provision of power. Figure 6.23 adds contracted availability of power capacity for application 1 (light blue) to the archetype parallel stacking (see Figure 6.19).

Application 1 could be a symmetric application, which pays for positive as well as negative power provision. It also pays for the mere availability of power capacity (light blue) and the actual provision of power (dark blue). In this scenario, the 20 MW system is paid for the availability of ±10 MW in periods 1, 2, and 4. It is also paid for providing active power of +10 MW to application 1 and application 2 in period 2. In period 3, the system recharges. This means the system only contracted applications with obligations that can always be met.

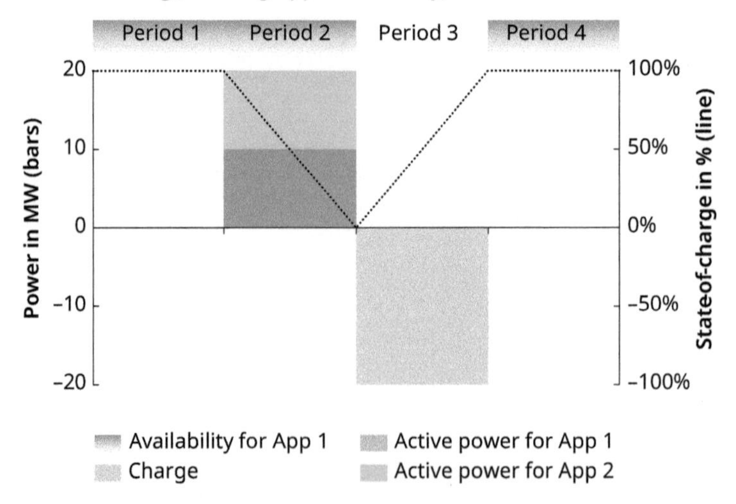

Figure 6.23 Revenue stacking through parallel provision of multiple applications. Light blue reflects availability for application 1. Dark blue and orange reflect active power provision to application 1 and 2 respectively. An example for application 1 could be frequency response. An example for application 2 could be wholesale market arbitrage.

Alternatively, a storage system could indicate its availability to multiple applications and receive respective payments from each. This is shown in Figure 6.24, which adds the paid availability of power capacity for application 1 and 2 to the archetype of sequential stacking (see Figure 6.20).

The storage system contracts its power capacity availability simultaneously to application 1 and 2. This reflects the archetype overlapped stacking for power capacity *availability*. As a result, the system gets paid for reserving its power capacity to applications 1 and 2 in periods 1, 2, and 4. In period 2 it is also paid for actively providing power to application 1. It does not need to be available for application 1 in period 3 and is only paid for reserving its power capacity for application 2 while it charges. In period 4 it then provides active power to application 4.

In this particular scenario, the system can meet its contractual obligations. This is a lucky coincidence, however. If application 2 were called at +20 MW in period 3, the storage system could not provide it (SoC was at 0). Similarly, if application 1 were called again in period 4 with +20 MW (simultaneously with application 2), the system would not have enough energy to provide for both applications. Therefore, this type of revenue stacking should not be called overlapped, but rather 'overbooked stacking'.

For applications where actual activation of active power is subject to a tendering process 'overbooked stacking' may be manageable. High-priced bids would mean the storage system is not called into activation although being paid for availability. However, in most cases

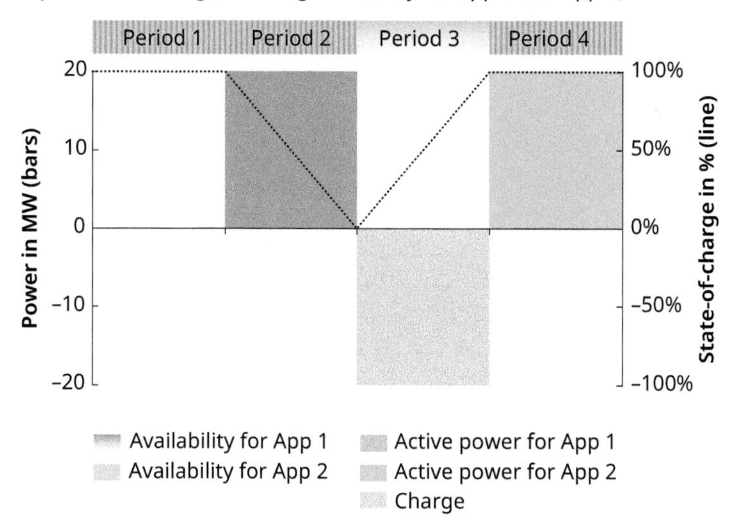

Figure 6.24 Revenue stacking through simultaneous reservation of power capacity availability and sequential active provision of power capacity to two applications. Light blue and orange reflect availability for applications 1 and 2 respectively. Dark blue and dark orange reflect active power provision to application 1 and 2 respectively. An example for application 1 could be frequency response. An example for application 2 could be frequency regulation.

this type of revenue stacking would likely be illegal and result in penalties for failure to deliver. Also, the high level of transparency in system-relevant applications through obligatory real-time communication of system operation parameters may mean that customers will uncover overbooked stacking quickly.

When planning for revenue stacking, there are six key topics to consider:

1. **Technology**: The practical ability of the storage system to meet the operational requirements for multiple applications in the same timeframe must be assessed carefully (e.g. ability to manage state-of-charge) to avoid penalties when contractual obligations cannot be met.
2. **Lifetime**: Additional charge–discharge cycles of the storage system when serving additional applications may degrade the system disproportionately. This can significantly shorten its lifetime and eliminate the economic benefits of the additional revenue streams.
3. **Timelines**: Different applications may have different tendering timelines and contract lengths, which must be managed to ensure contractual obligations can always be met.
4. **Regulation**: Regulators may devise rules that forbid the provision of multiple applications within the same time frame or specify conditions for doing so. A prominent example is the California Public Utilities Commission, which published a set of rules in 2018 for utilities to procure electricity storage services.[37] These define at which network level applications can be provided, which applications have priority over others, and which information must be communicated by storage operators to utilities.
5. **Counterparties**: Different applications may have different counterparties with different contractual and technical interfaces that must be managed.
6. **Connection**: Different applications may require the storage system to be connected at the transmission or distribution network, or at the customer site. If multiple applications are targeted, project developers need to ensure that they can be targeted from the chosen connection point.

The value of revenue stacking is not quantified here as it is very case-specific, driven by the extent to which technology and regulations permit or prevent services from being delivered together.[19] It should be expected that the value of stacking two services will be less than the sum of their individual values due to partial blocking or masking between them but will still be greater than the value of providing one service alone.[19]

In practice, project developers and operators run optimization algorithms that consider the operational feasibility of providing multiple applications and optimize the operational schedule (long term and short term) to maximize revenues.

6.6 Financial appraisal

Before energy storage projects get built, their profitability is assessed. This is usually part of the project development phase (see Chapter 2). It follows the same logic as any financial modelling of investments. However, there are some special considerations for energy storage projects, which will be highlighted in this section.

There are two approaches to choose an energy storage technology and obtain an initial view on a project's profitability. In markets where storage systems target only one application and revenues are known, this can be done based on the application-specific lifetime cost (see Chapter 5). In markets where energy storage systems provide multiple applications and stack revenues, this assessment is done through a combination of revenue modelling and financial modelling. In both cases, a financial model is the basis for the final investment decision.

1. **Revenue modelling**: An optimization model with a high temporal resolution (e.g. half-hour). It uses technology performance parameters and revenue data (e.g. auction results) to optimize storage operation for maximum annual revenues. This can be done for one or multiple applications. The resulting revenues are a key input to the financial model.
2. **Financial modelling**: A discounted cash flow model with a low temporal resolution (e.g. yearly) that determines the returns of an investment in an energy storage system and is the basis for investors to take an investment decision. It accounts for the technology cost and performance parameters, the application requirements and revenues, taxation, subsidies, and financing parameters like equity vs debt ratio and debt interest rates.

KEY INSIGHT

The standard profitability metrics are key outputs from the financial modelling of energy storage projects: net present value (NPV), internal rate of return (IRR), and payback period of the investment.

One particular consideration to be aware of for energy storage projects is that annual revenues are usually treated with respect to power capacity (e.g. USD/kW-year). The reason is twofold. First, for applications that pay for delivered energy, this metric includes asset utilization. Second, in many markets grid connections are difficult and costly to obtain, which is an incentive to optimize revenues for it.

Taxation is determined by governments and therefore highly market specific. However, there are key elements that persist across markets and should be considered in the financial modelling of storage systems. These are

- **Corporate income & business tax**: Taxation of the income and/or property value of businesses (e.g. energy storage operators/owners) and payable to federal governments and/or local authorities. These need to be deducted from revenues.
- **Value added tax & import duties**: Taxes on the purchase of goods in general (e.g. energy storage technology) and/or based on the import location. These need to be added to cost quotes of technology providers.
- **Power tax & levies**: Taxes that usually come on top of electricity wholesale prices and are part of the electricity purchase price for storage systems. In some markets, however, storage systems may be exempt from power tax and levies to support their deployment and avoid double taxation.[42,43]

While the first two tax types are relevant to any business, power taxes and levies are particularly important to electricity storage systems due to the high consumption of electricity when charging.

A particular issue to be aware of is double taxation, which can occur when electricity storage systems are classified as consumers. In this case, they pay power taxes (and levies) when they consume electricity. However, these taxes and levies are paid again after the electricity is discharged and then consumed by the end-customer. Here, the electricity that was stored in the storage system is taxed twice.[44] Similarly, if electricity storage systems are classified as consumers and generators since they both consume and deliver electricity, they themselves may be subject to taxes and levies that apply to consumed and delivered electricity. This issue can be resolved through a clear definition of electricity storage systems (e.g. consumer, generator, separate category), including the taxes and levies that must be paid.

With respect to taxation, the depreciation period for electricity storage projects should also be considered.[45] Similar to other technology investments, electricity storage investment cost can be depreciated over time and thereby reduce corporate income tax for this period.

FREQUENTLY ASKED QUESTION

I heard insurance cost can significantly affect profitability of energy storage projects. Is this true?
Insurance cost for energy storage projects must be accounted for in the O&M cost. The key insurance is fire, theft, and damage. Since energy storage systems are still perceived as relatively novel technology, only a handful of insurers are willing to underwrite these risks for storage, even in mature energy storage markets like Great Britain as of 2022. Insurers typically charged > 5,000 USD/MW per year for standard systems and > 3,000 USD/MW per year for systems that are optimized to minimize these risks and meet respective regulations (e.g. UL9540A). These premiums are likely to fall as insurers gain experience with energy storage technologies. The extent to which insurance affects profitability of an energy storage project must therefore be carefully assessed during the financial appraisal.

Energy storage is supported by many types of subsidies, including:

- **Grants**: Provide direct cash to reduce total investment cost. Example: Great Britain's Longer Duration Energy Storage Demonstration program.[46]
- **Fixed/variable tariffs or premiums**: May increase revenues and provide revenue certainty to facilitate project financing. Example: Germany's innovation tender scheme.[47]
- **Loans with low interest rates**: Reduce financing cost. Example: Australian sustainable household scheme.[48]
- **Tax credits**: Reduce taxation that would otherwise fully apply to the energy storage project. Example: US investment tax credit.[49]

In line with project finance in general, the financing parameter debt/equity ratio (or gearing) is critically important. This determines how much of the investment in energy storage projects can be financed with lower-cost debt, for example loans from banks, and how much must be provided as equity from investors directly. As lenders tend to be more risk-averse, the ratio strongly depends on the maturity of the energy storage technology and the reliability of future revenues (see the FAQ below). A higher share of debt will usually increase the return for investors because they need to provide less capital upfront. This, however, depends on the interest rate that needs to be paid on the debt.

FREQUENTLY ASKED QUESTION

How can merchant electricity storage projects provide certainty on future revenues in the absence of PPAs, fixed tariffs, or premiums?
Selected operators of electricity storage projects are offering price floors on revenue (e.g. for 7 years), which transfers the downside risk of the project from lenders to operators and thereby improves the project's risk profile from the lender's perspective. However, price floors come at a cost, which the operator will deduct from revenues. This will reduce overall profitability of the energy storage project, but from the equity investor's perspective, the higher debt/equity ratio may outweigh this reduction.

Table 6.2 lists exemplary values for the parameters that are relevant for the financial appraisal of electricity storage projects.

Table 6.2 Parameters for the financial appraisal of electricity storage projects. In a financial model, these need to be considered together with technology cost and performance parameters and application requirements.

Parameter	Unit	Value	Consideration for energy storage projects
Revenues			
Initial annual revenue	USD/kW-year	400	Should consider all applications provided, usually given in USD/kW-year
Revenue escalator	% p.a.	1%	Should account for future development of considered revenue streams and potential new ones
Taxation			
Value added tax	%	15%	Must be added to technology provider quotes on investment cost
Corporate income tax	%	10%	No specific considerations
Investment cost depreciation period	years	10	Should correspond to legally defined depreciation period for electricity storage systems if applicable

(*Continued*)

Table 6.2 (*Continued*)

Parameter	Unit	Value	Consideration for energy storage projects
Subsidies			
Investment cost subsidy	%	10%	Sample type of electricity storage subsidy—if applicable, deduct from investment cost
Financing			
Equity share	%	70%	Usually high for energy storage projects (> 50%) due to uncertainty of future revenues
Debt share	%	30%	Usually low for energy storage projects (< 50%) due to uncertainty of future revenues
Debt interest rate	% p.a.	3%	No specific considerations
Loan duration	years	10	No specific considerations

Figure 6.25 shows the typical annual and cumulative cash flows from a discounted cash flow model if these parameters are applied to the project finances of the exemplary vanadium redox-flow battery system providing peak capacity services introduced in Chapter 5. Panel a) shows the cash flows for the project in general. Panel b) shows the cash flows to the equity provider.

The project would have an NPV of ~USD 5 million (m) and a simple and discounted payback period of 9 or 16 years respectively. The project IRR would be ~10.5%, which is at the lower end of the threshold required by private sector investors in the energy storage industry.

However, only 70% of the investment cost is provided as equity. Therefore, to the equity provider the NPV would be ~USD 8 m and simple and discounted payback period would be 8 or 13 years respectively. At ~12.5%, the equity IRR would then be within the threshold required by private sector investors and make the project investable. That is because the equity provider needs to provide only 70% of the investment cost upfront and can pay 30% as debt service during the loan period. Debt service payments during the loan period are discounted. At the given interest rate of 3% per year, the sum of 70% upfront investment and 30% discounted debt service is lower than 100% upfront investment.

Figure 6.26 shows the results of a sensitivity analysis of key technology cost, application requirement and financial appraisal parameters on a) project IRR and b) equity IRR. In both cases, absolute changes of ±20% in annual revenue, investment cost or annual charge-discharge cycles affect IRR by more than 10% (relative). Taxation and subsidies have a relatively small impact on IRR of less than 5% (relative). The impact of financing parameters is more

(a) **Project cash flows**

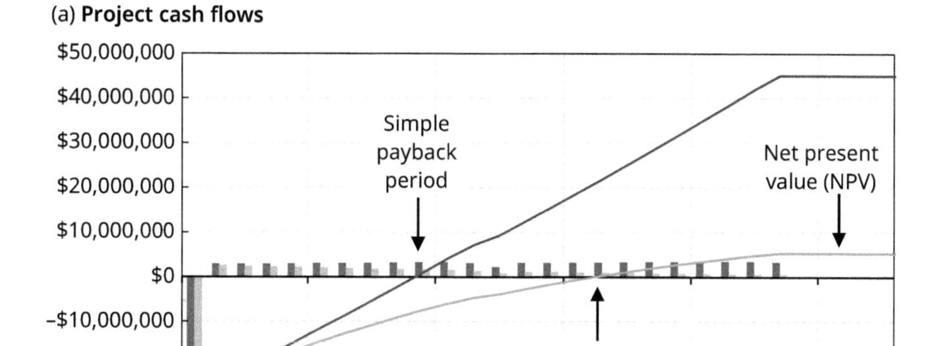

(b) **Equity cash flows**

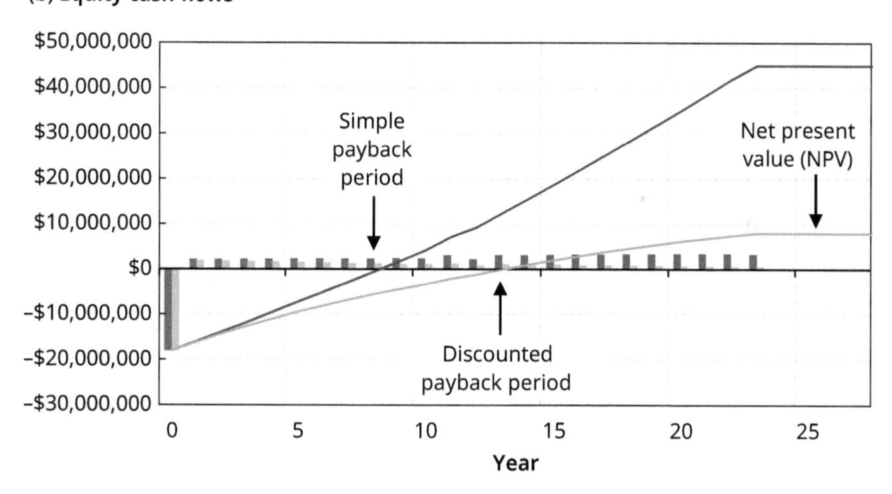

Figure 6.25 Annual cash flows for exemplary vanadium flow battery project in a peak capacity application. (a) Cash flows on project level, (b) cash flows to equity holder. Simple payback period highlights the year when cumulative cash flows turn positive. Discounted payback period highlights the year when discounted cumulative cash flows turn positive. Net present value (NPV) is equal to the final value of discounted cumulative cash flows. The project and equity IRR is the discount rate at which respective NPV would become zero.

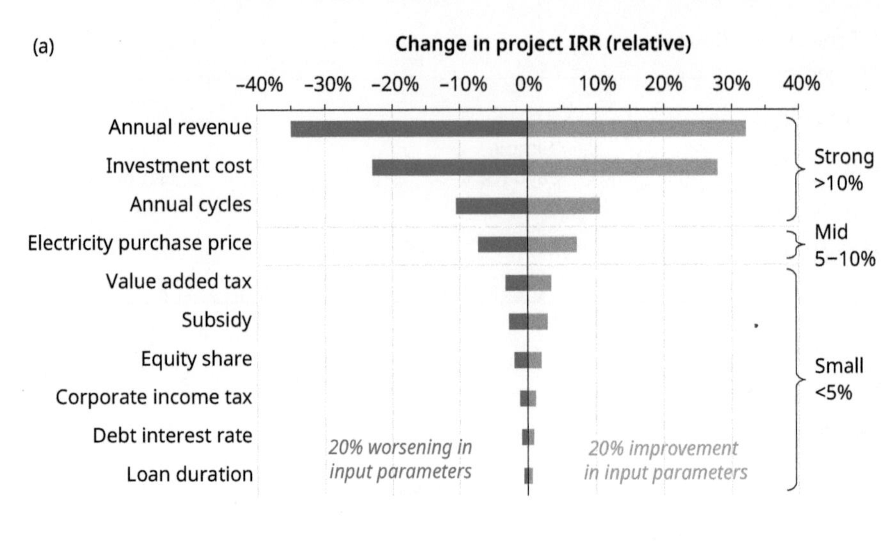

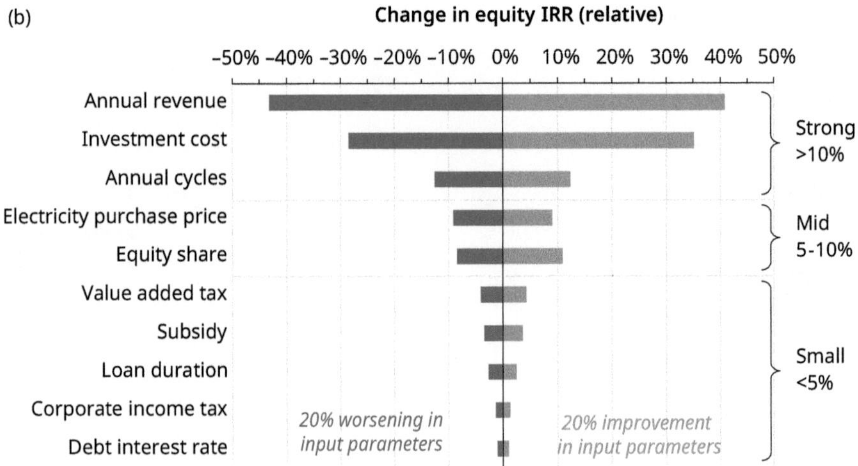

Figure 6.26 Sensitivity analysis of project profitability. Impact of ±20% change in key input parameters on (a) project IRR (relative) and (b) equity IRR (relative).

pronounced for equity IRR, where the share of equity has a moderate impact of 5–10% on IRR similar to the electricity purchase price. Also, loan duration and debt interest rate have a slightly stronger impact than on the project IRR although still less than 5%.

KEY INSIGHT

The two parameters which most strongly affect the profitability of energy storage projects are the annual revenue and investment cost.

6.7 Discussion

6.7.1 No universal business model

If one thing is evident from this evaluation of storage services it is the substantial variability in revenue: across services, across markets, and across different studies. Key drivers include changes to the services or regulations (e.g. the pricing rules, products being created or withdrawn), evolution of the underlying market conditions (e.g. the amount of renewables, conventional capacity, and interconnection relative to peak demand), and the amount of storage deployed (due to its revenue cannibalization effect).

This implies that there is not a generic business model that storage developers can follow. If a market proves to be particularly profitable in one country, that is no indication that it will be in another part of the world, or that it will still offer good investment opportunities in subsequent years. This is seen in previous assessments of the most attractive markets for storage for frequency regulation or response, which tend to change from year to year.[50,51] When new pockets of profitability open up, they are quickly exploited and exhausted, so investors and developers must be agile.

The assessment of profitability performed in section 6.3 allows the value of services to be harmonized according to the key service requirements (duration and cycling frequency). This enables a structured comparison to technology lifetime cost (from Chapter 5) to reveal the operating space in which storage could be profitable. This allows several markets and technologies to be screened quickly and easily, yet in a systematic way.

6.7.2 Substantial variability in arbitrage

As with other services, high variability in the profits that can be gained from arbitrage across markets, years and technologies has been identified. The same system operating in the same market could expect around 25% variation in its profits from year to year. This volatility impacts both the ability and cost of raising capital for investment. The P90 cashflow (the annual profit that would be exceeded in 90% of years) is a common benchmark used to size the debt a project may take on,[52] and for energy storage providing arbitrage in the markets we study, this will be around one-third lower than the central expectation (i.e. P50 profit).

Technology parameters influence the profit that is available, with more efficient and longer duration storage able to command the highest premium. That said, the marginal benefit of increasing either of these must be carefully weighed up against the additional cost of the technology (see the Worked Example in section 6.8).

These findings relate to energy-only markets, which remunerate storage based on the energy it delivers, and so expose it to the volatility of prices during the specific times it operates. Markets which also include capacity payments (a payment per kW of capacity for being available at specific times)[12] will provide greater certainty over revenue between technologies and with respect to performance parameters. They may also offer lower volatility from year to year, depending on how the capacity payments are determined (e.g. auctioned vs administered).

6.7.3 A little storage can go a long way

Storage will cannibalize its own revenue in the same way as currently seen with renewable generators, although more strongly. Each additional storage system will lower the profitability of those already operating in a market as well as its own. From the developers' perspective, this means any investment decision must be mindful of potential future installations in their market reducing available profits. From the policy-maker's and system planner's perspective, this means less storage may be installed than would be desired for meeting system or environmental objectives. Forms of remuneration other than directly from energy arbitrage energy would help to counteract this, such as capacity payments.

6.7.4 Limitations

The assessment of revenue and profitability in markets is based on a survey of existing studies and market results. It is inevitable that the individual studies will have differences in time frame, scope, and the definition of available services. These differences will all contribute to the observed variability in anticipated revenue. The results should therefore be taken as a guide to what value might be available rather than a bespoke assessment upon which business decisions should be made. In practice, investors who conduct such a survey for a specific market could apply the methodology demonstrated here of mapping revenue onto a landscape plot to identify the combination of cycling frequency and storage duration which yields the greatest revenue or profitability.

The operating schedules and profitability calculated for storage arbitrage are a best-case scenario as they are a retrospective assessment that uses full knowledge of all past and future power prices throughout the year. Real-world profitability will be lower as imperfect forecasts of future power prices will lead to deviations from the 'perfect' dispatch that could come from having perfect foresight. Academic estimates suggest that price forecasting errors could be expected to reduce storage revenues by 10–20% from those obtained with perfect foresight.[9,53–55] Operating schedules and profits will also be influenced by dynamic effects within the storage device. Round-trip efficiency is not necessarily constant, and for some technologies it varies with cell temperature (and thus with the recent operating history). High rates of self-discharge would reduce profitability, and change the optimal operating schedule towards having shorter times between charging and discharging. Cycling-induced degradation will influence operating decisions for certain technologies, as they must make sufficient profit to exceed the cost of depreciating the remaining value of the system. This can be included indirectly via a marginal cost for charging and/or discharging. If the impact on lifetime of discharging one MWh of energy can be quantified, it can then be converted into a financial cost and added to the requirement for profit.

The analysis in this chapter does not quantify the opportunities for revenue stacking, where one device can provide multiple services at the same time and thus access multiple revenue streams. This is because the value that can be accessed through this approach is highly case-specific. Revenue stacking has proven critical in the development of financially viable storage projects to date and that should be an important consideration for developers.

6.8 Worked example

In this worked example we will analyse the financial viability of providing arbitrage, thereby replicating some of the analysis from section 6.4. Imagine a developer wishes to build a 10 MW battery system to provide arbitrage in Germany. They can choose between two technologies which are 78% and 82% efficient and wish to assess which is the better investment.

1. Open <www.EnergyStorage.ninja> and go to the 'Arbitrage' tab
2. Model the operation and profitability of providing arbitrage in Germany by:

   a. Choose 'Germany' as country and 2018 as reference year for wholesale power prices

      Power prices:

      Choose a country:

      Germany ▾

      Choose a year:

      2018

   b. Choose the technology parameters to represent a 10 MW system with 6 hour duration

      Storage parameters:

      Power capacity in MW

      10

      Discharge duration in hours

      6

   c. Set round trip efficiency to 78% and marginal cost of operation to 0 USD/MWh

      Round-trip efficiency in %

      78

      Marginal cost of operation in $/MWh

      0

   d. Click 'Calculate' to perform the dispatch optimization

After a few seconds you will see a visualization of Germany's day-ahead power prices during 2018, including dispatch of your chosen storage system (top), and the state of charge (bottom):

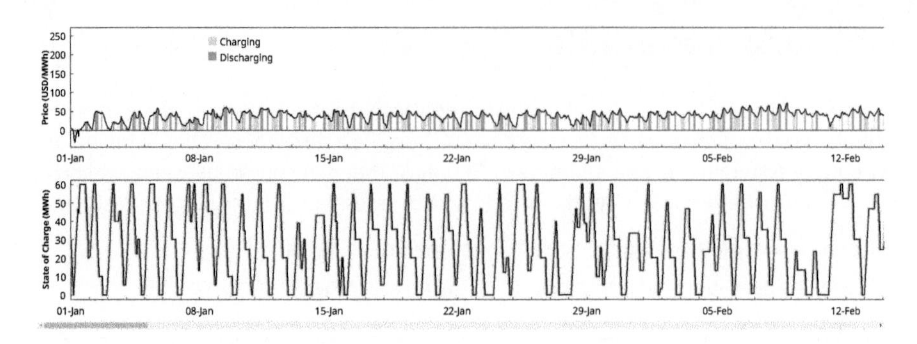

It is not possible to clearly visualize 8,760 hours of data on a single screen, so the chart shows the first few weeks of data and can be scrolled horizontally to show the rest of the year. Scrolling to the right reveals a week of very high prices in late November, which the storage system capitalizes upon:

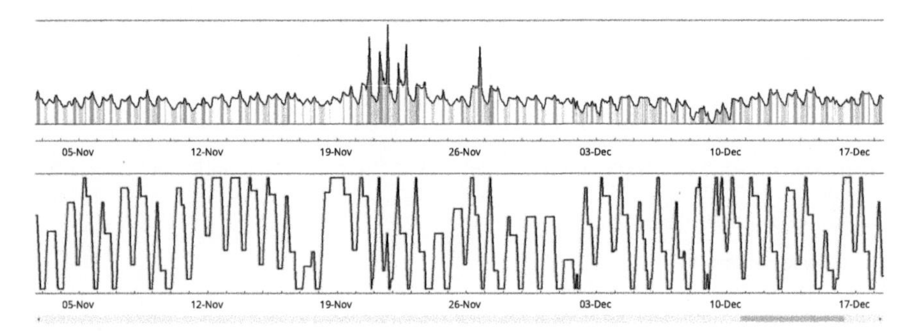

Over the whole year, the system generates USD 285,556 profit from arbitrage, giving a specific profit of 28.56 USD/kW-year. Repeating this calculation with a round-trip efficiency of 82% yields a specific profit of 33.62 USD/kW-year (18% higher).

This profit estimate can then be combined with financial information on the potential projects to assess their viability. If this project was to be financed at 8% with a 20-year lifetime, for example, these profits equate to a present value of 280 USD/kW (for 78% efficiency) or 330 USD/kW (for 82% efficiency), as given in Equations (1) and (2). Thus, if all other elements of the project are held constant (e.g. O&M cost), then it would be better to purchase the more efficient system if its total investment cost is not more than 18% higher than the less efficient one.

$$PV\left(8\%\text{ rate, } 20\text{ years, } \$28.56\text{ annually}\right) = \sum_{t=1}^{20} \frac{\$28.56}{\left(1+8\%\right)^{t}} = \$280.41 \tag{1}$$

$$PV\left(8\%\text{ rate, } 20\text{ years, } \$33.62\text{ annually}\right) = \sum_{t=1}^{20} \frac{\$33.62}{\left(1+8\%\right)^{t}} = \$330.09 \tag{2}$$

You've done it!

6.9 References

1. Battke B and Schmidt TS. 'Cost-efficient demand-pull policies for multi-purpose technologies—The case of stationary electricity storage' (2015) 155 *Applied Energy* 334–48.
2. Wood Mackenzie. *Global Energy Storage Outlook H1 2021* (Edinburgh: Wood Mackenzie, 2021).
3. King D, Browne J, Layard R, O'Donnell G, *et al. A Global Apollo Programme to Combat Climate Change* (London: The London School of Economics and Political Science (LSE), 2015).
4. Denholm P, Cole W, Frazier AW, Podkaminer K, *et al. The Four Phases of Storage Deployment: A Framework for the Expanding Role of Storage in the U.S. Power System* (Golden: National Renewable Energy Laboratory (NREL), 2021).
5. Green R. 'Electricity and Markets' (2005) 21 *Oxford Review of Economic Policy* 67–87.
6. Pollitt M. *Business Models for Future Energy Systems* (Oxford: University of Cambridge Energy Policy Research Group, 2016). Available at: <https://www.eprg.group.cam.ac.uk/wp-content/uploads/2016/10/M.-Pollitt-_BMs-BIEE_22Sep16.pdf>.
7. Mayer K and Trück S. 'Electricity Markets around the World' (2018) 9 *Journal of Commodity Markets* 77–100.
8. Carbon Trust. *Can Storage Help Reduce the Cost of a Future UK Electricity System?* (London: The Carbon Trust, 2016).
9. Staffell I and Rustomji M. 'Maximising the Value of Electricity Storage' (2016) 8 *Journal of Energy Storage* 212–25.
10. Schäfer C. Batteriespeicher dominieren den PRL-Markt. *Regelleistung-Online* (2022). Available at: <https://www.regelleistung-online.de/batteriespeicher-dominieren-den-prl-markt/> accessed on 5 November 2022.
11. Figgener J, Stenzel P, Kairies KP, Linßen J, *et al.* 'The Development of Stationary Battery Storage Systems in Germany—Status 2020' (2021) 33 *Journal of Energy Storage* 101982.
12. Glachant J-M, Joskow PL, and Pollitt MG. *Handbook on Electricity Markets* (Cheltenham: Edward Elgar, 2021).
13. Cramton P. 'Electricity Market Design: The Good, the Bad, and the Ugly' in Ralph H. Sprague, Jr. (ed.), *Proceedings of the 36th Annual Hawaii International Conference on System Sciences, HICSS 2003* vol. 3 (Los Alamitos: IEEE Computer Society, 2003), 54b.
14. Balducci PJ, Alam MJE, Hardy TD, and Wu D. 'Assigning Value to Energy Storage Systems at Multiple Points in an Electrical Grid' (2018) 11 *Energy & Environmental Science* 1926–44.
15. Housden J. *An Evaluation of Energy Storage Profitability in the United Kingdom, Germany, Australia and the United States* (London: Imperial College London, 2019).
16. Stephan A, Battke B, Beuse MD, Clausdeinken JH, *et al.* 'Limiting the Public Cost of Stationary Battery Deployment by Combining Applications' (2016) 1 *Nature Energy* 16079.
17. Teng F, Aunedi M, Moreira R, Strbac G, *et al.* 'Business Case for Distributed Energy Storage' (2017) *CIRED—Open Access Proceedings Journal* 1605–8.
18. Braeuer F, Rominger J, McKenna R, and Fichtner W. 'Battery Storage Systems: An Economic Model-Based Analysis of Parallel Revenue Streams and General Implications for Industry' (2019) 239 *Applied Energy* 1424–40.
19. Gardiner D, Schmidt O, Heptonstall P, Gross R, *et al.* 'Quantifying the Impact of Policy on the Investment Case for Residential Electricity Storage in the UK' (2020) 27 *Journal of Energy Storage* 101140.

20. Akhil A, Huff G, Currier AB, Kaun BC, *et al. DOE/EPRI 2015 Electricity Storage Handbook in Collaboration with NRECA* (Albuquerque: Sandia National Laboratories, 2015).
21. Miller S. *Enhanced Frequency Response* (London: National Grid, 2016).
22. Next Kraftwerke. Primärreserve & Primärregelleistung—Was ist das? (2019). Available at: <https://www.next-kraftwerke.de/wissen/primaerreserve-primaerregelleistung> accessed 5 November 2022.
23. Joos M and Staffell I. 'Short-term Integration Costs of Variable Renewable Energy: Wind Curtailment and Balancing in Britain and Germany' (2018) 86 *Renewable and Sustainable Energy Reviews*, 45–65.
24. Staffell I. 'Measuring the Progress and Impacts of Decarbonising British Electricity' (2017) 102 *Energy Policy*, 463–475.
25. McGrath G and Comstock O. Battery Systems on the U.S. Power Grid are Increasingly Used to Respond to Price (U.S. Energy Information Administration, 2022). Available at: <https://www.eia.gov/todayinenergy/detail.php?id=53199> accessed 5 November 2022.
26. BloombergNEF. *New Energy Outlook* (London: BloombergNEF, 2021).
27. Grams CM, Beerli R, Pfenninger S, Staffell I *et al*. 'Balancing Europe's Wind-Power Output Through Spatial Deployment Informed by Weather Regimes' (2017) 7 *Nature Climate Change* 557–62.
28. Heptonstall PJ and Gross RJK. 'A Systematic Review of the Costs and Impacts of Integrating Variable Renewables into Power Grids' (2020) 6(1) *Nature Energy* 72–83.
29. López Prol J, Steininger KW, and Zilberman D. 'The Cannibalization Effect of Wind and Solar in the California Wholesale Electricity Market' (2020) 85 *Energy Economics* 104552.
30. Musker T. *Wholesale Power Price Cannibalisation* (Norwich: Cornwall Insight, 2018).
31. Halttunen K, Staffell I, Slade R, Green R, *et al*. 'Global Assessment of the Merit-Order Effect and Revenue Cannibalisation for Variable Renewable Energy' (2020) *SSRN*. Available at: <https://dx.doi.org/10.2139/ssrn.3741232>.
32. Ward KR and Staffell I. 'Simulating Price-aware Electricity Storage Without Linear Optimisation' (2018) 20 *Journal of Energy Storage* 78–91.
33. López Prol J and Schill WP. 'The Economics of Variable Renewable Energy and Electricity Storage' (2021) 13 *Annual Review of Resource Economics* 443–67.
34. Staffell I and Green R. 'Is There Still Merit in the Merit Order Stack? The Impact of Dynamic Constraints on Optimal Plant Mix' (2016) 31 *IEEE Transactions on Power Systems* 43–53.
35. Green R and Staffell I. 'The Contribution of Taxes, Subsidies, and Regulations to British Electricity Decarbonization' (2021) 5 *Joule* 2625–45.
36. Everoze Partners. *Cracking the Code: A Guide to Energy Storage Revenue Strewams and How to Derisk Them* (Bristol: Everoze Partners, 2016).
37. Bowen T, Chernyakhovskiy I, and Denholm P. *Grid-Scale Battery Storage—Frequently Asked Questions* (Golden: NREL, 2019).
38. Ofgem. *Targeted Charging Review: Decision and Impact Assessment* (London: The Office of Gas and Electricity Markets, 2019).
39. Colthorpe A. Software: The Driving Force Putting Batteries at the Heart of the Energy Transition (2021). Available at: <https://www.energy-storage.news/software-the-driving-force-putting-batteries-at-the-heart-of-the-energy-transition/> accessed 5 November 2022.

40. Grundy A. Unlocking Stacking of BOAs with Frequency Response Services (2019). Available at: <https://www.energy-storage.news/significant-moment-for-uk-storage-as-regulator-bids-to-end-double-charging-with-formal-definition/> accessed 5 November 2022.
41. National Grid ESO. *Capacity Market Stress Event Guide* (2020).
42. Diermann R. Bundesregierung will mit EnWG-Novelle Doppelbelastung von Energiespeichern beenden (2021). Available at: <https://www.pv-magazine.de/2021/02/10/bundesregierung-will-mit-enwg-novelle-doppelbelastung-von-energiespeichern-beenden/> accessed 5 November 2022.
43. Grundy A. '"Significant Moment" for UK Storage as Regulator Bids to End Double Charging with Formal Definition' (2019) *Energy Storage News*. Available at: <https://www.current-news.co.uk/news/ofgem-unveils-formal-storage-definition-in-bid-to-end-double-charging> accessed 5 November 2022.
44. EASE. Conclusions on EASE Reply to the European Commission Public Consultation on the Revision of the Energy Taxation Directive (Brussels: European Association for Storage of Energy, 2020).
45. Seltmann T. Steuerliche Behandlung von Batteriespeichern präzisiert (2020). Available at: <https://www.pv-magazine.de/2020/03/30/steuerliche-behandlung-von-batteriespeichern-praezisiert/> accessed 5 November 2022.
46. BEIS. Longer Duration Energy Storage Demonstration Programme: Successful Projects (2022). Available at: <https://www.gov.uk/government/publications/longer-duration-energy-storage-demonstration-programme-successful-projects> accessed 5 November 2022.
47. Bundesnetzagentur. Innovationsausschreibungen (2022). Available at: <https://www.bundesnetzagentur.de/DE/Fachthemen/ElektrizitaetundGas/Ausschreibungen/Innovation/start.html> accessed 5 November 2022.
48. ACT Government. Sustainable Household Scheme—Climate Choices (2021). Available at: <https://www.climatechoices.act.gov.au/policy-programs/sustainable-household-scheme> accessed 5 November 2022.
49. U.S. Senate. *Inflation Reduction Act* (2022). Available at: <https://www.epw.senate.gov/public/index.cfm/inflation-reduction-act-of-2022>.
50. Pang A and Khodabakhsh S. Energy Storage: Opportunities, Key Trends and Market Drivers. *White & Case LLP* (2018). Available at: <https://www.whitecase.com/publications/insight/energy-storage-opportunities-key-trends-and-market-drivers> accessed 5 November 2022.
51. Mayr F. The Pockets of Energy Storage Opportunity (Berlin: Apricum GmbH, 2020). Available at: <https://apricum-group.com/the-pockets-of-energy-storage-opportunity/> accessed 5 November 2022.
52. Ostrovnaya A, Staffell I, Donovan C, and Gross R. 'The High Cost of Electricity Price Uncertainty' (2020) *SSRN Electronic Journal*. Available at: <https://dx.doi.org/10.2139/ssrn.3588288>.
53. Sioshansi R, Denholm P, Jenkin T, and Weiss J. 'Estimating the Value of Electricity Storage in PJM: Arbitrage and Some Welfare Effects' (2009) 31 *Energy Economics* 269–77.
54. McConnell D, Forcey T, and Sandiford M. 'Estimating the Value of Electricity Storage in an Energy-Only Wholesale Market' (2015) 159 *Applied Energy* 422–32.
55. Connolly D, Lund H, Finn P, Mathiesen BV *et al*. 'Practical Operation Strategies for Pumped Hydroelectric Energy Storage (PHES) Utilising Electricity Price Arbitrage' (2011) 39 *Energy Policy* 4189–96.

7 System value
Making sense

KEY INSIGHT	WHAT IT MEANS
Total system cost accounts for the lifetime cost of a technology as well as the financial implications associated with its impact on the reliability and operability of a specific power system.	This considers the value of a technology not in isolation, but from a system's perspective. It is therefore relevant for system planners and operators.
Electricity storage can reduce total system cost compared to conventional flexibility technologies in systems with high penetration of variable renewable energy.	This is shown by a wide range of studies, and is driven by a reduction in the operating costs to balance supply and demand.
The marginal value of additional electricity storage capacity reduces with increasing deployment of flexibility technologies.	This reflects the 'cannibalization' effect observed with arbitrage revenues, but it affects the value to society rather than to private investors.

The flexibility capacity required to balance systems with high shares of variable renewable energy (VRE) can be approximated as:

VRE generation *(relative to electricity demand)*	**Energy capacity of storage *(relative to annual demand)***	**Power capacity of storage *(relative to peak demand)***
50% VRE share	< 0.02%	< 20%
80% VRE share	0.03–0.1%	20–50%
90% VRE share	0.05–0.2%	25–75%

This shorthand provides an initial estimate for the amount of flexibility capacity a power system may require to ensure reliable operation with increasing VRE shares.

The requirement for electricity storage capacity in a power system increases exponentially with the share of energy coming from variable renewable sources. This trend is observed across dozens of independent studies covering four major regions.	As the share of VRE increases, it becomes increasingly difficult to integrate each additional wind or solar farm while ensuring that supply and demand balance at all times. The requirements for both energy and power capacity of storage therefore increase exponentially.

(Continued)

Monetizing Energy Storage. Oliver Schmidt & Iain Staffell, Oxford University Press. © Oliver Schmidt & Iain Staffell (2023).
DOI: 10.1093/oso/9780192888174.003.0007

KEY INSIGHT	WHAT IT MEANS
Deploying inflexible low-carbon generators like nuclear power in parallel with variable generators like wind and solar further increases the need for flexibility capacity (such as electricity storage), rather than reduces it.	While small amounts of nuclear power may help to balance supply and demand in systems with variable renewables, large capacities will lead to situations where excess electricity supply must be curtailed. This drives the need for electricity storage in cost-optimal systems.
Short-duration storage with < 4 hours discharge is as effective as medium- or long-duration storage at balancing wind and solar with demand up to 80% VRE share. Medium-duration storage with < 16 hours is as effective as long-duration storage up until 90% VRE share.	Long-duration electricity storage, with multiple days or weeks of discharge capacity, will only add value to systems with very high shares of wind and solar generation.
The amount of fossil fuels held by countries as strategic energy reserves would amount to multiple thousand terawatt hours (TWhs) if converted to electricity storage energy capacity.	This showcases the physical challenge of fully decarbonizing energy systems. Significant volumes of fossil fuel storage systems need to be converted or replaced with systems that store low-carbon fuels or electricity. These systems will only be cycled once per year, giving very limited revenues from energy-only arbitrage.

7.1 Total system cost

The increasing penetration of low-carbon generation capacity requires more power system flexibility. LCOE for generation or LCOS for electricity storage technologies are an intuitive metric for technology-specific cost comparisons. From a system perspective, however, both metrics are ambiguous, because they do not account for output variability or the impact of a technology's operation on the electricity system in terms of reliability and operability.[1] The concept of system value determines the value of a technology to the power system as a whole as the difference in total system cost (TSC) caused by the deployment of a technology.[2] The concept therefore explicitly accounts for lifetime cost and the impact on power system reliability and operability, but it requires comprehensive energy system models to determine this value. The value itself can be given as the absolute difference in TSC (%), normalized per annual energy demand (USD/MWh_{el}), normalized per installed capacity of the technology (USD/kWh_{cap}), or normalized and annuitized per installed capacity (USD/kW-year).

KEY INSIGHT

Total system cost accounts for the lifetime cost of a technology as well as the financial implications associated with its impact on the reliability and operability of a specific power system.

A range of studies analyse the TSC of low-carbon power systems with variable renewable and flexibility technologies compared to systems with conventional, dispatchable generators. For the US, a detailed grid simulation model of the balancing areas in Colorado and Missouri found a system value of 145 USD_{2011}/kW-year for electricity storage with 8 hours discharge duration, highlighting that the reduction of operational costs is more significant when the device provides capacity (i.e. frequency regulation) rather than energy services (i.e. following reserve).[3] These results are in line with a similar model for the Texas power system, which found system values of 55–85, 120–200, and 160–270 USD_{2017}/kW-year for 1, 4, and 8 hour discharge duration respectively.[4] More broadly, a capacity expansion model of the same power system identified a 7 to 12% reduction in electricity generation investment and operation cost for 90% emission reduction as a result of electricity storage deployment. Cost savings were achieved through increased utilization of installed resources and greater penetration of lowest cost low-carbon resources.[5] This translated to a value range of 286–572 and 103–257 USD_{2016}/kWh of installed electricity storage capacity for the first 10 GW of a 2 or 10 hour discharge duration technology respectively.

In a set of European power supply scenarios, similar TSC was found for a system with 0% variable renewable electricity and one with 85% combined with electricity storage power capacity at 23% of peak demand.[6] An integrated assessment model of 24 European countries quantified the system value of electricity storage as a 3–5 USD_{2016}/MWh reduction in the integration cost of variable renewable electricity.[7] A more simplified power system model based on long-term meteorological and load data and a 90% share of variable renewable electricity found long-term storage with an energy capacity equivalent to 168 average load hours to reduce total system cost by 10%, while more efficient short-term storage equivalent to 4 load hours achieves 20%.[8]

KEY INSIGHT

Electricity storage can reduce total system cost compared to conventional flexibility technologies in systems with high penetration of variable renewable energy.

Table 7.1 lists studies that model Great Britain's power system up to 2050 and include electricity storage. The penetration of variable renewables ranges from 0 to 87% of annual electricity demand and 0 to 91% generation capacity. Electricity storage capacity is included at 0.004 to 0.027% of annual demand (energy) or 3 to 57% of peak demand (power). The storage technologies modelled range from one generic proxy for all technologies to a full suite of seven different technologies. While all studies consider storage and interconnection, other flexibility options like demand-side response (DSR) and hydropower are not always included.

In terms of system value, BloombergNEF finds a 2% reduction in total system cost by 2030 in a scenario where an additional 4.8 GW of electricity storage capacity is deployed.[10] This translates to a reduction of 0.7 GBP/MWh produced in the system. In comparing two scenarios with and without electricity storage, Heuberger et al. identify a reduction of 15%.[2] However, it is highlighted that the first GW of storage already leads to a reduction of 13%, thereby putting the result in line with the previous study (15% – 13% = 2%). The analyses by the Committee on Climate Change and the Carbon Trust determine annual TSC savings for

Table 7.1 Overview of studies investigating the system impact of electricity storage and other flexibility options on a low-carbon power system in Great Britain. System value in some studies reported explicitly for electricity storage. Other studies report system value for deployment of flexibility options in general. VRE—Variable renewable energy.

Study, Year (Institution)	Time horizon	VRE share*	Storage capacity**	System value	Flexibility options	Storage options
BEIS, 2017 (Government)[9]	2015–2035	25–55% 32–62%	0.007–0.023% 5–18%	–	Storage, Interconnection	–
BNEF, 2018 (Industry)[10]	2030, 2040	43–75% 56–85%	0.011–0.027% 12–57%	2% TSC reduction or 0.7 GBP/ MWh_{el} (67% VRE, 16.5 GW vs 11.7 GW electricity storage)	OCGT, Storage Interconnection, Hydropower, DSR, Other	Pumped storage, Small-scale batteries, Utility-scale batteries
Carbon Trust, 2016 (Government)[11]	2020, 2030, 2050	25–34% 37–48%	0.005–0.020% 4–23%	1.4–2.4 GBPbn p.a. (net) (100g_{CO2}/ kWh_{el} target, deployment of flexibility options)	OCGT, Storage Interconnection, DSR	Pumped storage, Bulk storage, Distributed storage
CCC, 2015 (Government)[12]	2030	0–83% 0–75%	0.004–0.024% 3–20%	3–3.8 GBPbn p.a. (gross) (100g_{CO2}/ kWh_{el} target, deployment of flexibility options)	OCGT, Storage Interconnection, Hydropower	Pumped storage, Other dedicated storage
Edmunds, 2014 (Academic)[13]	2020–2030	16–39% 24–54%	0.009–0.027% 4–6%	–	Storage, Interconnection, Hydropower	Pumped storage
Heuberger, 2017 (Academic)[2]	2035	– 49–60%	– 0–10%	15% TSC reduction or 515 GBP/ kW_{cap} (70 GBP/t_{CO2}, 9.5 GW vs 0 GW electricity storage)	OCGT, Storage Interconnection	Compressed air

(*Continued*)

Table 7.1 (*Continued*)

Study, Year (Institution)	Time horizon	VRE share*	Storage capacity**	System value	Flexibility options	Storage options
Heuberger, 2018 (Academic)[14]	2015–2050	14–76% 27–86%	0.007–0.023% 5–18%	–	OCGT, Storage Interconnection	Pumped storage, Battery
National Grid, 2018 (Industry)[15]	2020–2050	26–63% 35–74%	0.007–0.019% 10–38%	–	Storage Interconnection, Hydropower, DSR	Pumped storage, Decentral battery, Grid-scale battery, Fuel cells, Liquid air, Vehicle to grid, Compressed air
Pfenninger, 2015 (Academic)[16]	–	– 85%	– 5-25%	50–130 GBP/MWh_{el} (90% VRE, scenarios with vs without storage)	OCGT, Storage Interconnection, Hydropower, Tidal	Pumped storage, Grid-scale batteries
Price, 2018 (Academic)[17]	2050	52–87% 68–85%	0.006–0.024% 6–24%	–	OCGT, Storage Interconnection,	Pumped hydro, Sodium sulphur
Zeyringer, 2018 (Academic)[18]	2050	49, 76% 73, 91%	0.009, 0.015% 8,14%	–	OCGT, Storage Interconnection	Pumped hydro, Sodium sulphur

**Upper value: Energy generation, Lower value: Power capacity*
***Upper value: Energy capacity relative to annual electricity demand, Lower value: Power capacity relative to peak demand*

a power system with a carbon intensity of $100g_{CO2}$/kWh with flexibility technologies at GBP 1.4–2.4 or 3–3.8 billion in 2030 compared to no flexibility options. While the former refers to net savings, which include investment cost of the flexibility technologies, the latter uses gross savings, which does not.[11,12] At 90% VRE penetration, the value of electricity storage specifically is identified by Pfenninger et al. at 50–130 GBP/MWh_{el} system-wide electricity cost (32–35% of respective TSC), when adding electricity storage at a cost of 350 GBP/kWh_{cap} to a range of scenarios.[16]

KEY INSIGHT

The marginal value of additional electricity storage capacity reduces with increasing deployment of flexibility technologies.

7.2 The driver for system value

Not all studies explicitly quantify the system value of electricity storage. System value originates from the ability of storage to increase the utilization of power system assets like variable or inflexible generators and thereby increase their penetration.[5] Therefore, some studies only explore this capability without quantifying its financial value.

Figure 7.1 compares the findings of 30 studies across the US, EU, Germany, and Great Britain (GB), regarding the required electricity storage energy capacity and power capacity in low-carbon power systems with increasing VRE shares. The energy capacity and power capacity requirements are displayed relative to annual electricity and peak power demand respectively. Most studies appear to agree that for up to a VRE penetration of 50%, a power system requires less than 0.02% energy storage capacity and 20% power capacity. Taking Great Britain as an example with ~50 GW peak and 300 TWh annual demand, this would amount to 10 GW and 60 GWh of electricity storage capacity.

Storage requirements increase exponentially at higher levels of VRE penetration. Moving to 80% and 90% penetration, the energy capacity requirement increases to 0.03–0.1% and 0.05–0.2% respectively (60–300 GWh and 150–600 GWh for Great Britain's power system). Power capacity requirements increase to 20–50% and 25–75% (10–25 GW and 12.5–37.5 GW). There is substantial variation between the findings of different studies, especially in terms of energy storage capacity, as the studies in Figure 7.1 cover several geographies and make their own varied assumptions about the mix of technologies which provide flexibility (i.e. storage vs interconnection, flexible generation, and DSR).

KEY INSIGHT

The flexibility capacity required to balance systems with high shares of variable renewable energy (VRE) can be approximated as:

VRE generation ***(relative to electricity demand)***	**Energy capacity of storage** ***(relative to annual demand)***	**Power capacity of storage** ***(relative to peak demand)***
50% VRE share	< 0.02%	< 20%
80% VRE share	0.03–0.1%	20–50%
90% VRE share	0.05–0.2%	25–75%

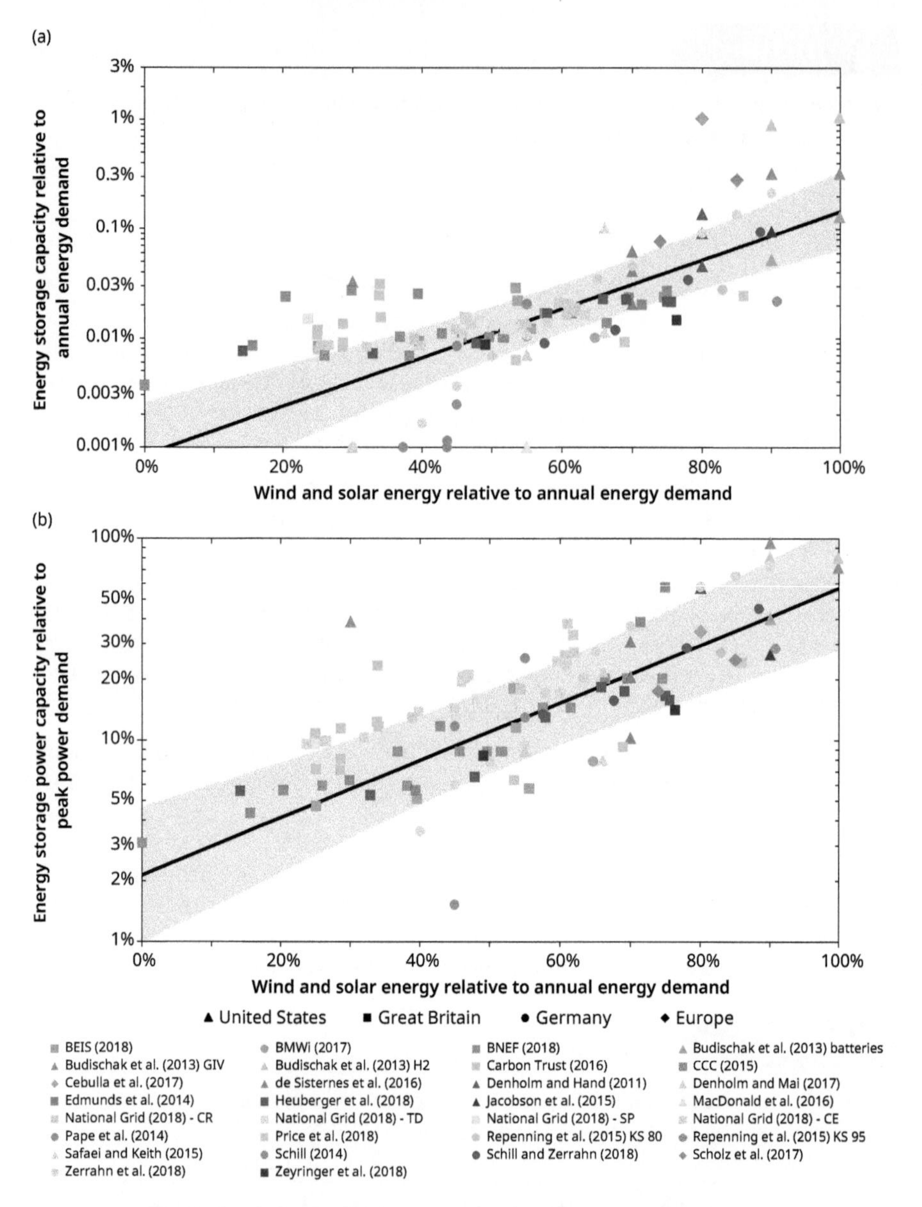

Figure 7.1 (a) Electricity storage energy capacity and (b) power capacity requirements as a function of variable renewable energy penetration. Capacity requirements are displayed relative to annual electricity or peak power demand respectively. Data based on a literature review of 30 studies modelling electricity storage requirements in low-carbon power systems in the US, Great Britain, Germany, and the EU.[4–6,9–15,17,18,20–31] Black lines show log-linear regressions and shaded areas the confidence intervals. Budischak scenarios: GIV—Grid-integrated vehicles, National Grid scenarios: CR—Community Renewables, TD—Two Degrees, SP—Slow Progress, CE—Consumer Evolution. Repenning scenarios: KS 80/95–80%/95% emission reduction.

KEY INSIGHT

The requirement for electricity storage capacity in a power system increases exponentially with the share of energy coming from variable renewable sources. This trend is observed across dozens of independent studies covering four major regions.

The trends used to summarize the results in Figure 7.1 suggest that both electricity storage power and energy capacity requirements increase exponentially with the penetration of VRE, as has been highlighted in other studies.[19] Incorporating low shares of VRE is relatively easy as their variability can be accommodated by slight alterations in the dispatch of conventional power stations. As VRE share increases further it becomes increasingly difficult to manage the balance between supply and demand. Exponentially increasing power capacity is needed to consume the increasing amounts of excess renewable energy for later discharge in most cost-efficient low-carbon energy systems. The combined impact of additional variable power capacity and the time it is supposed to generate electricity means that energy capacity requirements increase at an even higher exponential rate to ensure sufficient electricity is available at all times. Not only is more backup power needed but it is also needed for longer periods of time.

Selected studies argue that energy storage power capacity increases linearly rather than exponentially with increasing variable renewable energy penetration.[20] That conclusion is drawn from 15 independent studies (compared to 30 studies in this book) which show notable scatter around the trend, so rather than the exact functional form of the relationship, the important conclusion is that other studies support the insight that the rate at which storage power capacity increases is smaller than for energy capacity.

7.3 Case study for Great Britain

This section analyses electricity storage power capacity and energy capacity requirements for Great Britain's (GB) power system as a case in point. In this power system, nuclear is projected to play a significant role. The same is true for other power markets (e.g. France, China). Given its economic incentive for constant power output due to low fuel and high investment cost, nuclear's impact as 'inflexible' technology on flexibility requirements should be considered as well, similar to variable solar and wind.

FREQUENTLY ASKED QUESTION

Isn't low-carbon baseload power like nuclear an alternative to energy storage or other flexibility options?

Unfortunately, baseload generators are usually inflexible. That means flexible operation would either not be possible technically or would negatively impact project economics. Hence, there may be situations where variable generators already meet peak power demand and baseload power generators produce excess

(Continued)

electricity. Thus, for cost-optimal power systems increasing shares of variable generation should be met by increasing flexibility capacity. Flexible generators can produce when variable generators fall short of demand and stop generating when they overproduce. It is like a puzzle where different pieces need to match for completion.

Figure 7.2 plots the relationship between electricity storage power capacity and nuclear power capacity from several studies of the future GB power system (see Table 7.1). It shows that regardless of more variable renewables, storage capacity increases in line with nuclear capacity, by 0.2% for each 1% increase of nuclear power share, when considering all studies.

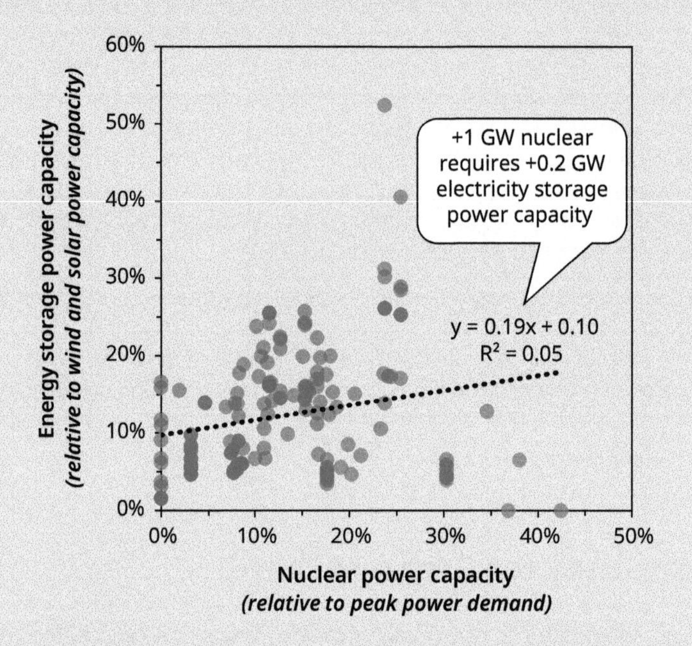

Figure 7.2 Relationship between electricity storage power capacity normalized by total wind and solar power capacity, and nuclear power capacity relative to peak power demand for all investigated studies on scenarios for the future GB power system (see Table 7.1).

KEY INSIGHT

Deploying inflexible low-carbon generators like nuclear power in parallel to variable generators like wind and solar further increases the need for flexibility capacity (such as electricity storage), rather than reduces it.

Figure 7.3 displays the electricity storage requirements as modelled by various studies for low-carbon scenarios of the future GB power system (see Table 7.1). The required absolute storage power capacity in a system with up to 90% wind, solar, and nuclear generation capacity could remain at the 4 GW installed in 2021 or increase to 35 GW (panel a). A similarly

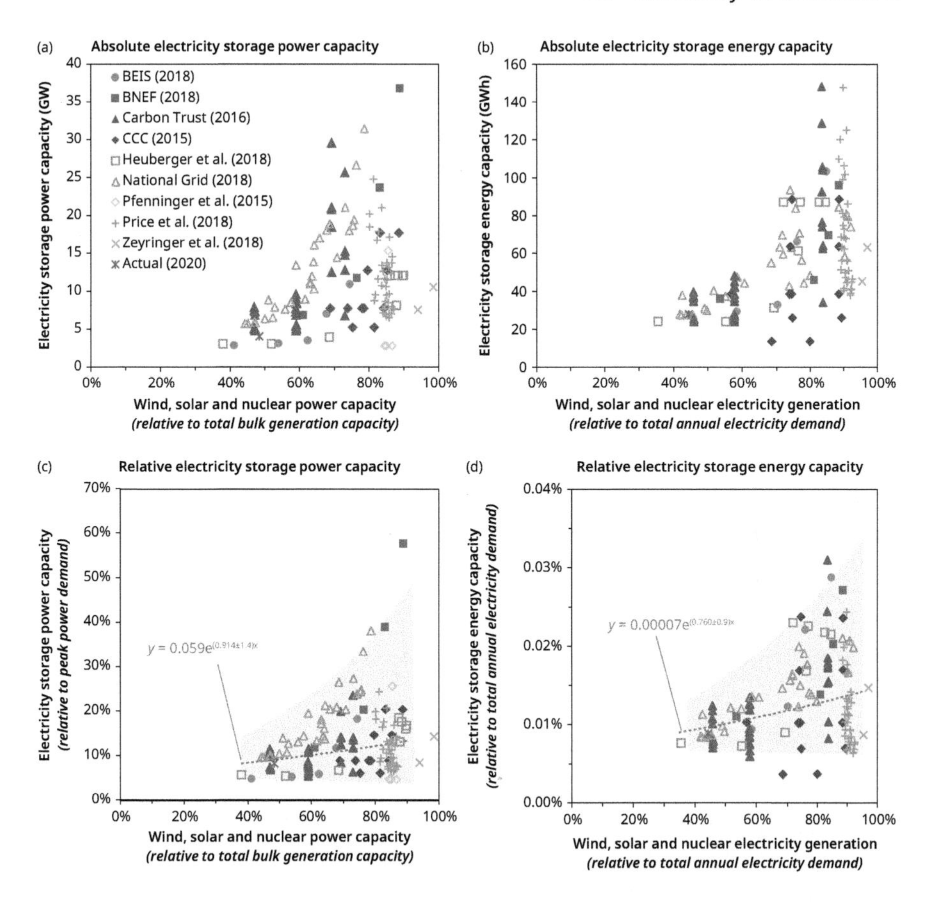

Figure 7.3 Electricity storage capacity requirements as a function of wind, solar, and nuclear penetration based on various studies modelling low-carbon generation scenarios in the GB power system.[9–18] Panels give storage capacity in absolute (a and b) and relative (c and d) terms, considering power (a and c) and energy (b and d) capacity. Bulk generation capacity consists of wind, solar, nuclear, coal, gas (combined cycle), biomass, geothermal, waste, and wave. Dotted lines represent exponential fits to the complete data set. Shaded areas represent indicate respective uncertainty ranges excluding outliers. The formulas for exponential fits and uncertainty ranges are given.

wide range is observed for electricity storage energy capacity. At 90% penetration of wind, solar, and nuclear energy generation (relative to total annual electricity demand), required storage capacity could be as low as today at 30 GWh or up to 140 GWh (panel b).

The electricity storage requirement range becomes more defined when accounting for differing peak power and annual electricity demand assumptions. The current power system at 48% penetration of wind, solar, and nuclear power capacity has 8% electricity storage power capacity relative to peak demand. According to the studies, it could range from 5–20% at 60% and 5–40% at 80% penetration (panel c). These values are equivalent to 2.5–10 GW (60%) and 2.5–20 GW (80%) electricity storage power capacity at the peak demand of 50 GW. Similarly, electricity storage energy capacity relative to annual electricity demand could remain at the current level of 0.01% or increase to 0.015% or 0.025% at 60% or 80% wind, solar, and nuclear energy penetration respectively, depending on the scenario (panel d).

Both of these results for the required electricity storage power capacity and energy capacity in GB are at the lower end of the ranges identified by the wider selection of studies and geographies in the previous section.

The shaded areas in Figure 7.3 identify possible maximum and minimum deployment levels for electricity storage subject to the assumptions in the various studies and scenarios. They identify the possible range of electricity storage capacity requirements to enable low-carbon power systems with up to 90% wind, solar, and nuclear power and energy share in Great Britain's power system.

However, the wide ranges also highlight that low-carbon power systems do not necessarily depend on electricity storage; rather, it can be a valuable enabler under certain conditions. This insight is supported by the fact that the observed variations in electricity storage requirements are not study-specific but vary by individual scenario. For example, Pfenninger et al. feature one scenario with 15 GW storage (25% relative to peak power demand) at 85% wind, solar, and nuclear power share, compared to 3 GW (5% relative to peak power demand) in all other scenarios.[16] In these other scenarios alternative flexibility technologies like interconnectors, demand-side response, gas (open cycle), hydropower, and oil and diesel generators balance variable and inflexible generation with demand.

FREQUENTLY ASKED QUESTION

So, is energy storage a must have in low-carbon power systems with variable renewable generators?
No, cost-efficient flexibility is a must-have. There are four major flexibility options:

- Flexible generation, e.g. gas plants with carbon capture, hydropower
- Interconnection, e.g. between power systems with different weather and demand patterns
- Demand response, e.g. ability to flexibly adjust demand without incurring major cost on the demand side
- Energy storage

Energy storage is only one flexibility option. Due to strong cost reductions and the wide range of services that storage systems can provide to power producers, network operators and consumers, it is the option that appears to be in public focus at times. However, this does not rule out other flexibility options. System planners should always choose the most sustainable, reliable and cost-efficient options first.

A more comprehensive approach to assessing the system value potential of electricity storage in integrating low-carbon electricity is to analyse the overall flexibility capacity requirements, regardless of which technology provides them (see Figure 7.4). It shows that up to 40% wind, solar, and nuclear power share, less than 20% flexibility capacity relative to annual peak power demand is required. This increases to a range of 40–100% above 80% wind, solar, and nuclear power. These findings are in line with the findings in Figure 7.1.

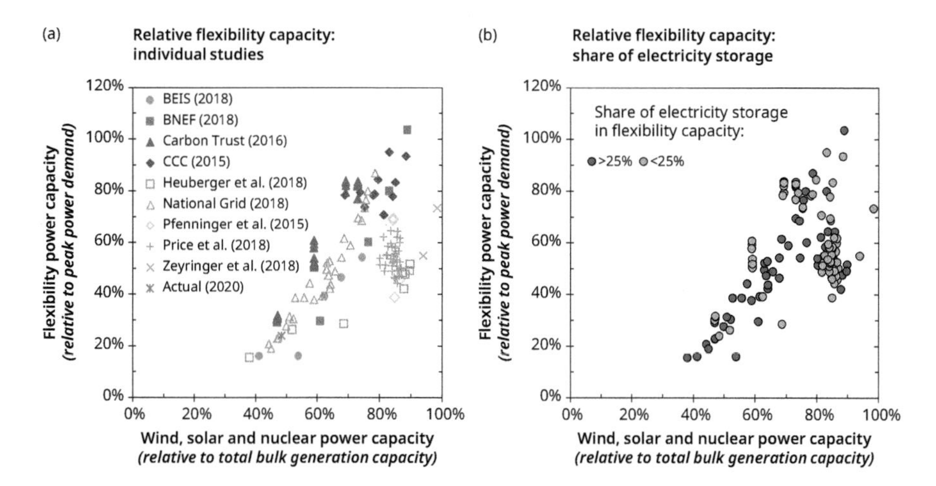

Figure 7.4 Flexibility power capacity requirements relative to peak power demand as a function of wind, solar and nuclear power penetration for the GB power system. (a) Results from individual studies. (b) Differentiation of studies by the share of flexibility power capacity provided by electricity storage. Bulk generation capacity consists of wind, solar, nuclear, coal, gas (combined cycle), biomass, geothermal, waste, and wave. Flexibility capacity consists of electricity storage, interconnectors, demand-side response, gas (open cycle), hydropower, oil, and diesel.

Figure 7.4 b shows that the share of electricity storage has no impact on the required flexibility capacity. This confirms that flexibility technologies can be used relatively interchangeably. The analysis therefore shows the maximum theoretical potential for electricity storage. The practical share of each technology is determined by technology constraints (e.g. CO_2 emissions for flexible generation, limited interconnection possibilities) and their economic market value (e.g. electricity storage vs demand-side response).

This analysis reveals three insights. First, the requirement for flexibility power capacity appears to increase linearly with increasing shares of wind, solar, and nuclear power capacity. Figure 7.5 a displays linear regression trendlines as visual guides for studies with more than two data points and low variability.

Second, there appears to be a flexibility baseline at 20% capacity of peak demand. This holds from a wind, solar, and nuclear penetration of 0% up to 40% in all studies. It indicates that nearly half of power system electricity can come from relatively inflexible or variable sources before there are additional needs for flexibility.

Third, there seem to be two schools of thought in terms of modelling flexibility capacity requirements. This becomes evident when comparing studies from industry and government with those from academics (see Figure 7.5 b). Academic studies suggest half as much impact from adding low-carbon generation: an extra 0.8% for each 1% share of wind, solar and nuclear, versus an extra 1.6% from industry and government studies. This equates to an average of 60% versus 100% flexibility capacity at 90% VRE and nuclear share. This is likely the result of modelling different system margins for firm reliable capacity (i.e. all capacity except wind and solar). For most academic studies, the margin is less than 20% above peak demand (see Figure 7.6). Hence, this is the optimistic school of thought. Industry

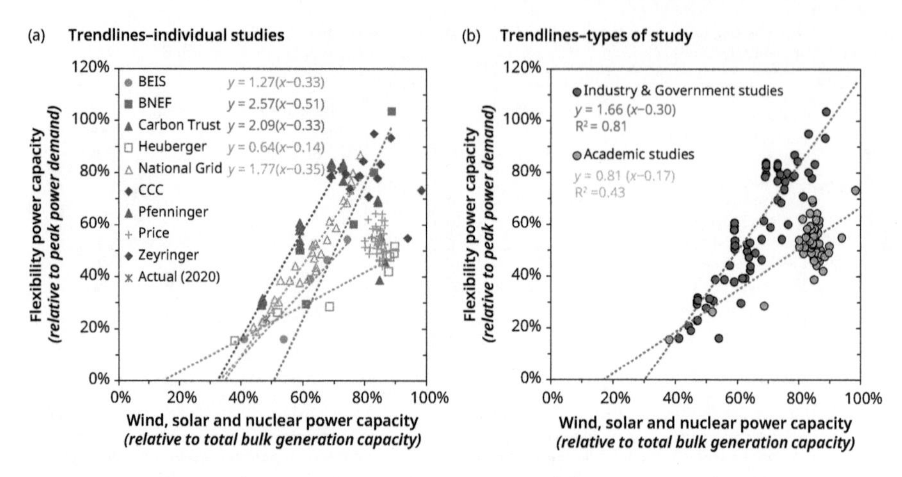

Figure 7.5 Trendlines for flexibility power capacity requirements relative to peak power demand as a function of wind, solar and nuclear power penetration for the GB power system. Trendlines in (a) are given for data series with more than two data points and coefficient of determination of $R^2 \geq 0.85$. Trendline formulas are rearranged so the negative offset gives the x-axis intercept (i.e. the share of capacity above which flexibility is needed). Bulk generation capacity consists of wind, solar, nuclear, coal, gas (combined cycle), biomass, geothermal, waste, and wave. Flexibility capacity consists of electricity storage, interconnectors, demand-side response, gas (open cycle), hydropower, oil, and diesel. Grouping used in (b): Industry and Government: BEIS (2018), BNEF (2018), Carbon Trust (2016), CCC (2015), National Grid (2018). Academia: Heuberger et al. (2018), Pfenninger et al. (2015), Price et al. (2018), Zeyringer et al. (2018).

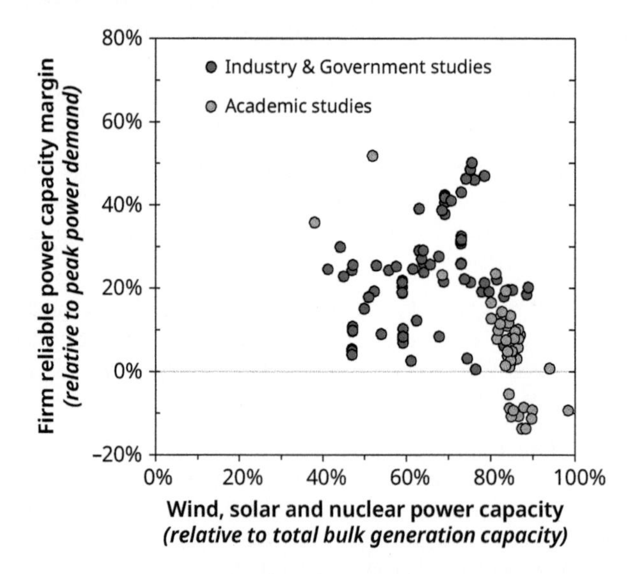

Figure 7.6 Firm reliable power capacity margin relative to peak power demand as a function of wind, solar and nuclear power penetration for the GB power system. Firm reliable capacity consists of all power generation capacity except wind, solar, and wave. Data differentiated according to commissioning institution. Industry and Government: BEIS (2018), BNEF (2018), Carbon Trust (2016), CCC (2015), National Grid (2018). Academia: Heuberger et al. (2018), Pfenninger et al. (2015), Price et al. (2018), Zeyringer et al. (2018).

and government studies model more than 20%, making those the conservative school of thought. Section 7.6 assesses possible motivations for both approaches.

The two schools of thought can be used for two different approaches to plan low-carbon power systems. The more conservative one suggests that flexibility capacity is only needed once wind, solar, and nuclear make up 30% of the generation portfolio. It will then increase by 1.66% relative to peak power demand with each additional 1% of low-carbon power capacity (see Figure 7.5 b). According to this approach, a power system based only on wind, solar, and nuclear power would need flexibility capacity at 115% of peak demand.

The less conservative approach in academic studies suggests that no flexibility capacity is needed below 17% wind, solar, and nuclear power penetration. It will increase by 0.81% of peak power demand for each additional 1% low-carbon power capacity in the power mix. According to this approach, a power system based only on wind, solar, and nuclear power would require flexibility power capacity at 65% of peak power demand.

7.4 Insights for the global power system

The two approaches outlined in the previous section can be useful in planning low-carbon power systems to assess flexibility power capacity requirements that could be fulfilled with electricity storage. Figure 7.7 uses these approaches in a thought experiment on the amount of flexibility required globally if the power generation mix changes in line with projections made in the IPCC 1.5 °C report to keep global average temperature increase below 2 °C.[32-34]

In 2020, wind, solar, and nuclear power penetration relative to total bulk generation capacity was just over 20%, and would require nearly no flexibility capacity. The 1,300 GW hydro and oil-based capacity and around ~400 GW electricity storage, demand-side response and interconnection that is deployed amounts to nearly 40% of the global noncoincident peak power demand of 4,800 GW (see Chapter 8 for details).

As the power penetration of wind, solar, and nuclear increases to 76% by 2050, the flexibility capacity is projected to decrease from 40% to 20%. This is because peak demand increases faster than projected additions for hydro- and oil-based generation, the only flexibility technologies explicitly listed in the database behind the IPCC reports.[33,34] This reveals an increasing gap between flexibility requirements and installed flexibility capacity from the early 2030s (conservative) or late 2030s (optimistic approach) onwards. The conservative approach requires additional 400 GW flexibility already by 2035. For 2050, the optimistic approach suggests 2,100 GW additional flexibility capacity and the conservative another 2,600 GW on top. At even higher penetration of low-carbon generators, Jacobson et al.'s roadmap for a 100% renewable energy system in 139 countries in 2050 finds flexibility capacity requirements at 60% of peak power demand, which is closely aligned to the suggestion of the optimistic approach.[35]

This 'thought experiment' shows that, notwithstanding the limitations discussed in section 7.6, the outlined approaches can be used for high-level approximations of maximum and minimum flexibility power capacity requirements for low-carbon power systems. These two

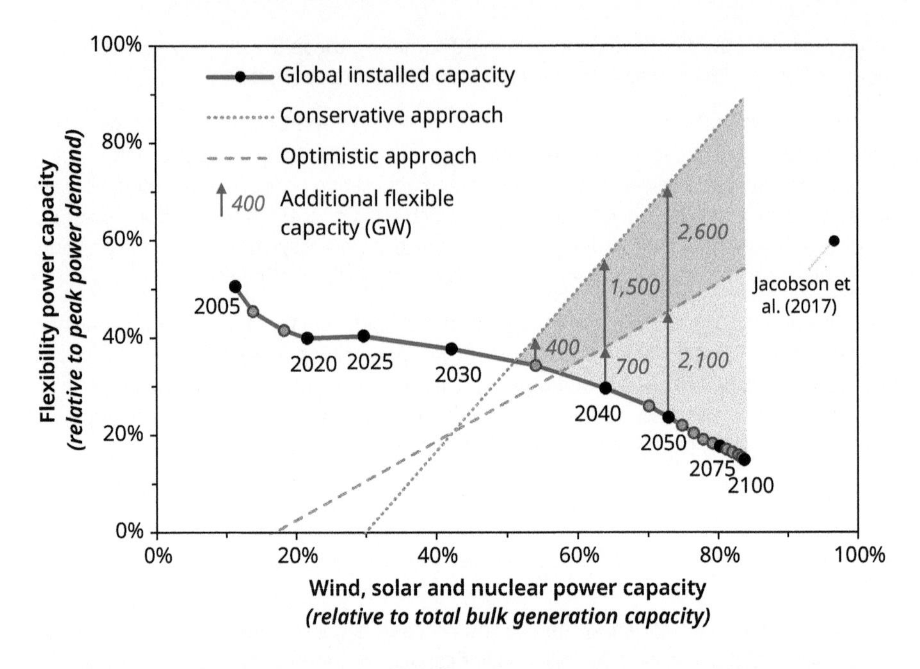

Figure 7.7 'Thought experiment' on global flexibility power capacity requirements. Red line shows the evolution of global installed power capacity based on the median across IPCC scenarios with 50% probability of keeping global average temperature increase below 2 °C.[32–34] Conservative and optimistic approaches reflect flexibility capacity requirements as identified in Figure 7.5. Red numbers indicate additional flexibility capacity required on top of projected capacities for hydro and oil-based power capacity and 2020 capacity levels for electricity storage (175 GW), interconnection (180 GW) and demand-side response (40 GW).[36–38] In 2050, global annual electricity demand is modelled at 48,000 TWh, non-coincidental peak demand at 10,000 GW, total power generation capacity at 15,700 GW with 2,000 GW hydro and oil-based generation. For comparison, 2020 values are 23,300 TWh (annual demand), 4,800 GW (peak), 6,300 GW (total capacity), 1,300 GW (hydro, oil). The result from Jacobson et al. for a 100% wind, solar, and hydropower based energy system for 139 countries is displayed for comparison (peak: 11,800 GW; hydropower and other flexibility power capacity: 7,060 GW).[35]

approaches outline the theoretical need for electricity storage, form the basis to assess its total financial system value, and can guide future power system planning to ensure sufficient flexibility capacity.

FREQUENTLY ASKED QUESTION

Why is the demand for flexible capacity so high in low-carbon power systems?
The increase in flexibility requirements is a direct result of the low capacity credit of wind and solar. When aggregated, the studies for the GB power system suggest a credit of only 10%, in other words each additional GW of wind and solar displaces 0.1 GW of firm reliable incumbent capacity (see Figure 7.8). This means there is a need for nearly 1:1 backup of variable capacity with flexibility capacity.

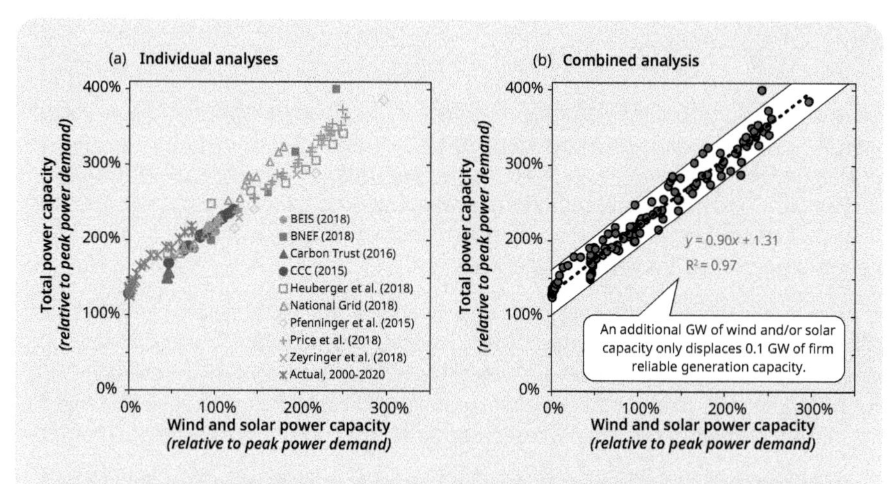

Figure 7.8 Total installed power capacity relative to peak power demand as a function of wind and solar capacity relative to peak power demand. (a) Individual GB power system studies. (b) Combined trend across all studies. Shaded area represents the linear regression and its prediction interval for the data.

7.5 Electricity vs energy perspective

Sections 7.3 and 7.4 analysed the potential electricity storage power capacity requirements to balance increasing shares of variable and/or inflexible low-carbon power supply. However, Figure 7.1 also shows that the required energy capacity increases with rising shares of low-carbon power supply, and that it increases exponentially rather than linearly.

Two perspectives are important when determining the required energy storage capacities:[39]

- **Electricity sector**: This perspective focuses on electricity storage needs to balance low-carbon electricity supply in the electricity sector (also called Power-to-Power; for respective studies see Figure 7.1)
- **Energy sector**: This perspective is broader by focusing on energy storage needs to balance low-carbon energy supply in the energy sector overall. It includes coupling of the electricity to other energy sectors like heat or transport (also called Power-to-X).[40–42]

There is a wide range of studies that analyse the need for energy storage from both perspectives. Many of these studies implement very detailed assumptions on technology and market-specific parameters. The priority of this book is, however, to promote a fundamental understanding on the requirements for energy storage to decarbonize the electricity and overall energy sectors. The following analysis is therefore intentionally kept simple to focus on fundamental insights that are valid across energy storage technologies and power markets.

7.5.1 Electricity sector

This section systematically investigates the impact of different electricity storage energy capacities and efficiencies on integrating low-carbon electricity generation. It focuses on variable generation from wind and solar power and uses Great Britain's power system as an example. The analysis is based on meteorological wind and solar data, assumes that electricity can flow unconstrained from generation to consumption and that all generation capacity beyond wind and solar is fully flexible.

An initial step is to identify the ideal ratio of wind and solar in the generation mix that best matches hourly demand to minimize the need for electricity storage. Hourly wind and solar generation data are scaled to meet a certain share of total electricity demand overall. Hourly electricity demand is then subtracted from the scaled wind and solar generation. This reveals how much of this generation actually meets demand and is consumed, rather than curtailed.

Figure 7.9 shows that a wind:solar ratio of 85%:15% best coincides with Great Britain's hourly electricity demand pattern. If wind and solar generate 100% of total electricity demand at

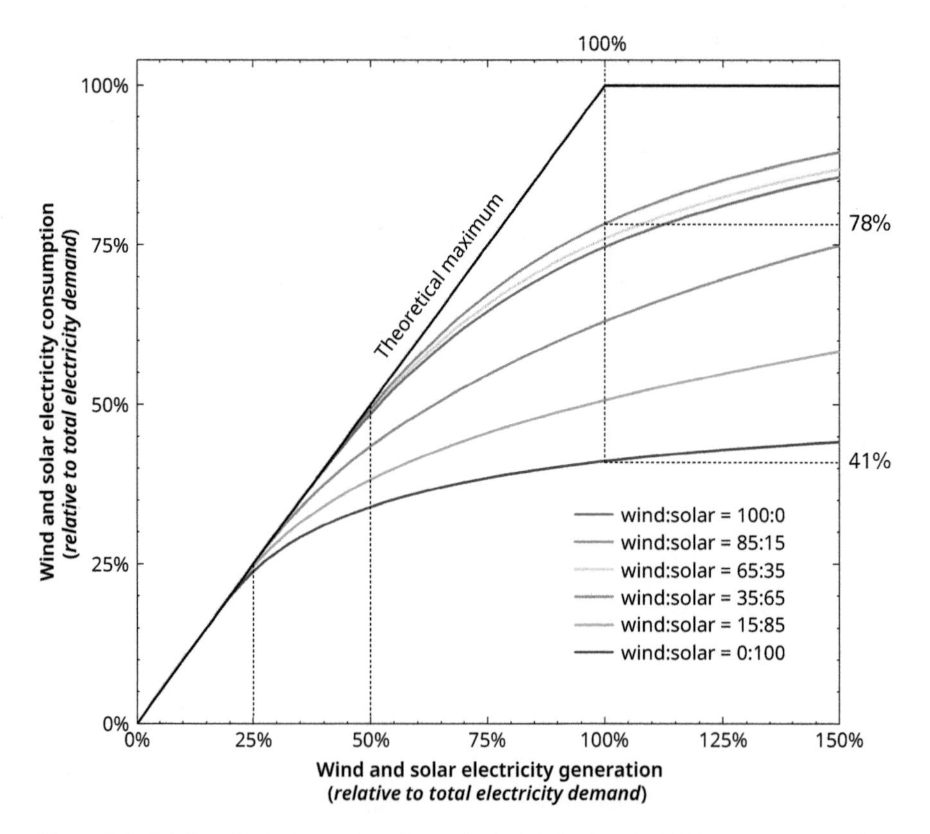

Figure 7.9 Relationship between the share of wind and solar electricity generation and the share which can be consumed (accounting for mismatch between supply and demand) without electricity storage. Different colours refer to different shares of wind versus solar generation. Based on meteorological and demand data for 1991–2019 for Great Britain. Average annual electricity demand is ~320 TWh.

this ratio, they actually provide 78% of hourly demand. In contrast, if the wind:solar ratio was 0%:100%, only 41% could be consumed. Other studies confirm an ideal wind:solar ratio of 85%:15% for Great Britain.[43,44]

The analysis also reveals that below 25% wind and solar generation there is only minimal mismatch (< 1%) with hourly demand for any wind:solar ratio. For the ideal wind:solar ratio this value increases to 50%. There would be no need for electricity storage to integrate excess wind and solar generation below these penetration levels.

The next step is to model the ability of storage energy capacity to integrate excess solar and wind generation. Figure 7.10 shows the results for different sizes of 100% efficient electricity storage. The capacities are given relative to the average hourly electricity demand (average load hours) and in absolute terms (TWh). Assuming 20% wind and solar overcapacity, such that total generation amounts to 120% of total electricity demand, 84% could be consumed without any storage (compared to 78% without overcapacity in Figure 7.9). Storage systems

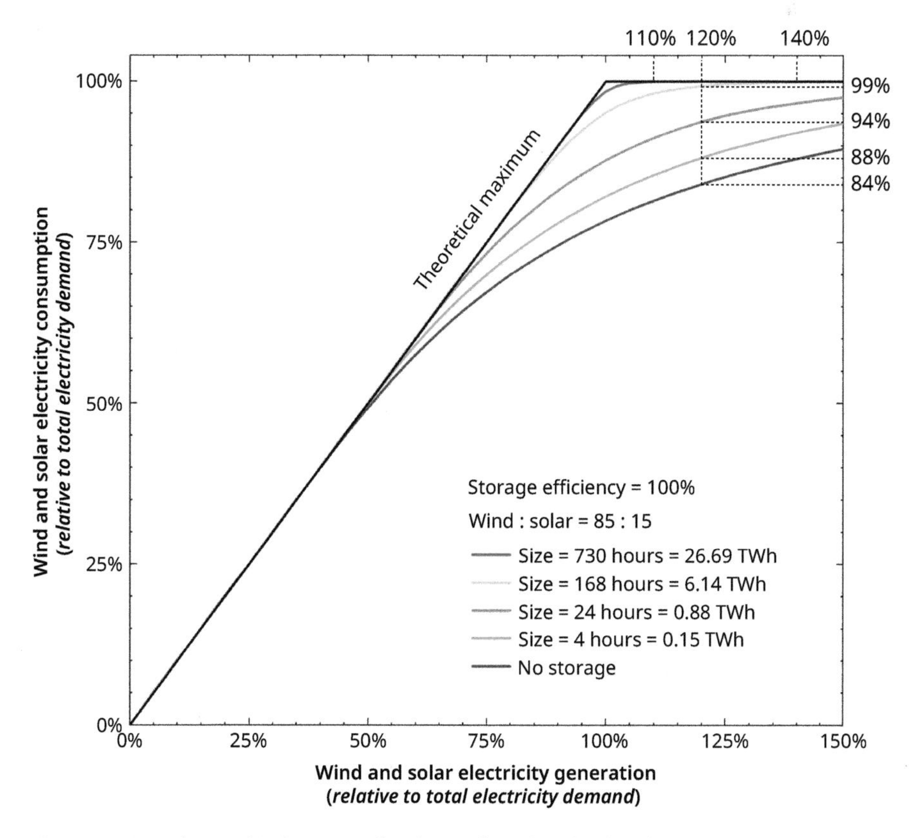

Figure 7.10 Relationship between the share of wind and solar electricity generation and the share which can be consumed (accounting for mismatch between supply and demand) with different energy storage capacities, assuming 100% round-trip efficiency. Different colours refer to different amounts of energy storage capacity. Based on meteorological and demand data for 1991–2019 for Great Britain. Average annual electricity demand is ~320 TWh. hours = Average load hour, indicating the average hourly electricity demand.

that could supply average demand for four hours, a day, a week or a month would increase that share to 88%, 94%, 99%, or 100% respectively.

A zero-carbon electricity system could be realized with ~110% solar and wind penetration and an electricity storage energy capacity of 27 TWh (average demand for one month). Alternatively, ~140% penetration and 6 TWh (one week) would suffice. This shows how wind and solar overcapacity can reduce the need for electricity storage energy capacity.

However, these values are significantly lower than the 67 TWh or 63 TWh identified in other studies.[43,44] The reason is that electricity storage systems are not 100% efficient. Those studies assume a round-trip efficiency of 40–45% for long-duration hydrogen storage. Applying a 40% efficiency to the 27 TWh would increase the required energy storage capacity to 67 TWh, putting this finding in line with those from other studies.

Figure 7.11 shows the impact of integrating wind and solar generation for energy storage capacities of different sizes and different round-trip efficiencies (RT). The three storage types could be classified as:

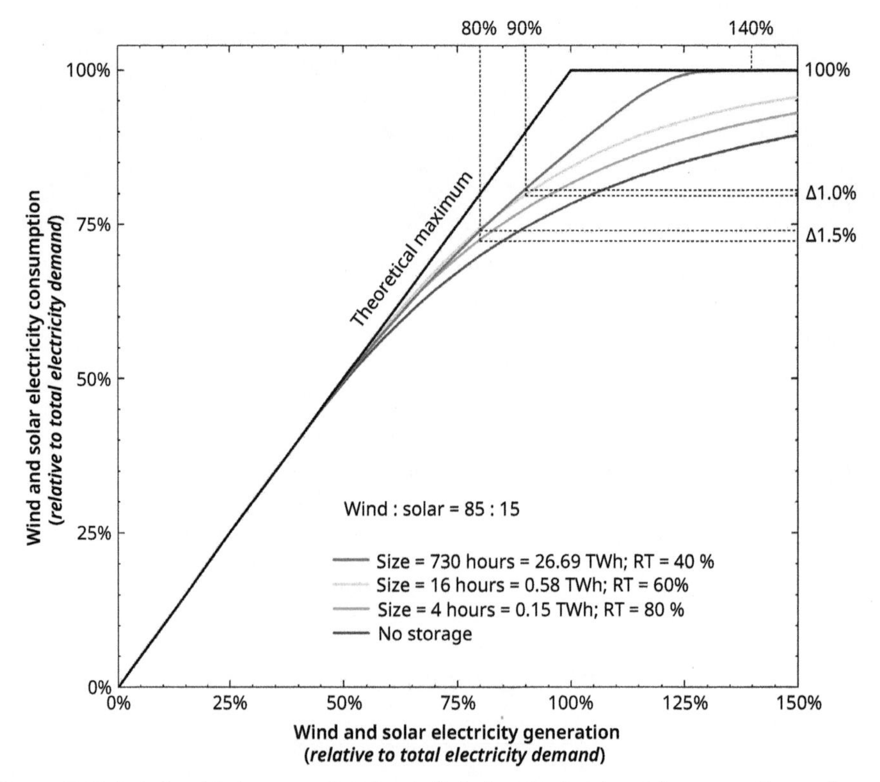

Figure 7.11 Relationship between the share of wind and solar electricity generation and the share which can be consumed (accounting for mismatch between supply and demand) with different energy storage capacities and round-trip efficiencies. Different colours refer to different sizes and round-trip efficiencies of energy storage capacity. Based on meteorological and demand data for 1991–2019 for Great Britain. Average annual electricity demand is ~320 TWh. hours = Average load hour, indicating the average hourly electricity demand. RT = Round-trip efficiency.

- **Small, short duration**: Energy storage capacity of ~0.05% of total annual electricity demand (0.15 TWh) that can supply average demand for 4 hours at 80% round-trip efficiency; Example: lithium-ion battery storage
- **Medium size, medium duration**: Energy storage capacity of ~0.18% of total annual electricity demand (0.58 TWh) that can supply average demand for 16 hours at 60% round-trip efficiency; Example: compressed air storage (efficiency averaged between adiabatic and diabatic type)
- **Large, long duration**: Energy storage capacity of ~20% of total annual electricity demand (~27 TWh) to supply average demand for 730 hours (1 month) at 40% round-trip efficiency; Example: hydrogen storage.

The analysis reveals that up to 80% wind and solar generation, the small, short-duration storage is as effective as medium- and long-duration storage (only 1.5% difference in the amount of wind and solar generation that can be used at the hourly level). Up to 90% penetration, medium-sized and medium-duration storage is as effective as long-duration storage (only 1% difference). However, to fully meet electricity demand with wind and solar generation, large, long-duration storage is needed along with over-building renewables. In this example, it would amount to ~27 TWh at 40% round-trip efficiency and 40% wind and solar overcapacity. Such large-scale electricity storage systems do not yet exist.

Table 7.2 summarizes the key insights on the role of electricity storage in integrating wind and solar generation for Great Britain's power system and other major power markets.

The key insights are:

- **There are optimal wind:solar ratios that minimize the mismatch between wind and solar generation and hourly demand**: For markets with high wind and limited solar resource (like Europe) it is > 80:20. For markets with a higher solar resource, it changes to ~70:30.
- **No storage is needed if the share of wind and solar generation below 25%:** At this share, wind and solar generation can always be fully integrated with demand (*condition: the remaining generation capacity is fully flexible*)
- **No storage is needed if the optimal wind:solar ratio is applied and their generation share is below 45%**: At this share and ratio, wind and solar generation can always be fully integrated with demand (*condition: the remaining generation capacity is fully flexible)*
- **Small, short-duration storage is sufficient to integrate wind and solar generation if its share is below 80%**: The difference in integrating this share with hourly demand compared to medium-sized and -duration storage systems is < 3%
- **Medium-sized and medium-duration storage is sufficient to integrate wind and solar generation if its share is below 90%**: The difference in integrating this share with demand compared to large, long-duration systems is < 3%
- **Large, long-duration storage systems and wind and solar overcapacity are required to meet 100% of demand with wind and solar generation alone**: All markets can achieve full integration of wind and solar generation with a storage system sized at ~20% of total annual electricity demand and ~140% wind and solar generation capacity relative to total annual electricity demand.

- **Wind and solar overcapacity can significantly reduce the required electricity storage energy capacity**: A 100% efficient store sized to ~20% of annual electricity demand could ensure demand is fully met with 110% wind and solar generation; at 140% wind and solar generation this would reduce to a size of ~2%.

KEY INSIGHT

Short-duration storage with < 4 hours discharge is as effective as medium- or long-duration storage at balancing wind and solar with demand up to 80% VRE share. Medium-duration storage with < 16 hours is as effective as long-duration storage up until 90% VRE share.

Table 7.2 Key insights on wind and solar ratios and the impact of electricity storage in integrating wind and solar generation across different power markets. QLD—Queensland, PJM—Regional transmission organization in the north-east of the US.

	Great Britain	Germany	France	Australia (QLD)	Japan	US (PJM)
Best wind:solar ratio, i.e. least mismatch between solar and wind generation and hourly demand	85:15	83:17	86:14	69:31	68:32	68:32
Share of wind and solar generation from which mismatch to hourly demand > 1% (worst wind:solar ratio)	25%	27.5%	25%	30%	27.5%	27.5%
Share of wind and solar generation from which mismatch to hourly demand > 1% (best wind:solar ratio)	50%	47.5%	47.5%	45%	47.5%	42.5%
Δ in solar and wind integration between short- and medium-duration storage (at 80% generation share)	1.5%	1.7%	1.7%	2.7%	2.7%	2.8%
Δ in solar and wind integration between medium- and long-duration storage (at 90% generation share)	1.0%	2.2%	2.0%	0.5%	1.2%	0.8%
Required solar and wind generation share for its consumption to reach 100% with long-duration storage	140%	140%	132.5%	140%	137.5%	142.5%
Size of required long-duration storage at 40% round-trip efficiency in TWh	27	40	40	4	75	23

Table 7.3 Comparison of insights from systematic analysis in this section and from review of academic studies in Figure 7.1.

Wind and solar penetration	Insights from academic studies shown in Figure 7.1	Insights from systematic analysis
< 50%	Less than 0.02% of electricity storage energy capacity relative to annual electricity demand needed	No storage may be needed if wind:solar ratio is optimized
< 80%	Electricity storage energy capacity of 0.02%–0.1% relative to annual electricity demand is needed	Small, short-duration storage at 0.05% of annual electricity demand is as effective as larger storage systems
< 90%	Electricity storage energy capacity of 0.02%–1% relative to annual electricity demand is needed	Medium-size,-duration storage at 0.2% of annual electricity demand is as effective as larger storage systems

These insights are in line with other studies and confirm the findings from Figure 7.1 on required electricity storage energy capacity as shown in Table 7.3.[39,43]

This analysis is kept deliberately simple and high-level to transparently derive fundamental insights that hold true across power markets and scenarios. Further detail is provided in other studies, for example on:[8,43–45]

- Cost optimization between generation overcapacity and electricity storage
- Impact on total electricity storage capacity and cost when combining electricity storage systems of different sizes and efficiency
- Impact of changing demand (absolute size, hourly variation) due to electrification of heat and transport demands
- Impact of baseload generation on the required electricity storage energy capacities (e.g. nuclear)

FREQUENTLY ASKED QUESTION

Do you see storage capacity being provided by large centralized systems or rather by local city or residential storage?

There is most likely going to be a mix of central large-scale systems and smaller decentral communal/commercial and residential systems. In fact, most studies project a ~50:50 mix between the two going forward.[46,47] It is worthwhile differentiating by application though to understand better where centralized and where decentralized systems may be deployed. This is shown in Figure 3.23. Applications on the generation side tend to be larger and on the consumption side smaller. Network services could be provided by both.

7.5.2 Energy sector

The previous section highlighted the need to deploy TWh-scale electricity storage to enable 100% wind and solar supplied electricity sectors. However, the challenge becomes even greater when expanding the perspective to the entire energy sector.

The scale of this challenge can be understood by looking at current strategic energy reserves in fossil fuels. For example, the US has energy storage reserves of nearly 5,000 $TWh_{calorific}$ (based on the heating value of the fuels) for petroleum, crude oil, motor fuels, heating and other oils, and natural gas. Assuming these fuels can be converted to electricity with an efficiency of 40%, this would amount to 2,000 $TWh_{electric}$.[48] This is several orders of magnitudes larger than the electricity storage capacities identified for decarbonization of the electricity sector. Not all of these strategic reserves are for use in the energy sector; however, those that are used as industry feedstocks may also have to be replaced with electricity-derived chemical fuels like green hydrogen in future.

KEY INSIGHT

The amount of fossil fuels held by countries as strategic energy reserves would amount to multiple thousand terawatt hours (TWhs) if converted to electricity storage energy capacity.

For a more detailed analysis on energy storage requirements to decarbonize the energy sector, Table 7.4 and Figure 7.12 focus on natural gas storage in the EU and the US. They confirm the significant storage volumes required (> 1,000 $TWh_{calorific}$) and show the distinct seasonality in charge and discharge of the storage volumes (once per year), in line with the regional heating periods. The respective maximum charge/discharge capacities reveal that the gas storage systems have minimum discharge durations in the region of 2-3 months (energy to power ratio of > 1,300 hours).

These facts show the sheer scale of energy storage that societies rely upon. The 'equivalent' amount of electricity storage required could be notably lower though, as the technologies

Table 7.4 Total volume, maximum charge/discharge capacities and respective discharge duration for natural gas storage in the EU and the US.[49,50]

	European Union	United States
Natural gas storage volume [$TWh_{calorific}$]	1,120	1,260
Maximum charge/discharge [$GW_{calorific}$]	840	650
Minimum discharge duration [hours]	1,350	1,940
Annual charge-discharge cycle [#]	1	1

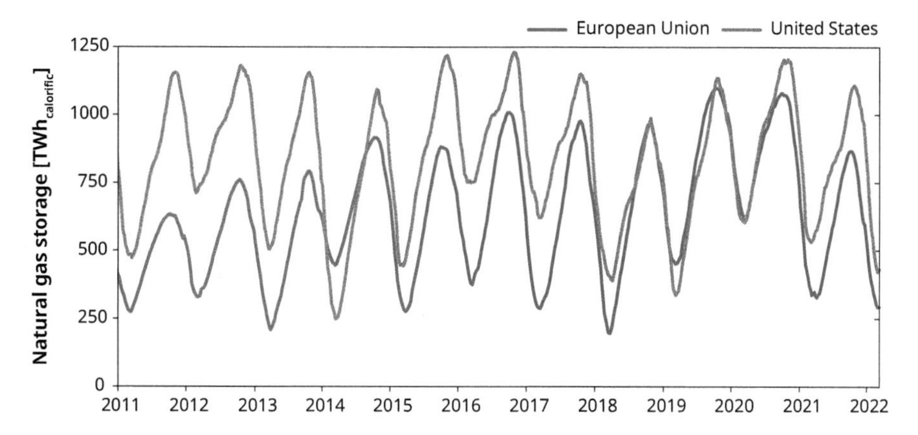

Figure 7.12 Natural gas stored in the European Union and the United States between 2011 and 2022. Total volumes for the European Union grow over time due to new facilities coming online and changing reporting standards.[51]

for end-use conversion are more efficient. If heating is electrified via heat pumps, the amount of storage needed for heating demand could be a factor of 3 to 4 lower due to their high coefficient of performance (i.e. 1 unit of electricity is converted to 3 to 4 units of heat).[52] However, this still leaves a requirement for hundreds of $TWh_{electric}$. A similar analysis could be conducted for energy storage reserves for transport, yielding similar orders of magnitude storage requirement.

This cursory analysis provides a snapshot of the challenge with decarbonizing the whole energy sector. The significant volumes of fossil fuel storage systems need to be converted or replaced with systems that store low-carbon based fuels/electricity. More importantly, these stores are only charged and discharged once per year, which also creates an economic challenge.

7.6 Discussion

7.6.1 Two schools of thought

The analysis revealed a more conservative approach in industry and government than in academia for planning the future low-carbon power system. It should be noted that the academic studies focus on accurate representation of wind and solar variability using resource data from multiple years and high spatial and temporal resolution.[14,16–18] Academic studies may also consider improvements in ancillary service provision which allows more services to be provided with less capacity.[11,53,54] Both rationales may justify system adequacy without the need for excessive capacity margins. However, temporal resolution is always above one hour, which may underestimate flexibility needs for short-term system balancing. On the other side, industry studies were conducted by direct stakeholders of the power system like the system operator, National Grid. This could mean that these studies use more realistic assumptions due to better industry insights and/or account for larger safety margins due to real-world liabilities of the stakeholders.

7.6.2 Role of nuclear power

The role of nuclear in low-carbon power systems is highly debated. One reason is that base-load nuclear power increases the ability to meet peak power demand, which should reduce demand for flexible power capacity. However, the investigated studies optimize for most cost-efficient energy systems. Balancing renewable supply and consumer demand cost-effectively requires the ability to quickly increase or reduce power output in light of the variability of renewables and demand, which is not technically feasible or would make nuclear uneconomic.[55] Otherwise, there may be situations where renewables meet demand and nuclear produces excess electricity. Naturally, renewable generation could be curtailed in these situations and all studies assume this to a certain degree of it, but excessive curtailment will not yield cost-optimal solutions.

This contrast between providing for peak power demand but not adjusting power output reflects the wider debate about the future role of nuclear in low-carbon power systems.[56,57] Recent findings suggest value in a limited amount of nuclear to decarbonize power systems with low overall flexibility, but highlight the preference to meet peak demand with flexibility capacity.[53] The more detailed analysis of the investigated studies confirms that additional nuclear capacity in the presence of variable renewables is projected to actually increase the need for flexibility capacity.

7.6.3 Comparison to other flexibility options

Figure 7.4 shows no difference in flexibility capacity requirement between studies with more electricity storage capacity than others. It could therefore be argued that flexibility options can be used interchangeably. However, this conclusion neglects possible technology constraints in providing flexibility at certain times and durations. For example, flexible power provision through electricity storage is limited by its discharge duration, whereas provision through interconnection is limited by the spatial correlation of weather and demand patterns.[58] These limitations have direct financial implications. For example, the UK energy markets regulator 'de-rates' the capacity contribution electricity storage can make in capacity market auctions based on discharge duration.[59] As of 2021 storage systems with 0.5 hours discharge duration only receive remuneration for 13% of their power capacity when bidding into the one-year-ahead capacity market, while systems above 5 hours receive 95%.[60] This de-rating is supposed to reflect the equivalent firm capacity at the time and duration of peak net demand (gross demand minus output from variable renewables).[59] However, similar limitations apply to other flexibility options. Interconnector capacity is de-rated at between 10% and 97%,[60] DSR at 79% and flexible generation capacity (e.g. open cycle gas turbine (OCGT), oil, diesel, hydro) at 91–95% based on historical station availability.

For the present analysis this has two implications. First, the modelled flexibility capacity appears to refer to equivalent firm capacity, because flexibility options are not de-rated. Therefore, absolute flexibility capacity requirements may be higher than identified here. Second, the low de-rating of electricity storage with more than 5 hours discharge duration (95%) implies that this flexibility option is considered to have highest system value in meeting peak demand for the GB power system with 40–50% of solar, wind and nuclear energy share.

FREQUENTLY ASKED QUESTION

In light of the millions of batteries in EVs that could be used to support matching power supply and demand, do we need stationary storage at all?
Adapting the rate and time of charging EVs appears like a low-hanging fruit to support matching power supply and demand. However, supplying electricity to the grid may be more complex as it may affect driving schedules (the primary purpose of EVs) and lifetime/warranty of the battery. In addition, the evolution of transport must be considered where car sharing or autonomous driving may become more common and vehicles have less 'idle time' in general. As such, EVs will play a role in providing power system flexibility, but the degree is still highly uncertain and it appears unlikely they will fully replace the need for stationary storage.

7.6.4 Long-duration storage economics

These cursory analyses of the required energy storage capacities to decarbonize the electricity and overall energy sectors, reveal the *TWh-challenge* of converting existing fossil fuel or deploying new storage systems. However, the economic case for these long-duration storage systems may be the hardest challenge to overcome. Chapter 5 shows lifetime cost of > 500 $USD/MWh_{electric}$ for an exemplary seasonal electricity storage system by 2030-40. The European gas network discharges up to 800 TWh over winter, which equates to approximately 320 TWh of electricity. Given this cost of storage, society would have to pay USD 160 billion per year to move electricity between summer surplus and winter shortfall, which will be challenging to justify. Policy-makers, business professionals, and academics need to study whether flexible low-carbon alternatives like flexible generation (e.g. natural gas with carbon capture and storage, geothermal generation), demand-side response or network expansion could be more feasible and more economic. Alternatively, regulatory changes and incentive mechanisms need to be established to give the required energy storage capacity a viable business model.

7.6.5 Limitations

The normalization of flexibility requirements to peak or annual demand or overall power capacity and energy capacity are supposed to make the identified insights applicable to all power systems. In reality, however, differences in flexibility requirements are likely to result from three dimensions:

- Type of renewable resource—for example, flexibility capacity requirements have been found to be larger in solar power dominated systems[20,61]
- Temporal distribution of resource availability—for example, in systems that span multiple time zones, maximum solar production at noon in one zone might coincide with the afternoon demand peak in another, reducing overall flexibility needs[25]
- Existing power system assets—for example, a more flexible power plant portfolio (more combined cycle gas turbines (CCGTs) rather than coal or nuclear) could reduce flexibility needs

These three dimensions should serve to qualify results based on the requirements identified in this chapter to guide power system planning and ensure sufficient flexibility capacity.

The analyses on required energy storage capacities to decarbonize the electricity sector and the energy sector as a whole are intentionally kept simple. However, the key limitations that should be mentioned are:

- Usage of meteorological wind and solar data instead of actual generation data
- Assumption of unconstrained power flows within markets
- Assumption of fully flexible generation capacity beyond wind and solar
- Usage of historical demand data instead of projection of future demands
- Lack of differentiation between fossil fuel reserves required as fuel in the energy sector and as feedstock in industry

In more detailed studies that are supposed to enable direct policy or investment decisions, these limitations must be omitted.

FREQUENTLY ASKED QUESTION

These analyses focus on 'developed' electricity systems in Western and APAC regions. What role could storage play in the 'developing world'?
In regions without a fully developed centralized electricity system, energy storage could provide energy access through the deployment of solar home systems, nano-grids, or microgrids. These could empower more than ~1.2 billion people that currently have no access to electricity.[62,63] These regions could potentially 'leapfrog' the development of centralized and often fossil-fuel based electricity systems and directly move to a network of interconnected mini-grids powered by decentral renewables and electricity storage systems.

7.7 Worked example

This chapter has looked into the value that electricity storage technologies (and flexibility technologies more generally) provide by integrating variable renewable and other inflexible low-carbon electricity generators. While every power system is different based on available renewable resources, temporal distribution of resource availability, and existing power plant portfolio, some general trends were identified that can help to derive initial estimates of electricity storage/flexibility capacity needs to integrate low-carbon energy sources. This worked example will guide you through the derivation of these needs.

A question:
The government in Germany plans to increase the share of renewable generated electricity from ~45% in 2020 to ~80% in 2030.[64] How much electricity storage will be needed to enable that?

How to answer:

First, we need to identify how much of the renewable electricity will be variable. In 2020, 32% of electricity came from wind and solar (variable) and 13% from dispatchable hydro, biomass, waste incineration, and geothermal.[65] Assuming that the share of the latter will stay constant, the share of wind and solar must increase to 67% to hit the 80% renewables target by 2030. Second, we need to identify total annual electricity and peak power demand. These stood at 570 TWh and 80 GW.[65,66]

1. Open <www.EnergyStorage.ninja> and go to the 'System need' tab
2. Enter the parameters of the German power system for 2020

Peak power demand GW 80 | Annual electricity demand TWh 570 | Share of intermittent renewable electricity 0% 67% 100%

You will see that in 2020 the German power system needed an estimated ~6 GW and ~35 GWh of electricity storage capacity. Pumped hydro capacity stood at 6.3 GW and 37 GWh.[67] The capacity of large-scale battery storage was ~0.5 GW and ~0.6 GWh.[68] Thus, there was sufficient storage capacity to integrate variable renewables in 2020.

3. Now, enter the parameters of the German power system for 2030. Let's assume peak power demand and annual energy demand will stay constant (although they are more likely to increase due to electrification of transport and heat).

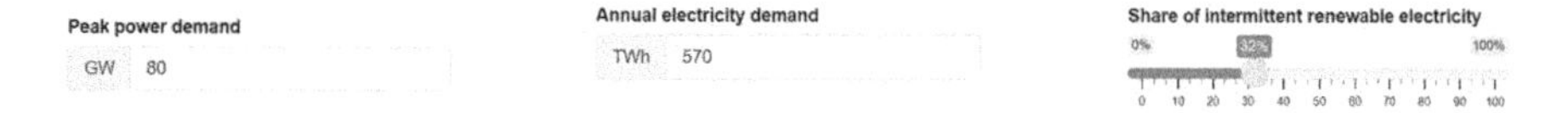

This shows that electricity storage capacity would have to increase to 17 GW (14–26 GW) and 160 GWh (range: 120–410 GWh).

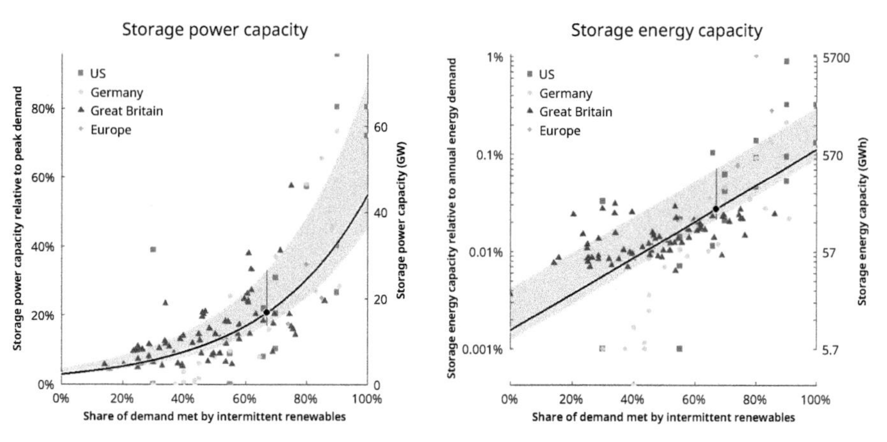

You've done it!

Congratulations. You have derived a first-cut estimate of the electricity storage capacity required for the German power system to increase variable renewable electricity generation to 67% (and total renewable generation to 80%). You have identified that current capacity (mostly pumped hydro) must triple in terms of power capacity and more than quadruple in terms of energy capacity. Given that suitable spots for pumped hydro in this densely populated country are already taken, this additional capacity is likely to come from battery or other novel technologies used for stationary electricity storage.

However, please be aware that these values need to be taken with caution. While the studies underlying these results look at electricity storage energy capacity needs only, the power capacity needs may also be met by other flexibility technologies (see section 7.3). The need for energy capacity highly depends on the availability of alternative generation technologies (e.g. gas peaking plants) and potential imports from neighbouring countries.

7.8 References

1. Larsson S, Fantazzini D, Davidsson S, Kullander S, *et al.* 'Reviewing Electricity Production Cost Assessments' (2014) 30 *Renewable and Sustainable Energy Reviews* 170–83.
2. Heuberger CF, Staffell I, Shah N, and MacDowell N. 'A Systems Approach to Quantifying the Value of Power Generation and Energy Storage Technologies in Future Electricity Networks' (2017) 107 *Computers & Chemical Engineering* 247–56.
3. Denholm P, Jorgenson J, Hummon M, Jenkin T, *et al. The Value of Energy Storage for Grid Applications* (Golden: National Renewable Energy Laboratory, 2013).
4. Denholm P and Mai T. *Timescales of Energy Storage Needed for Reducing Renewable Energy Curtailment* (Golden: National Renewable Energy Laboratory, 2017).
5. de Sisternes FJ, Jenkins JD, and Botterud A. 'The Value of Energy Storage in Decarbonizing the Electricity Sector' (2016) 175 *Applied Energy*, 368–379.
6. Scholz Y, Gils HC, and Pietzcker RC. 'Application of a High-Detail Energy System Model to Derive Power Sector Characteristics at High Wind and Solar Shares' (2017) 64 *Energy Economics* 568–82.
7. Després J, Mima S, Kitous A, Criqui P, *et al.* 'Storage as a Flexibility Option in Power Systems with High Shares of Variable Renewable Energy Sources: A POLES-based S 64 *Energy Economics* 638–50.
8. Weitemeyer S, Kleinhans D, Wienholt L, Vogt T, *et al.* 'A European Perspective: Potential of Grid and Storage for Balancing Renewable Power Systems' (2016) 4 *Energy Technology* 114–22.
9. BEIS. *Updated Energy and Emissions Projections 2017* (London: Department for Business, Energy & Industrial Strategy, 2018).
10. Marquina D, Rooze J, Cheung A, and Berryman I. *Flexibility Solutions for High-Renewable Energy Systems* (London: BloombergNEF, 2018).
11. Sanders D, Hart A, Brunert J, Strbac G, *et al. An Analysis of Electricity System Flexibility for Great Britain* (London: The Carbon Trust, 2016).
12. Strbac G, Aunedi M, Pudjianto D, Teng F, *et al. Value of Flexibility in a Decarbonised Grid and System Externalities of Low-Carbon Generation Technologies* (London: Imperial College London and NERA Economic Consulting, 2015).
13. Edmunds RK, Cockerill TT, Foxon TJ, Ingham DB, *et al.* 'Technical Benefits of Energy Storage and Electricity Interconnections in Future British Power Systems' (2014) 70 *Energy* 577–87.

14. Heuberger CF, Staffell I, Shah N, and Mac Dowell N. 'Impact of myopic decision making and disruptive events in power systems planning' (2018) 3 *Nature Energy* 634–40.
15. National Grid. *Future Energy Scenarios* (London: National Grid, 2018).
16. Pfenninger S and Keirstead J. 'Renewables, Nuclear, or Fossil Fuels? Scenarios for Great Britain's Power System Considering Costs, Emissions and Energy Security' (2015) 152 *Applied Energy* 83–93.
17. Price J, Zeyringer M, Konadu D, Sobral Mourão Z, *et al.* 'Low Carbon Electricity Systems for Great Britain In 2050: An Energy-Land-Water Perspective' (2018) 228 *Applied Energy* 928–41.
18. Zeyringer M, Price J, Fais B, Li P-H, *et al.* 'Designing Low-Carbon Power Systems for Great Britain in 2050 that are Robust to the Spatiotemporal and Inter-Annual Variability of Weather' (2018) 3 *Nature Energy* 395–403.
19. Zerrahn A, Schill W-P, and Kemfert C. 'On the Economics of Electrical Storage for Variable Renewable Energy Sources' (2018) 108 *European Economic Review* 259–79.
20. Cebulla F, Haas J, Eichman J, Nowak W, *et al.* 'How Much Electrical Energy Storage Do We Need? A Synthesis for the U.S., Europe, and Germany' (2018) 181 *Journal of Cleaner Production* 449–59.
21. Bernath C, Bossmann T, Deac G, Elsland R, *et al. Langfristszenarien für die Transformation des Energiesystems in Deutschland* (Karlsruhe, Aachen, Heidelberg: Fraunhofer ISI, Consentec GmbH, ifeu - Institut für Energie- und Umweltforschung Heidelberg GmbH, 2017).
22. Budischak C, Sewell D, Thomson H, Mach L, *et al.* 'Cost-minimized Combinations of Wind Power, Solar Power and Electrochemical Storage, Powering the Grid up to 99.9% of the Time' (2013) 225 *Journal of Power Sources* 60–74.
23. Denholm P and Hand M. 'Grid Flexibility and Storage Required to Achieve Very High Penetration of Variable Renewable Electricity' (2011) 39 *Energy Policy* 1817–30.
24. Jacobson MZ, Delucchi MA, Cameron MA, and Frewa BA. 'Low-cost Solution to the Grid Reliability Problem with 100% Penetration Of Intermittent Wind, Water, and Solar For All Purposes' (2015) 112 *PNAS*, 15060–15065.
25. MacDonald AE, Clack CTM, Alexander A, Dunbar A, *et al.* 'Future Cost-Competitive Electricity Systems and Their Impact on US CO2 Emissions' (2016) 6 *Nature Climate Change* 526–31.
26. Pape C, Gerhardt N, Härtel P, Scholz A, *et al. Roadmap Speicher* (in German) (Kassel, Aachen, Würzburg: Stiftung Umweltenergiereicht, IAEW, Fraunhofer IWES, 2014).
27. Repenning J, Emele L, Blanck R, Böttcher H, *et al. Klimaschutzszenario 2050* (in German) (Berlin, Karlsruhe: Öko-Institut, Fraunhofer ISI, 2015).
28. Safaei H and Keith DW. 'How Much Bulk Energy Storage is Needed to Decarbonize Electricity?' (2015) 8 *Energy & Environmental Science* 3409–17.
29. Schill W-P. 'Residual Load, Renewable Surplus Generation and Storage Requirements in Germany' (2014) 73 *Energy Policy* 65–79.
30. Schill W-P and Zerrahn A. 'Long-run Power Storage Requirements for High Shares of Renewables: Results and Sensitivities' (2018) 83 *Renewable and Sustainable Energy Reviews* 156–71.
31. Zerrahn A and Schill W-P. 'Long-run Power Storage Requirements for High Shares of Renewables: Review and a New Model' (2017) 79 *Renewable and Sustainable Energy Reviews* 1518–34.
32. Masson-Delmotte V, Zhai P, Pörtner H-O, Roberts D, *et al. IPCC Special Report: Global Warming of 1.5 C* (Paris: Intergovernmental Panel on Climate Change, 2018).
33. Huppmann D, Kriegler E, Krey V, Riahi K, *et al.* IAMC 1.5°C Scenario Explorer and Data hosted by IIASA (2019). Available at: <https://iiasa.ac.at/models-and-data/iamc-15degc-scenario-explorer> accessed 5 November 2022.

34. Huppmann D, Rogelj J, Kriegler E, Krey V, *et al.* 'A New Scenario Resource for Integrated 1.5 °C Research' (2018) 8 *Nature Climate Change* 1027–30.

35. Jacobson MZ, Delucchi MA, Bauer ZAF, Goodman SC, *et al.* '100% Clean and Renewable Wind, Water, and Sunlight All-Sector Energy Roadmaps for 139 Countries of the World' (2017) 1 *Joule* 108–21.

36. International Energy Agency. *World Energy Outlook* (Paris: IEA, 2018).

37. International Energy Agency. *Energy Storage* (Paris: IEA, 2022).

38. International Hydropower Association (2022). Available at: <https://www.hydropower.org/> accessed 5 November 2022.

39. Schill W-P. 'Electricity Storage and the Renewable Energy Transition' (2020) 4 *Joule* 2047–64.

40. Brown T, Schlachtberger D, Kies A, Schramm S, *et al.* 'Synergies of Sector Coupling and Transmission Reinforcement in a Cost-Optimised, Highly Renewable European Energy System' (2018) 160 *Energy* 720–39.

41. Mathiesen BV, Lund H, Connolly D, Wenzel H, *et al.* 'Smart Energy Systems for Coherent 100% Renewable Energy and Transport Solutions' (2015) 145 *Applied Energy* 139–54.

42. Acatech, Leopoldina and Akademienunion. *Coupling the Different Energy Sectors—Options for the Next Phase of the Energy Transition* (Munich, Halle, Mainz: 2018).

43. Cárdenas B, Swinfen-styles L, Rouse J, and Garvey SD. 'Short-, Medium-, and Long-Duration Energy Storage in a 100% Renewable Electricity Grid: A UK Case Study' (2021) 14 *Energies* 8524.

44. Cosgrove P and Roulstone T. *Royal Society Working Paper on Energy Storage Multi-Year Studies* (London: Royal Society, 2020).

45. Weitemeyer S, Kleinhans D, Siemer L, and Agert C. 'Optimal Combination of Energy Storages for Prospective Power Supply Systems Based on Renewable Energy Sources' (2018) 20 *Journal of Energy Storage* 581–9.

46. BloombergNEF. Aggregation Paves the Way for Behind-The-Meter Storage (2017). Available at: <https://about.bnef.com/blog/aggregation-paves-way-behind-meter-storage/> accessed 5 November 2022.

47. Wood Mackenzie. *Global Energy Storage Outlook H1 2021* (Edinburgh: Wood Mackenzie, 2021).

48. Kittner N, Castellanos S, Hidalgo-Gonzalez P, Kammen DM *et al.* 'Cross-sector Storage and Modeling Needed for Deep Decarbonization' (2021) 5 *Joule* 2529–34.

49. U.S. Energy Information Administration (EIA). Weekly Natural Gas Storage Report (2022). Available at: <https://ir.eia.gov/ngs/ngs.html> accessed 5 November 2022.

50. AGSI. Gas Infrastructure Europe (2022). Available at: <https://agsi.gie.eu/> accessed 5 November 2022.

51. AGSI. Data Definitions (2022). Available at: <https://agsi.gie.eu/data-definition> accessed 5 November 2022.

52. Staffell I, Brett D, Brandon N, and Hawkes A. 'A Review of Domestic Heat Pumps' (2012) 5 *Energy & Environmental Science* 9291–306.

53. Strbac G, Aunedi M, and Pudjianto D. *Value of Baseload Capacity in Low-carbon GB Electricity System Report prepared for Ofgem Imperial College Project Team* (London: Ofgem, 2018).

54. Teng F, Aunedi M, Moreira R, Strbac G, *et al.* 'Business Case for Distributed Energy Storage' (2017) *CIRED—Open Access Proceedings Journal* 1605–8.

55. Lokhov A. *Technical and Economic Aspects of Load Following with Nuclear Power Plants* (Paris: OECD & NEA, 2011).

56. Joskow PL and Parsons JE. 'The Economic Future of Nuclear Power' (2009) 138 *Daedalus* 45–59.

57. Morris C. Can Nuclear and Renewables Coexist? (2018). Available at: <https://energytransition.org/2018/03/can-nuclear-and-renewables-coexist/> accessed 5 November 2022.

58. Grams CM, Beerli R, Pfenninger S, Staffell I, *et al.* 'Balancing Europe's Wind-Power Output through Spatial Deployment Informed by Weather Regimes' (2017) 7 *Nature Climate Change* 557–62.

59. Ofgem. *The Capacity Market (Amendment) Rules* (London: The Office of Gas and Electricity Markets, 2018).

60. National Grid ESO. *Electricity Capacity Report* (London: National Grid, 2021).

61. Pietzcker RC, Ueckerdt F, Carrara S, Sytze H, *et al.* 'System Integration of Wind and Solar Power in Integrated Assessment Models: A Cross-Model Evaluation of New Approaches' (2017) 64 *Energy Economics* 583–99.

62. Few S, Schmidt O, Gambhir A, Stephenson E, *et al. Energy Storage Trends for Off-Grid Services in Emerging Markets Insights from Social Enterprises* (London: Shell Foundation, 2018).

63. Few S, Schmidt O, and Gambhir A. 'Energy Access Through Electricity Storage: Insights from Technology Providers and Market Enablers' (2019) 48 *Energy for Sustainable Development* 1–10.

64. Koalitionsvertrag SPD Gruene FDP. Mehr Fortschritt wagen (2021). Available at: <https://www.gruene.de/artikel/koalitionsvertrag-mehr-fortschritt-wagen> accessed 5 November 2022.

65. Statistisches Bundesamt. Bruttostromerzeugung in Deutschland (2022). Available at: <https://www.destatis.de/DE/Themen/Branchen-Unternehmen/Energie/Erzeugung/Tabellen/bruttostromerzeugung.html> accessed 5 November 2022.

66. BDEW Bundesverband der Energie- und Wasserwirtschaft. *Energy Market Germany 2020* (2020).

67. Heimerl S and Kohler B. 'Aktueller Stand der Pumpspeicherkraftwerke in Deutschland Bestehende Pumpspeicherkraftwerke' (2017) 10 *Wasserwirtschaft*, 77–79.

68. Figgener J, Stenzel P, Kairies KP, Linßen J, *et al.* 'The Development of Stationary Battery Storage Systems in Germany—Status 2020' (2021) 33 *Journal of Energy Storage* 101982.

PART III: Methods to assess Energy Storage

8 Methods
Doing it yourself

8.1 Introduction

Chapter 1 introduces the required transformation of the energy sector and the available flexibility options to balance supply and demand in low-carbon power system. It also determines the amount of energy stored in fossil fuels in the UK in 2020. This is done by identifying the respective stock levels of coal, crude oil, petroleum products, and natural gas in the UK in 2020.[1] The respective amounts in tonnes or cubic metres are converted to calorific energy in TWh by using the lower heating values for each fuel.[2] The calorific energy values are multiplied by 40% conversion efficiency to determine the respective electric energy that could be generated when burning these fuels. 40% is a high-level average of coal, oil, and gas-fired power stations.[3]

For the transport sector, it is assumed that the 32 million vehicles in the UK have a 60-litre petrol tank. This is compared to 200,000 battery-electric vehicles and 200,000 plug-in hybrid electric vehicles with battery capacities of 50 kWh and 10 kWh respectively.[4]

8.2 Metrics

Chapter 2 provides indicative cost contributions for key upstream and downstream value chain steps. These are derived in Table 8.1, including the contributions of individual components if applicable.

Table 8.1 Cost components of a lithium-ion battery system, including indicative cost contributions.[5–8]

Product scope	Cost contribution
Cell	**35%**
Electrodes	~45%
Electrolyte	~15%
Separators	~15%

(Continued)

Monetizing Energy Storage. Oliver Schmidt & Iain Staffell, Oxford University Press. © Oliver Schmidt & Iain Staffell (2023).
DOI: 10.1093/oso/9780192888174.003.0008

Table 8.1 (*Continued*)

Product scope	Cost contribution
Current Collectors	~20%
Terminals & cell container	~5%
Pack	**15%**
Temperature control	~10%
Wiring and connectors	~20%
Housing	~20%
Battery Management System	~50%
Balance-of system	**10%**
Container	*Depends on project requirements*
Monitors, controls	
Thermal control	
Fire suppression	
Power control system	**10%**
Inverter/converter	~65%
Energy management system	~20%
Data management	~15%
System integration	**5%**
Assembly of components to system	*Depends on project requirements*
Tailoring of system to application	

(*Continued*)

Table 8.1 (*Continued*)

Product scope	Cost contribution
Project development	**10%**
Permits	
Land acquisition	*Highly project specific*
Financial and technical studies	
Installation	**15%**
Engineering studies	
Procurement of system	*Highly project specific*
Construction	
Commissioning	
Total installed system	**100%**

8.3 Technologies and applications

Chapter 3 compiles values for 17 cost and performance parameters for nine electricity storage technologies. Special focus is placed on using industry-validated sources that are based on manufacturer quotes and have a track record for realistic data. The resulting values were cross-checked with multiple industry experts via e-mail exchanges and direct phone calls.

The selection of sample companies active in the individual upstream product value chain steps for the nine electricity storage technologies is based on manufacturing of the products listed in Table 8.2. The companies were identified through news articles on specific energy storage projects or as market leaders in industry reports on the respective technology.

The market size values for global installed electricity storage systems by 2020 are determined using a range of data sources (see Table 8.3). The data were double-checked to ensure only operational storage systems were included.

The annual market for lead-acid batteries in 2020 was estimated at 415 GWh, which would correspond to 3,300 GW at energy-to-power ratios of 1:8 (typical for engine starter batteries).[14] However, most of these batteries are used for automotive applications or specialized stationary use cases like telecom tower power provision or uninterrupted power supply. The market size estimate therefore only contains the stationary battery systems that are connected to energy networks and listed in the Department of Energy Global Energy Storage Database.[12]

Table 8.2 Sample products of the upstream product value chain steps for the nine focus electricity storage technologies.

	Materials	Components	Storage unit(s)	System integration
Pumed hydro	Iron ore	Turbine blades	Pelton turbine	Turbine + Pump + Pipes + Dam
Compressed air	Copper	Electromagn. coils	Generator	Compressor + Generator + Cavern
Flywheel	Carbon fibres	Rotor	Flywheel	Flywheel + PCS + BoP
Lead acid	Lead	Electrode	Pack	Pack + PCS + BoP
Lithium ion	Nickel	Electrode	Cell, Pack	Pack + PCS + BoP
Sodium sulphur	Sodium	Solid electrolyte	Cell, Pack	Pack + PCS + BoP
Redox flow	Vanadium	Electrode	Cell stack	Cell stack + PCS + BoP
Supercapacitor	Activated carbon	Electrode	Supercapacitor	Supercapacitor + PCS + BoP
Hydrogen	Polymers	Membrane	Electrolyser	Electrolyser + PCS + BoP

Table 8.3 Operational utility-scale electricity storage capacity in energy system applications by 2020.

Technology	Capacity	Comment	Data source
Pumped hydro	158 GW	n/a	IHA[9]
Lithium ion	15 GW	n/a	WoodMac, IEA[10,11]
Flow battery	< 0.1 GW	Only utility-scale systems in energy system	DoE database[12]
Supercapacitors	< 0.1 GW	Only utility-scale systems in energy system	DoE database[12]
Hydrogen	< 0.1 GW	Only utility-scale systems in energy system	DoE database[12]
Compressed air	0.4 GW	n/a	DoE database[12]
Flywheel	< 0.1 GW	Only utility-scale systems in energy system	DoE database[12]
Lead acid	< 0.1 GW	Only utility-scale systems in energy system	DoE database[12]
Sodium sulphur	0.6 GW	n/a	NGK Insulators[13]

Chapter 3 also identifies 23 common electricity storage applications. These are allocated to 13 archetypal applications for which lifetime cost are modelled in Chapter 5. Allocation of common to archetypal applications is based on similar technical requirements, that is, discharge duration and annual cycles. The final requirements of the archetypal applications are chosen such that a broad spectrum of discharge duration and annual cycles requirements is covered, but the values are still within or close to the ranges given in the literature.

Table 8.4 shows the technical requirements for the 23 common electricity applications as given in the literature and the final archetypal values (bold).

Table 8.4 Discharge duration and annual cycle requirements for the 23 common electricity storage applications.[15-17] Final values for archetypal applications are given in bold. Comments justify choice of values where necessary.

Application	Discharge [hours]	Annual cycles [#]	Response time [s]	Comment
Frequency regulation	0.25–1	250–10,000	< 10	High cycle values may not refer to *full equivalent charge-discharge cycles*, but individual cycles that do not exploit full depth-of-discharge
Renewables smoothing	0.25–4	>4,000	< 10	
Power quality	< 0.25	10–200	< 10	
Frequency regulation	**0.5**	**300**	**< 10**	
Black Start	0.25–1.0	10–20	> 10	No comment
Black start	**1**	**10**	**> 10**	
Peak capacity	2–6	5–500	> 10	No comment
Peak capacity	**4**	**300**	**> 10**	
Congestion relief	1–6	50–500	> 10	
Load following	0.25–1	100–1,000	> 10	
Congestion mgmt.	**1**	**1,000**	**> 10**	
Ramping reserve	0.25–1	20–50	< 10	No comment
Contingency reserve	0.25–1	20–50	< 10	
Inertial services	0.25–1	50–100	< 10	
Frequency response	**0.5**	**50**	**< 10**	

(*Continued*)

Table 8.4 (*Continued*)

Application	Discharge [hours]	Annual cycles [#]	Response time [s]	Comment
Seasonal storage	n/a	n/a	> 10	No parameters for seasonal storage in the three studies
Seasonal storage	**700**	**3**	**> 10**	
Backup power	1–4	10–100	> 10	No comment
Power reliability	4–10	< 50	> 10	
Power reliability	**4**	**10**	**> 10**	
Renewables firming	2–10	5–500	> 10	No comment
Renewables integration	**8**	**300**	**> 10**	
Transmission deferral	2–8	10–500	> 10	No comment
Distribution deferral	1–6	50–500	> 10	
T&D deferral	**8**	**100**	**> 10**	
Demand charge mgmt.	1–4	50–500	> 10	No comment
Demand charge mgmt.	**1**	**100**	**> 10**	
RE self-consumption	2–6	150–400	> 10	100 cycles to ensure differentiation to peak capacity; RE - Renewables
Self-consumption	**4**	**100**	**> 10**	
Renewables arbitrage	< 1	> 250	> 10	Parameters are highly dependent on power price volatility; ToU - Time-of-use tariffs
Wholesale arbitrage	2–10	300–400	> 10	
Retail arbitrage	2–4	50–250	> 10	
ToU bill management	3–6	400–1500	> 10	
Energy arbitrage	**4**	**700**	**> 10**	

(*Continued*)

Table 8.4 (*Continued*)

Application	Discharge [hours]	Annual cycles [#]	Response time [s]	Comment
Voltage support	–	–	< 10	No active power drawn from device

8.4 Investment cost

In Chapter 4, experience curve analysis is used to project future investment cost for electricity storage technologies. This is possible because experience curves identify the relationship between historical technology prices and cumulative capacity additions and can be used to extrapolate observed trends to the future.

8.4.1 Theory

Learning curves depict the improvement of a technology parameter (e.g. cost, size) as a function of experience (e.g. cumulative capacity or number of systems). More specifically, learning curves based on Wright describe the development of manufacturing cost in relation to increased cumulative production.[18] They have been described as the most objective method to project future cost of technologies.[19] Instead of manufacturing cost, experience curves depict product price development (i.e. investment cost) to account for all cost factors (e.g. research and development (R&D), sales, depreciation) and, while more uncertain than learning curves, are also suitable to explore future cost.[20,21] The rate at which product prices change is termed the experience rate. Cumulative production has been identified as the predictor of technology cost that performs best compared to other variables.[22]

There are a range of milestones in the development of the experience-curve methodology and its application in the energy sector, which are listed below:

- 1936: Theodore Wright describes the effect of learning on manufacturing cost in the aircraft industry and proposes a mathematical model.[18]
- 1962: Kenneth Arrow finds that the model holds true for whole 'capital goods industry' (i.e. industrial sector) and coins 'learning-by-doing' as the specific cost reduction in the manufacturing stage.[23]
- 1968: The Boston Consulting Group extends the cost inputs to the model to include all manufacturing inputs as well as any other costs required to deliver the product to the end user and coins the term 'experience curve'.[20]
- 2000: The International Energy Agency publishes learning curves for the most prominent energy generation technologies.[24]
- 2000+: A rich body of literature evolves on learning curves for energy technologies, with individual curves, reviews, and comparisons.[24-28]

Strengths of experience curve analysis

The experience curve model is appealing for several reasons. The idea that firms learn from experience in the past seems intuitive, models can be easily tested with empirical datasets, the high goodness-of-fit shows it works, and by reducing the complex process of innovation into a single parameter, the model is simple.[29]

The underlying reasons for cost reductions as a result of learning-by-doing in manufacturing are identified as spreading overhead cost over larger volumes (economies of scale), reducing inventory cost, cutting labour cost with process improvements, achieving greater division of labour, and improving efficiency through greater familiarity with the process.[30] Similar models were developed to account for these underlying reasons more specifically, like Moore's Law (power law of time); Goddard's Law (power law of annual production); Sinclair, Klepper, and Cohen's Law (power law of annual and cumulative production), and Nordhaus' Law (power law of time and cumulative production).[22]

Weaknesses of experience curve analysis

A common critique of experience curve analysis is the lack of causation and accountancy to the various cost reducing factors. Experience curves show how cost may reduce over time, but provide no explanation for the underlying reasons beyond its relationship to cumulative output (in the case of one-factor curves).[31] Additional cost reducing factors are R&D expenditures (learning-by-searching), improvement of product characteristics via user feedback (learning-by-using) and network relationships between research laboratories, industry, end-users, and political decision-makers that can lead to spill-over effects (learning-by-interacting).[27,32] Some authors suggest that experience curves largely reflect economies of scale.[33] While it has been shown that cumulative production is a suitable predictor for costs of solar PV modules, other factors than experience in manufacturing are responsible for this, namely plant size (economies of scale), efficiency improvements, and commodity costs.[34] It has also been argued that individual experience effects of single component improvements together may explain an aggregated form of experience for a product.[35] Component-specific experience rates could account for these individual effects, but separate production and cost data are difficult to obtain.[21]

Two-factor experience curves aim to disentangle two important learning factors: cumulative output (learning-by-doing) and knowledge stock (learning-by-searching).[36] They have been used to explain cost reductions for wind, solar, and conventional generation technologies as well as the 2017–2020 plunge in lithium-ion battery prices.[27,37,38] However, it has been argued that this approach is less robust than the proven concept of one-factor experience curves due to challenges in resolving the collinearity between cumulative output and knowledge stock, in choosing a proxy for knowledge stock (e.g. patents, R&D investment), and in obtaining the necessary data (e.g. private R&D investment).[21,36]

The use of price data as a proxy to reflect all cost input factors makes the analysis sensitive to pricing policies (i.e. the rationale behind determining product prices within a company based on product costs, customer demand, and market competition).[30]

High data variance can lead to significant variations of experience rates across studies and datasets. Depending on the spread of the data, it is possible to calculate different rates by changing the start and end point of the analysis and by including or excluding outliers.[21] In particular when price data are used, a period of at least 10 years' worth of historical data should be available for price trends to be reliably reflective of cost trends.[29]

Experience curves are incapable of predicting step-change innovations or accounting for product changes that might improve performance at the same cost.[30,34] It has been argued that radical product changes constitute new products that exhibit new experience rates.[21] Moreover, in situations with significant product changes, other indicators than the specific investment cost may be more appropriate to reflect learning outputs, such as product functionality or levelized cost of electricity for a power generation technology.[36,39]

The idea of experience improvement at a constant rate is also critiqued. Some argue that costs fall more rapidly during the R&D phase due to radical discontinuities (i.e. innovations leading to new technology features).[35,40] Others argue that learning might be stronger in the commercial phase due to competition.[41] What is clear though is that experience rates cannot project cost indefinitely and cost floors exist. Following the logic that relative cost shares of components with high experience rates decrease over time, a reduction of the aggregated rate for products over time appears feasible.[35] This can be represented in energy system models with 'kinked' (i.e. piece-wise linear) curves or experience rates that depreciate with time.[42-44]

Finally, a distinction between products that require extensive on-site construction and those mass-produced in centralized factories must be made, due to the often highly specific, custom-built nature of the former resulting in lower experience rates.[31]

8.4.2 Methodology

Experience rates (ER) are derived according to Wright's Law[18]

$$P(x) = A\,x^{b} \tag{1}$$

$$ER = 1 - 2^{b} \tag{2}$$

with $P(x)$ the product price, that is the total investment cost, at the cumulatively installed energy capacity or power capacity x of that technology. The normalisation factor A and experience factor b are obtained through a linear regression of the logarithms of the given price and capacity data. The trendline for the linear regression is the experience curve.

The normalisation factor A is obtained by determining the product price at the cumulative installed capacity of 1.

$$P(1) = A\,1^{b} = A \tag{3}$$

If $P(1)$ is not known, the intercept of the experience curve in logarithmic values with the y-axis can be determined since $\log(P(1)) = 0$. The respective formula for A then becomes

$$\log(P(1)) = \log(A) + b\,log(1) \tag{4}$$

$$A = 10^{\log(P(1))} \tag{5}$$

The experience factor *b* is the slope of the experience curve, that is the linear regression of the logarithmic values of product price and cumulative capacity data.

$$b = \frac{\log(P(x_2)) - \log(P(x_1))}{\log(x_2) - \log(x_1)} \quad (6)$$

Figure 8.1 exemplifies this approach for the example of lithium-ion EV battery pack price data and the respective product prices and cumulative capacity.

With respect to equation (6), it is important that the experience curve is based on a linear regression of the logarthmic values of product price and cumulative capacity. It should not be based on specific data points, for example the initial and latest point. Drawing a line between initial and latest data point in Figure 8.1 would yield a line with a reduced slope that would convert into an experience rate of 22% and therefore underestimate the actual cost-reduction trend.

Experience rate uncertainty reflects the 95% confidence interval based on the linear regression of the logarithmic values for product price and cumulative capacity. The confidence interval using the mean μ and standard error σ of the linear regression is calculated as:

$$\mu \pm 1.96\ \sigma \quad (7)$$

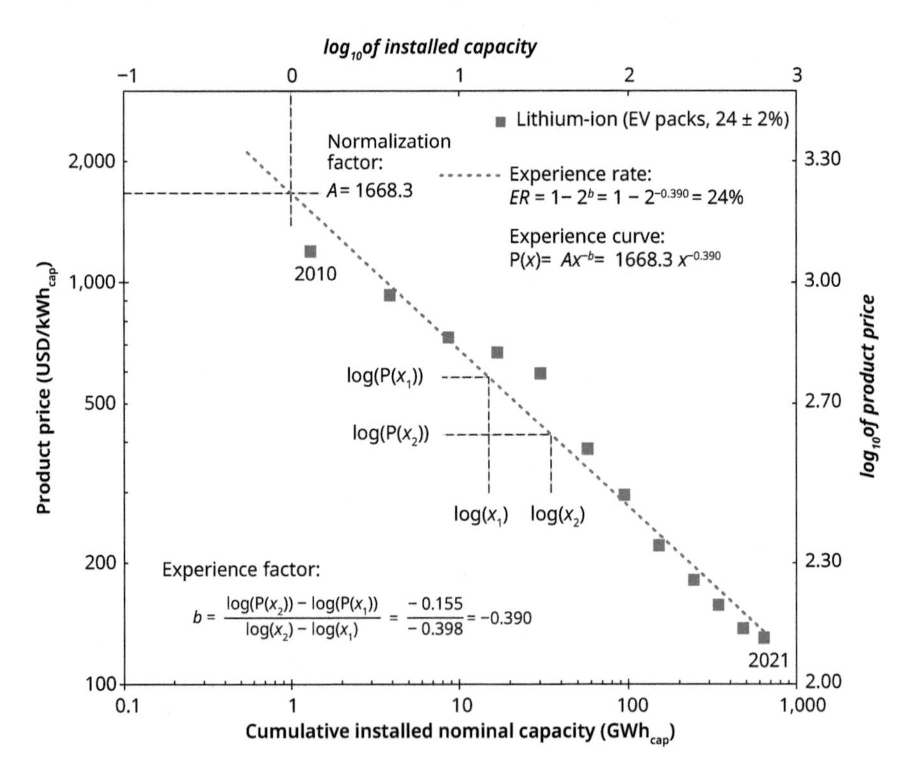

Figure 8.1 Experience curve for lithium-ion EV battery packs showing the methodology to derive experience parameters.

8.4.3 Data

Technology price and cumulative deployment data for the electricity storage technologies in this book are taken from peer-reviewed literature, research and industry reports, news items, electricity storage databases, and interviews with manufacturers (see Table 8.5). In the literature, learning rates (based on manufacturing cost) and experience rates (based on product price) are often used interchangeably. Therefore, the sources in the referenced literature are double-checked to ensure the use of actual product price data for consistency.

The geographic scope of the analysis is global. Where cumulative deployment data are available on company or country level, the data are scaled to global level with validated assumptions for the respective global market size. Regarding price data, it is assumed the global marketplace ensures that these are globally applicable.[36]

Product scope is differentiated into cell, pack, and installed system. Installed system includes the cost for project development, engineering and procurement, transportation, installation, and commissioning. Electrolyzers and fuel cells that are not yet equipped with other balance-of-system components are usually referred to as electrolyzer or fuel cell stacks. For simplicity, they are referred to as 'packs' in this book as the technology scope is comparable to battery packs. Additional information on the cost components included at each level can be found in Table 8.1.

Three application categories are distinguished in this analysis with subgroups to indicate technology size: portable (< 1 kWh), transport (hybrid electric vehicle: < 5 kWh; electric vehicle: > 25 kWh), and stationary (residential: < 30 kWh; utility: > 100 kWh).

Currency conversions are performed in two steps. First, historical prices are deflated in local currency with OECD Consumer Price Indices[45] and then converted to USD_{2020} with OECD exchange rates based on Purchasing Power Parities for gross domestic product (GDP).[46,47]

Conversions from energy-based to power-based data (USD/kWh, GWh vs USD/kW, GW) are performed using the reported power-to-energy ratios for each technology (see Table 8.5).

Technical and economic maturity assessments of electricity storage technologies in the literature are compared to the cumulative installed capacities in the present analysis.[48,49] It is found that those technologies termed 'Research & Development' or 'Developing' had just around 1 GWh installed (flow batteries, fuel cells), 'Demonstration & Deployment' or 'Developed' less than 100 GWh (sodium sulphur) and 'Commercialization' or 'Mature' more than 100 GWh (pumped hydro, lead acid, lithium ion, electrolysis) in 2020. If applicable, the economic maturity assessment is prioritized. Maturity categories are renamed to 'emerging', 'growing', and 'mature'.

8.4.4 Future investment cost

Future product prices of the analysed electricity storage technologies are determined using Wright's Law (equation (1)) with projected values for cumulative installed capacity.

Table 8.5 Overview of product price and cumulative installed capacity data sources, indicative power-to-energy ratio, and comments on individual technology dataset.

Pumped hydro	
Product scope	Installed system
Application	Stationary, utility-scale
Price data	2000–2020: Average price data for run-of-the-river and pumped hydro plants reported to IEA by OECD member states, South Africa, Brazil, and China.[50–53] 1980–2000: Based on mean price and experience rate for large and small hydro.[26] Scaling factor of 0.6 applied to all run-of-the-river data points to convert to pumped hydro cost. Factor is based on comparing run-of-the-river cost to pumped hydro cost.
Capacity data	Capacities and commissioning dates of plants listed in DoE database.[54] Validated with information of International Hydropower Association.[9]
Power-to-energy ratio	Power-to-energy ratio is 1:9.7, based on weighted average discharge duration of plants listed in DoE database,[54] excluding values > 100 hours.
Comment	The data are biased towards OECD countries, because majority of data points come from OECD member states. This could overestimate cost since construction of pumped hydro plants is labour-intense and labour cost in OECD countries is high.
Lead acid (pack)	
Product scope	Pack
Application	Multiple
Price data	US producer price index for lead-acid batteries larger than BCI dimensional group 8D (Applications: uninterruptable power supply, heavy duty vehicles, etc.).[55]
Capacity data	Lead end-use statistics in US, assuming 80% of consumption for battery production.[55] US data scaled-up to global data based on US share of 30% in global lead end-use.[56]
Power-to-energy ratio	Power-to-energy ratio is 8, given the vast majority is produced for the transport sector.[57]
Comment	Data taken from peer-reviewed study.[55] No economy-wide data on lead-acid battery production or sales available, so US lead consumption is used as proxy.
Lead acid (system)	
Product scope	Installed system
Application	Stationary, residential

Price data	Observed prices for systems < 30 kWh in German residential market 2013–2016.[58]
Capacity data	Based on information from German KfW incentive program.[59,60] Scaled to global market assuming German residential storage market is ~1/3 of global.[10]
Power-to-energy ratio	Power-to-energy ratio is 1:3, based on the average value for residential lead-acid batteries sold in Germany.[61,62]
Comment	Basically no lead-acid battery systems were deployed in residential application after 2016.[58]
Lithium ion (cell)	
Product scope	Cell
Application	Portable
Price data	Japanese Ministry for Economy, Trade and Industry statistics on lithium-ion consumer batteries 1995–2011,[63] Avicenne market reports for 2011–2016.[64]
Capacity data	Annual production data for consumer electronics lithium-ion batteries.[64–66]
Power-to-energy ratio	Power-to-energy ratio is 1:3.[64]
Comment	Dataset refers to cells used in portable consumer electronics—different from cells used in EV battery packs (see below).
Lithium ion (pack)	
Product scope	Pack
Application	Transport, electric vehicle
Price data	Annual battery pack price index for 2010–2021.[67]
Capacity data	Annual figures for EVs sold and respective battery pack size.[68]
Power-to-energy ratio	Power-to-energy ratio is assumed to be 1 (average between max. burst power output, max. charging power, and max. power output at most energy-intense steady-state driving mode).
Comment	No comment
Lithium ion (residential system)	
Product scope	Installed system
Application	Stationary, residential

(*Continued*)

Table 8.5 (*Continued*)

Price data	Observed prices for systems < 30 kWh in German residential market 2013–2021.[58]
Capacity data	Based on installed systems and average system size in Germany.[58] Scaled to global market assuming German residential storage market is ~1:3 of global.[10]
Power-to-energy ratio	Power-to-energy ratio is 1:3, based on average value for residential lithium-ion batteries sold in Germany.[69]
Comment	Dominant technology in Germany with market share of 99%, up from ~50% in 2013.[70]
Lithium ion (utility-scale system)	
Product scope	Installed system
Application	Stationary, utility-scale
Price data	Observed prices for stationary systems for 2010–2021 validated with industry experts and corrected for power-to-energy ratio where necessary.[8,71–73]
Capacity data	Systems listed in the DoE database until 2012 and industry reports on deployment figures until 2021.[10,74]
Power-to-energy ratio	Power-to-energy ratio is 1:4 based on latest data source and most likely average ratio for systems installed in the 2020s.[8]
Comment	–
Nickel-metal hydride	
Product scope	Pack
Application	Transport, hybrid electric vehicle
Price data	Modelled for 1997–2014 with annual car sales data and price projections for annual production levels.[75] Checked against official company statements on price reductions.
Capacity data	Toyota Prius sales figures[76] and battery specifications.[77] Toyota Prius sales make up ~50% of total global HEV sales.[78]
Power-to-energy ratio	Power-to-energy ratio is 15.6:1, based on Toyota Prius Generation III battery specification.[77]
Comment	Approach taken from peer-reviewed study.[79] Price data deflated in Japanese Yen and then converted to USD_{2020}. Price reduction of 75% from 1997–2010 in line with official Toyota statement from 2010.[80]

Sodium sulphur		
Product scope	Installed system	
Application	Stationary, utility-scale	
Price data	Industry reports, news items, and manufacturer interviews for 2007–2021.[15,81–87]	
Capacity data	Systems deployed by NGK.[87–90]	
Power-to-energy ratio	Power-to-energy ratio is 1:6 as specified by NGK.[15,85]	
Comment	NGK dominates market for sodium-sulphur systems.[54] No experience rate modelled due to very high standard error of underlying data.	
Vanadium redox flow		
Product scope	Installed system	
Application	Stationary, utility-scale	
Price data	Reported vanadium redox-flow system prices for 2008–2019.[85,91–93]	
Capacity data	Vanadium redox-flow systems listed as operational by 2017 in DoE database and annual market size for 2018 and 2019.[74,94]	
Power-to-energy ratio	Power-to-energy ratio is 1:4 based on weighted average of systems listed in DoE database.[74]	
Comment	Uncertain database due to high variety of sources for price and capacity data.	
	Electrolysis—Utility, Pack	**Fuel cells—Residential, Pack**
Product scope	Pack (stack)	Pack (stack)
Application	Stationary, utility-scale	Stationary, residential
Price data	1956–2002: Industry reports and academic publications.[95] 2002–2019: Manufacturer quotes.[96–99]	Japanese Enefarm-type systems,[100,101] i.e. Solid oxide (SOFC) and polymer electrolyte fuel cells (PEFC).
Capacity data	1956–2002: Industry reports and academic publications.[95] 2002–2019: Based on USD 100m market size.[102]	Sales numbers for Japanese Enefarm-type systems and average system size of 700 W.[100,101]
Power-to-energy ratio	Power-to-energy ratio is assumed at 1:10; appears feasible given average residential electricity consumption of 10 kWh/day and residential fuel cell system size of ~1 kW.	

(*Continued*)

Table 8.5 (*Continued*)

	Electrolysis—Utility, Pack	Fuel cells—Residential, Pack
Comment	Data based on alkaline technology—most mature electrolysis technology with longest record of price and capacity data.	Average of PEFC and SOFC systems.[101]
Solar PV Modules		
Price data	Solar PV module price data for 1980–2020.[103]	
Capacity data	Global cumulative capacity for 1980–2020.[103]	
Comment	–	
PV Inverters		
Price data	Inverter price data for 1990–2013.[104]	
Capacity data	Global cumulative PV shipments for 1990–2013.[104]	
Comment	PV inverters < $20kW_p$	

The normalization factor *A* and experience factor *b* were determined using historical data. Experience rate uncertainty is accounted for by using upper and lower values of the 95% confidence interval for the experience factor and the resulting normalization factor.

8.4.5 Raw material cost

The raw material cost for each storage technology is calculated by multiplying reported material inventories with commodity prices.[105–109] Commodity prices are drawn from peer-reviewed literature, the Bloomberg database, and a bottom-up engineering model.[5,55,110] For most commodities, monthly average price data were identified from 2010 to 2020. Based on this data, average, minimum, and maximum prices were determined. Monthly average price data represent supply contracts of battery manufacturers better than daily spot prices. They may even overestimate price fluctuations as raw material supply contracts can also cover multiple years. For those raw materials with insufficient data, only single price figures were used. Raw material cost uncertainty is based on variations in reported material inventories and commodity prices.

8.4.6 Time frame

In order to relate cost projections at future cumulative deployment levels to time, deployment curves as a function of time need to be produced. This is done using a logistic growth function for electricity storage application subgroups (i.e. consumer electronics, hybrid and battery electric vehicles, residential and utility storage).

$$A_n = \frac{A_{sat}}{1 + \frac{(A_{sat} - A_{base})}{A_{base}} e^{-r\,n}} \tag{8}$$

Where A_n is the annual market capacity in a particular year, A_{base} is the initial capacity, and A_{sat} the maximum annual market capacity that will be reached long-term, in other words the saturation capacity. *r* is the growth rate and *n* the number of periods after the start period. A_{base} and A_{sat} are based on the literature or own assumptions. *r* is fitted to historical and projected data points for annual market capacity from the literature through non-linear regression. The non-linear regression also yields the standard error of *r* to measure goodness-of-fit. Growth rate uncertainty is based on the maximum and minimum *r* determined in a Monte Carlo analysis of the non-linear regression.

The resulting annual market growth projections are used to relate future cumulative capacities to time to interpret projected cost reductions. That means it is assumed that each electricity storage technology obtains 100% market share in its respective application subgroup. On the one side, this shows the potential for cost reductions assuming 'the winner takes it all'. On the other side, it may lead to overly optimistic cost projections as not 100% of the market is captured by only one technology. Table 8.6 displays the resulting sigmoid curves (i.e. S-curves) for the modelled applications.

Table 8.6 Growth projections for various electricity storage application subgroups. The listed growth rates refer to the parameter 'r' of the logistic growth equation. Initial and saturation capacity refer to A_{base} and A_{sat} respectively.

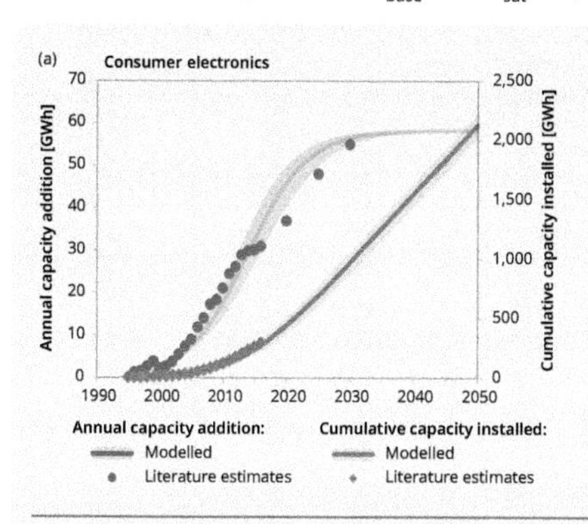

- Initial capacity: 1.0 GWh p.a. (1995)[63]
- Saturation: 60 GWh p.a. (2050)
- Growth rate: 0.214 (0.19–0.23)
- Literature estimates based on Avicenne and Citibank reports[63,66,111]

Comment: Saturation capacity determined assuming that 90% of a total of 10bn people have on average one portable device with a 20 Wh battery that lasts for 3 years.

(Continued)

Table 8.6 (*Continued*)

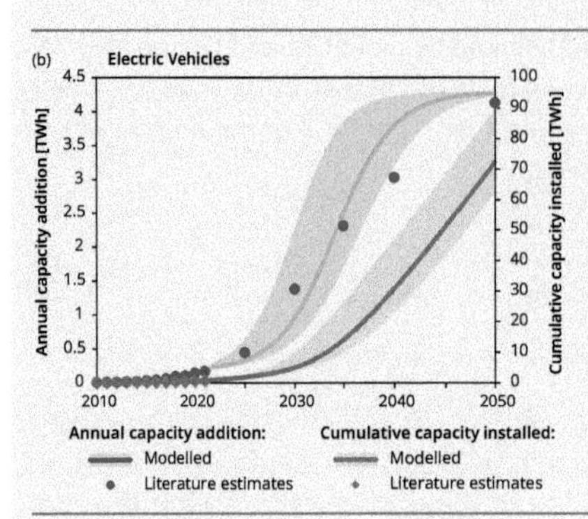

- Initial capacity: 1.0 GWh p.a. (2010)[112]
- Saturation: 4,000 GWh p.a. (2050)[113]
- Growth rate: 0.34 (0.31–0.42)
- Literature estimates based on BNEF and IEA global EV outlooks[68,114]

Comment: Saturation capacity based on assumption that all 75m cars that are sold per year (average in 2010–2020) are fully electric and have a 55 kWh battery in 2050

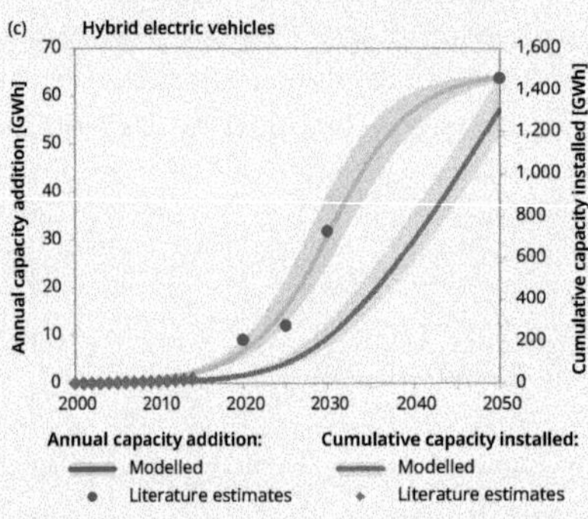

- Initial capacity: 0.06 GWh p.a. (1998)[76]
- Saturation: 65 GWh p.a. (2050)[113]
- Growth rate: 0.21 (0.20–0.22)
- Literature estimates based on Avicenne and Toyota[66,76,113]

Comment: Saturation capacity refers to alternative scenario where 50m hybrid electric cars are sold per year in 2050, all with a 1 kWh battery only

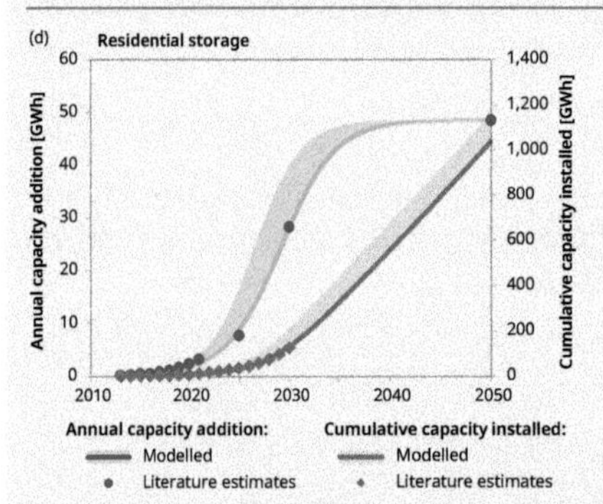

- Initial capacity: 0.13 GWh p.a. (2013)
- Saturation: 50 GWh p.a. (2050)
- Growth rate: 0.37 (0.36–0.44)
- Literature estimates based on WoodMac[10]

Comment: Saturation capacity based on assumption that 5% of all households globally (125m) will have a 6 kWh battery system with a 15-year lifetime by 2050

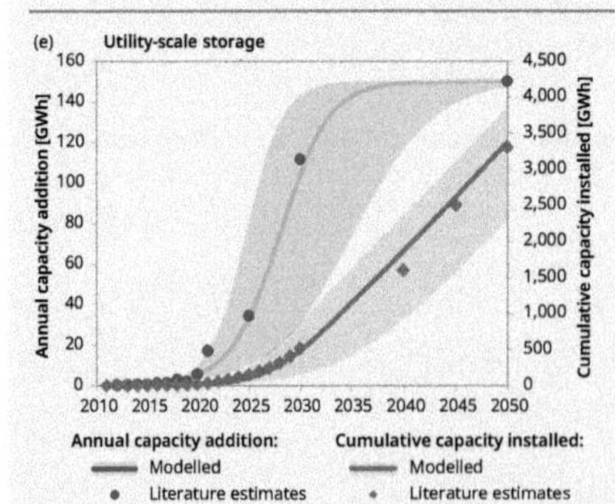

- Initial capacity: 0.10 GWh p.a. (2012)
- Saturation: 150 GWh p.a. (2050)[115]
- Growth rate: 0.41 (0.26–0.57)
- Literature estimates based on WoodMac and BNEF[10,115]

Comment: Rate of deployment highly dependent on rate of cost reductions as these unlock further applications for electricity storage.

8.4.7 Cumulative investment

The amount of cumulative investment required to deploy the projected volumes of cumulative capacity are calculated through the integral of equation (1). It determines the cumulative spend required to go from the current installed capacity x_1 to some future amount x_2, thus installing the amount $x_2 - x_1$ while product prices reduce:

$$\text{Cumulative Spend}\left(X\right)=\int_{X_1}^{X_2}\left(A\,X^{-b}\right)dx \tag{9}$$

Calculating this integral, while subtracting a target price (P_{target}, i.e. what consumers are willing to pay) from the product price, returns the cumulative subsidy required to deploy a defined amount of storage capacity at a subsidized target price.

Figure 8.2 visualizes the approach to calculate cumulative spend. The entire shaded area reflects the cumulative spend to deploy ~999 GWh of vanadium redox-flow battery systems and increase cumulative deployed capacity from ~1 GWh to 1,000 GWh. As a result, product price reduces from P_1 to P_{target}. If deployment of these electricity storage systems is subsidized at P_{target}, the cumulative spend of the subsidy provider is shown by the dark area and the cumulative spend of the subsidy receiver by the light area.

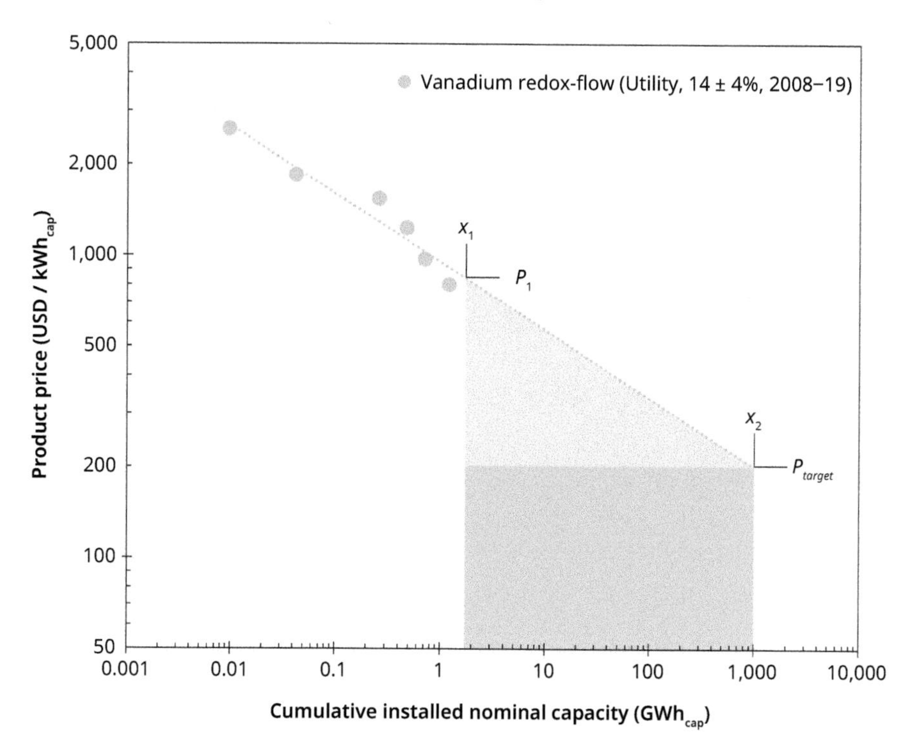

Figure 8.2 Visualization of the approach to calculate cumulative spend to reduce product price for vanadium flow battery systems from P_1 = 800 USD/kWh to P_2 = 200 USD/kWh. Dark-yellow area: cumulative spend of subsidy giver. Light-yellow area: Cumulative spend of subsidy receiver. Note that the size of the areas is not proportional to the total economic value due to the use of logarithmic axes, and the dark shaded area extends further below the axis to 0 USD/kWh.

8.4.8 Total cost of ownership

To assess the competitiveness of an energy storage technology in a specific application relative to existing alternatives, all cost and performance parameters relevant throughout the technology's lifetime must be considered.

For mobility applications, the metric total cost of ownership can be used. This metric sums all cost incurred by purchasing and operating the vehicle throughout its lifetime and divides it by the total mileage driven. It is based on the formula:[116]

$$TCO = \frac{\left(Capex_{Total} - \frac{RV}{(1+r)^N}\right) CRF + \frac{1}{N}\sum \frac{Opex}{(1+r)^n}}{Mileage} \quad (10)$$

with RV the residual value at the end of life, r the discount rate, N the lifetime in years, mileage the distance travelled per year, and CRF the capital recovery factor, itself a function of discount rate and lifetime:

$$CRF = \frac{r\,(1+r)^N}{(1+r)^N - 1} \quad (11)$$

The capital recovery factor converts a present value into a constant rate of cash flows over a given time frame (i.e. annuity), accounting for the discount factor r and the total payment periods N (lifetime in years in this case). In mathematical terms, it reflects the reciprocal of the annuity factor, itself the sum of the geometric series that constant, discounted cash flows represent.

By considering only fuel tank or battery pack and gasoline or power price, the formula for cost of ownership (CO) is specified as:

$$CO = \frac{\left(Capex_{Storage} - \frac{RV}{(1+r)^N}\right) CRF}{Mileage} + \frac{\left(\frac{P_{fuel}}{\eta} DoD\right)\frac{1}{CRF}}{\eta_{fuel} \sum_{n=1}^{N} (1 - Deg\ n)} \quad (12)$$

with P_{fuel} the gasoline or power price, η_{fuel} the fuel efficiency, η the round-trip efficiency of the energy storage device, DoD the depth of discharge, and Deg the annual degradation of the storage device, defined as the fraction of usable storage content lost per year. All inputs for these parameters can be found in Table 8.7.

The US is chosen for this example to complement studies that focus on electrification of personal vehicle transportation in this country.[117] A gasoline price of ~2 USD/gallon is chosen based on a reference crude oil price of around 40 USD/barrel, which was the average in 2020. This price, however, is at the lower bound of average prices over the last 20, 30, and 40 years (USD 54, USD 45, and USD 47).[118]

Lifetime cost for residential storage is calculated based on the formulas presented in section 8.5 with the input parameters in Table 8.7. Charging cost are modelled as the LCOE for a residential solar PV installation. The 2020 German retail power price is taken as reference

Table 8.7 Parameters and references used for lifetime cost analyses.

	Gasoline fuel tank	Lithium-ion battery pack	Residential lithium-ion system
Investment cost	USD 180	USD 4,500 (2020)*	USD 9,000 (2020) *
Capacity (distance)	17 gallons (510 mi)	30 kWh (107 mi)	–
Experience rate	0%[119]	24% ± 2%	13% ± 3%
Growth Rate	–	0.34 (0.31, 0.42)**	0.37 (0.36, 0.44) **
Residual Value	0% (of capital cost)	0% (of capital cost)	0% (of capital cost)
Warranted lifetime	100,000 miles/ 8 years	100,000 miles/ 8 years	3,500 cycles/ 10 years
Fuel Price	2.12 USD/gallon	0.13 USD/kWh[120]	<0.1 EUR/kWh (LCOE of solar PV)
Round-trip efficiency	100%	90%	92%
Depth-of-discharge	100%	80%	80%
Fuel efficiency	30 miles/gallon	4.46 miles/kWh***	–
Annual degradation	0% (of capacity)	2% (of capacity)	0.5% (of capacity)
Discount rate	5%	5%	5%
Retail power price	–	–	0.43 USD/kWh_e (2020)

* *Capacity × Price = 30 kWh × 150 USD/kWh (EV) or 6 kWh × 1,500 USD/kWh (residential)*
** *r parameter from equation (8)*
*** *Distance per capacity/(Capacity × Depth of discharge) = 107 miles/(30 kWh × 80%)*

power price up to 2040, with Germany chosen for this example, because recent growth in residential installations shows that it is a promising market for this application.[58]

In both applications (EV transportation, residential storage) recent deployment data show lithium ion as most common technology. Therefore, calculations are provided for this technology.[58,67]

8.4.9 Other cost projection methods

Chapter 4 compares projections based on experience curves to estimates made by experts. These estimates were obtained in structured discussions with experts. They rely on cognitive heuristics and are therefore subject to bias.[121]

Several methods recommended in the literature aim to minimize the use of heuristics and resulting biases in expert elicitations. While the comprehensive, visual presentation of all necessary evidence can minimize availability bias, asking for extreme values first and allowing for refining these before making a best guess can help to avoid anchoring bias.[121] It appears more challenging to minimize overconfidence bias. Here, neutrally formulated questions and diligent interview conduction with probing questions that allow the expert to justify estimates are useful tools.[121,122] Face-to-face interviews as opposed to telephone interviews or online surveys facilitate this evaluation of given probabilities and may ensure the elicitation is taken more seriously by experts.[123] These best-practice recommendations from the literature are implemented to obtain representative results and minimize cognitive heuristics and bias (see Table 8.8).

The estimates were obtained as follows (see Table 8.9). Before the interview, potential experts were contacted and, upon agreement of participation, an elicitation protocol was sent 2 weeks before the interview. The elicitation protocol outlines the motivation for the study, compiles background material on technological and economic aspects of the technology, describes the expert elicitation technique, and contains the elicitation questionnaire. Iterating this protocol with internal experts allowed for capturing the latest available and relevant

Table 8.8 Cognitive heuristics and bias and recommended countermeasures[121,124].

	Description	Countermeasure
Anchoring	Tendency to rely too heavily on a first piece of information (the 'anchor') and adjust relatively conservatively from this when making probabilistic decisions, rather than fully considering factors which may influence a quantity of interest, leading to overconfident estimates (i.e., too narrow ranges).	• Informing interviewee about heuristic • Asking for extreme estimates first (90th, 10th percentiles), then for median estimate (50th percentile) • Asking for reasons for estimates to lie outside of indicated range
Availability	Heuristic procedure of making a decision according to the ease with which one can imagine an event occurring, which may for example bias judgements towards recent trends or events.	• Informing interviewee about heuristic • Asking for reasons for estimates to lie outside of indicated range
Overconfidence	Heuristic procedure of making confidence intervals according to the span of ad-hoc imaginable outcomes that are too narrow due to limited information availability.	• Neutrally formulated questions • Probing questions allowing expert to justify estimates
Representativeness	Judgement based on insufficiently representative information.	Providing background material to compile latest data and research insights from multiple sources.

Table 8.9 Elicitation procedure.

Phase	Interactions with expert	Timeline/Duration
Before interview	1. Making initial contact 2. Sending elicitation protocol (background material, questionnaire)	- 2 weeks before interview
During interview	3. Discussing background material 4. Eliciting values of interest with questionnaire	1 hour during interview 1 hour during interview
After interview	5. Sending elicited values and possible implications for final approval	1 week after interview

information, phrasing unambiguous questions, and identifying academic and industry experts in the field.

During the interview, the first hour was spent discussing the background material to minimize any availability bias. The second hour was spent introducing the case studies to limit technical ambiguity and eliciting the values of interest. Experts were asked for 10th, 50th, and 90th percentile estimates with extreme values being identified first to minimize any anchoring bias.[121] Using probing questions, they were supported in critically assessing, refining, and verifying the given values. By eliciting distinct parameters (e.g. investment cost), instead of aggregate parameters that require implicit calculations (e.g. levelized cost), uncertainty was further minimized.[121,125] Audio recordings were made with the experts' permission to ensure all responses were captured correctly. After the interview, responses were transcribed into a spreadsheet and potential implications were derived based on the elicited values in a separate document. Both were sent to the expert to allow for adjustments, point out potential inconsistencies, ask for additional comments and receive final approval of the elicited values.

Eleven experts were interviewed on lithium-ion battery packs.[126] While 10 is a common number of experts to interview,[127] there is no one rule for the correct number of interviewees required. However, it is important to select a set of experts who adequately represent the diversity of expert opinion in the area.[123,128] As such, equal numbers of experts were selected from academia and industry. All experts come from the UK, which limits the represented perspective of that technology to the UK. The interviews took place between October 2015 and June 2016.

The resulting investment cost estimates for 2020 and 2030 are compared to projections based on previously identified experience curves for lithium-ion batteries. An experience rate of 16 ± 4% is used. This rate is based on price and cumulative deployment data from 2010 to 2015 and thereby reflects the available information at the time of expert elicitation interviews.[129]

Estimates were elicited in EUR (2016) or EUR (2015). The conversion to USD (2020) is performed using the Eurozone consumer price index inflation and the GDP-based power purchase parity exchange rate for 2020.[45,46]

8.5 Lifetime cost

In Chapter 5, the lifetime cost methodology for energy storage technologies is introduced and applied. This approach is appropriate to assess the economic case for energy storage technologies because lifetime cost account for all technical and economic parameters affecting the cost of delivering stored energy or power.

8.5.1 Methodology

Lifetime cost sum all cost incurred by a technology over its lifetime and divide those by the total output produced over its lifetime. They thereby reflect the minimum remuneration required for the project to have a net present value of zero. For energy storage technologies, there are two forms of lifetime cost. Levelized cost of storage (LCOS) quantifies the cost for discharging energy, in other words what you need to get paid for each MWh. Annuitized capacity cost (ACC) quantifies the cost of providing power capacity per year, that is, what you need to get paid for each kW in a year. Table 8.10 derives the equations for both metrics.

Table 8.10 Derivation of formulas for levelized cost of storage and annuitized capacity cost. NPV—net present value, $E_{out}(n)$—electricity discharged per year, $Cap_{p,nom}$—nominal power capacity, n—year, N—lifetime in years, r—discount rate. For simplicity, construction time is neglected.

Levelized cost of storage (LCOS)			Annuitized capacity cost (ACC)		
NPV	=	0	NPV	=	0
$NPV\ of\ cost$	=	$NPV\ of\ revenue$	$NPV\ of\ cost$	=	$NPV\ of\ revenue$
$\sum_{n}^{N}\frac{cost(n)}{(1+r)^n}$	=	$\sum_{n}^{N}\frac{revenue(n)}{(1+r)^n}$	$\sum_{n}^{N}\frac{cost(n)}{(1+r)^n}$	=	$\sum_{n}^{N}\frac{revenue(n)}{(1+r)^n}$
$\sum_{n}^{N}\frac{cost(n)}{(1+r)^n}$	=	$\sum_{n}^{N}\frac{E_{out}(n)\cdot LCOS}{(1+r)^n}$	$\sum_{n}^{N}\frac{cost(n)}{(1+r)^n}$	=	$\sum_{n}^{N}\frac{Cap_{p,nom}\cdot ACC}{(1+r)^n}$
$\sum_{n}^{N}\frac{cost(n)}{(1+r)^n}$	=	$LCOS\cdot\sum_{n}^{N}\frac{E_{out}(n)}{(1+r)^n}$	$\sum_{n}^{N}\frac{cost(n)}{(1+r)^n}$	=	$ACC\cdot\sum_{n}^{N}\frac{Cap_{p,nom}}{(1+r)^n}$
$LCOS$	=	$\frac{\sum_{n}^{N}\frac{cost(n)}{(1+r)^n}}{\sum_{n}^{N}\frac{E_{out}(n)}{(1+r)^n}}$	ACC	=	$\frac{\sum_{n}^{N}\frac{cost(n)}{(1+r)^n}}{\sum_{n}^{N}\frac{Cap_{p,nom}}{(1+r)^n}}$

Levelized cost of storage (LCOS) divides all cost incurred throughout the lifetime of the technology by the total discharged electricity. Annuitized capacity cost (ACC) divides all cost incurred throughout the lifetime of the technology by the power capacity and discounted lifetime. In both metrics, not only the cost but also the output need to be discounted as it represents future revenue. Equation (13) and equation (14) display the formulas for LCOS and ACC including all cost components. The unit for LCOS is USD/MWh and USD/kW-year for ACC.

$$LCOS = \frac{\sum_n^N \frac{Investment}{(1+r)^n} + \sum_n^N \frac{Replacement}{(1+r)^n} + \sum_n^N \frac{O\&M}{(1+r)^n} + \sum_n^N \frac{Charging}{(1+r)^n} + \frac{End\ of\ life}{(1+r)^{N+1}}}{\sum_n^N \frac{E_{out}(n)}{(1+r)^n}} \quad (13)$$

$$ACC = \frac{\sum_n^N \frac{Investment}{(1+r)^n} + \sum_n^N \frac{Replacement}{(1+r)^n} + \sum_n^N \frac{O\&M}{(1+r)^n} + \sum_n^N \frac{Charging}{(1+r)^n} + \frac{End\ of\ life}{(1+r)^{N+1}}}{\sum_n^N \frac{Cap_{p,nom}}{(1+r)^n}} \quad (14)$$

The equations incorporate all elements required to determine lifetime cost: Investment, replacement, operation and maintenance (O&M), charging, and end-of-life cost divided by electricity discharged (E_{out}) or power provided ($Cap_{p,nom}$) during the lifetime. They sum ongoing cost in each year (n) up to the lifetime (N), discounted by the discount rate (r).

For simplicity, equations (13) and (14) introduced N as the lifetime. This value differs depending on the component of the equation that is being calculated. It can refer to the construction time T_{con}, operational lifetime of the storage system N_{op} or the sum of two, the project lifetime $N_{project}$.

$$N_{project} = T_{con} + N_{op} \quad (15)$$

Project lifetime can be set to a pre-defined value representing an investment horizon (e.g. 20 years) or determined by calculating operational lifetime N_{op} based on the temporal (Deg_t) and cycle degradation (Deg_c) of the storage system and a pre-defined degradation threshold (EoL).

$$N_{op} = \frac{\log(EoL)}{\log(1-\mathrm{Deg}_t) + Cyc_{pa} \cdot \log(1-Deg_c)} \quad (16)$$

The unit for ACC has the subscript 'year' because power is the rate of work with respect to time. In equation (14) it is discounted over its lifetime given in years, which means ACC refer to an annual cost. In contrast, energy is the ability to do work, irrespective of time. LCOS therefore refer to the cost of a specific output unit (e.g. MWh).

Investment cost account for nominal power ($Cap_{p,nom}$) and energy capacity ($Cap_{e,nom}$) and specific power ($C_{p,inv}$) and energy storage capacity cost ($C_{e,inv}$). If construction of the technology takes multiple years (T_{con}), then the costs are discounted to reflect their present value with respect to the year in which they are spent. The assumption is made that equal shares of the total investment cost are spent in each construction year.

$$\sum_{n}^{N} \frac{Investment}{(1+r)^{n}} = \sum_{n=1}^{T_{con}} \frac{C_{p,inv} \cdot Cap_{p,nom} + C_{e,inv} \cdot Cap_{e,nom}}{(1+r)^{n-1}} \tag{17}$$

FREQUENTLY ASKED QUESTION

Does it matter if the power of the discount factor equation (1 + *r*) is set to *n* or (*n* – 1)?

Setting the power to (*n* – 1) slightly overestimates the present value of the investment cost. It determines the discount factor as of the beginning of each year. Example: in the initial year, there is no discounting since $(1 + r)^{(1-1)} = 1$. However, the value of money will reduce throughout the initial year. Alternatively, setting the power to *n* will reflect discount factor values at the end of each year. Example: in the initial year, there would already be discounting since $(1 + r)^{(1)} = (1 + r)$. This would then underestimate present values. The approach could be optimized by determining an average discount factor based on beginning of year and end of year values. For simplicity, the analyses in this book follow the approach of setting the power to (*n* – 1).

Replacement cost account for specific replacement cost for power ($C_{p,rep}$) and energy storage capacity ($C_{e,rep}$). There are two options to account for these.

Option 1: The cost could be accounted for in the specific year in which the replacement occurs. In this case, the replacement interval (T_{rep}) reflects the number of years after which replacement is necessary and is determined based on the number of cycles after which replacement (Cyc_{rep}) is required relative to annual cycles (Cyc_{pa}). The variable *k* reflects the replacement occurrence and the variable *R* the total number of replacements throughout the lifetime of the technology.

$$\sum_{n}^{N} \frac{Replacement}{(1+r)^{n}} = \sum_{k=1}^{R} \frac{C_{p,rep} \cdot Cap_{p,nom} + C_{e,rep} \cdot Cap_{e,nom}}{(1+r)^{T_{con}+k \cdot T_{rep}}} \tag{18}$$

with

$$T_{rep} = \frac{Cyc_{rep}}{Cyc_{pa}} \tag{19}$$

and

$$R = \frac{N_{op}}{T_{rep}} \tag{20}$$

Option 2: Replacement cost can be converted into an annual cost reflecting upfront payments into a savings account dedicated to future replacements to avoid large one-off expenses when actual replacements occur. In effect, the annuitized payment required to cover replacement cost through the lifetime of the system is charged each year. The NPV of this cost is given by (18) except that the constraint that R takes an integer number is relaxed (i.e. the total number of replacements can be a fractional number).

The analyses in this book determine replacement cost based on Option 2.

O&M cost account for power and energy specific operation and maintenance cost ($C_{p,om}$ and $C_{e,om}$) relative to nominal power capacity and annual charged electricity (E_{in}). Annual charged electricity accounts for the nominal energy capacity, the depth-of-discharge (*DoD*), round-trip efficiency (η_{RT}), cycle degradation (Deg_c), and time degradation (Deg_t).

$$\sum_n^N \frac{O\&M\ cost}{(1+r)^n} = \sum_{n=1}^{N_{op}} \frac{C_{p,om} \cdot Cap_{p,nom} + C_{e,om} \cdot E_{in}(n)}{(1+r)^{n+T_{con}-1}} \tag{21}$$

$$E_{in}(n) = \frac{(Cap_{e,nom} \cdot DoD \cdot Cyc_{pa})}{\eta_{RT}} \cdot (1-Deg_c)^{(n-1)\cdot Cyc_{pa}} \cdot (1-Deg_t)^{(n-1)} \tag{22}$$

Cycle degradation accounts for a predefined end-of-life threshold (*EoL*) relative to nominal energy storage capacity (e.g. 80%) and cycle lifetime ($Life_{cyc}$) of the technology. It can be derived as follows:

$$Cap_{e,nom} \cdot (1-Deg_c)^{Life_{cyc}} = EoL \cdot Cap_{e,nom} \tag{23}$$

$$Deg_c = 1 - EoL^{\frac{1}{Life_{cyc}}} \tag{24}$$

Similarly to discount factors, setting the power of the degradation factor equations to $(n-1)$ will slightly underestimate degradation and thereby overestimate charged electricity (see the earlier Frequently Asked Question).

Charging cost are determined by multiplying electricity price (P_{el}) with the annual charged electricity.

$$\sum_n^N \frac{Charging}{(1+r)^n} = \sum_{n=1}^{N_{op}} \frac{P_{el} \cdot E_{in}(n)}{(1+r)^{n+T_{con}-1}} \tag{25}$$

End-of-life cost consider specific cost (or residual value) for the power ($C_{p,eol}$) and energy components ($C_{e,eol}$) at their end-of-life. Nominal energy capacity needs to account for cycle and temporal degradation during operational lifetime. The costs are discounted towards the end of technology lifetime.

$$\sum_n^N \frac{End\ of\ life}{(1+r)^n} = \frac{C_{p,eol} \cdot Cap_{p,nom} + C_{e,eol} \cdot Cap_{e,nom} \cdot (1-Deg_t)^{N_{op}} \cdot (1-Deg_c)^{Cyc_{pa} \cdot N_{op}}}{(1+r)^{N_{project}+1}} \tag{26}$$

Discharged electricity (E_{out}) is determined by multiplying annual charged electricity with the technology's round-trip efficiency and self-discharge (η_{self}), and discounting the product over time.

$$\sum_n^N \frac{E_{out}(n)}{(1+r)^n} = \sum_{n=1}^{N_{op}} \frac{E_{in}(n) \cdot \eta_{RT} \cdot (1-\eta_{self})}{(1+r)^{n+T_{con}-1}} \tag{27}$$

The analysis uses proxy values for self-discharge that reflect the ability of the individual storage technologies to keep charge in idle state. The actual self-discharge depends on the usage of the technology in the modelled application and should be further specified in more detailed analyses. Since losses incurred through self-discharge depend on operation of the storage system, these are not accounted for in the given values for nominal energy capacity of the storage system.

Power provided is determined by discounting nominal power capacity over the lifetime.

$$\sum_{n}^{N} \frac{Cap_{p,nom}}{(1+r)^{n}} = \sum_{n=1}^{N} \frac{Cap_{p,nom}}{(1+r)^{n+T_{con}-1}} \quad (28)$$

8.5.2 Modelled technologies

In Chapter 5, the most cost-efficient electricity storage technologies for the chosen archetype applications are identified. However, not all technologies are modelled in each application. Pumped hydro and compressed air have response times of larger than 10 seconds and are usually built-in systems larger than 1 MW power capacity. They are therefore not modelled for fast response services: frequency regulation, frequency response, and high-cycle, and not for consumer-side services: self-consumption, demand reduction, and power reliability.

8.5.3 Investment cost assumptions

The modelled lifetime cost projections account for future investment cost projections. These are determined using experience curve analysis for the total investment cost of electricity storage systems (see Chapter 4). For those technologies where no experience rate is available, assumptions on future cost reduction have been made. The resulting relative investment cost reductions (see Table 8.11) are applied to the 2020 specific investment cost input parameters in Table 3.2.

8.5.4 Monte Carlo analysis

Monte Carlo simulations generate random samples from a given probability distribution to estimate or simulate expectations of mathematical functions under that distribution.[130] This method was first used systematically during the 1940s to investigate properties of neutron travel through radiation shielding as part of the Manhattan Project.[131] It is said to be named after the Monte Carlo Casino in Monaco, drawing a comparison between the random sampling for mathematical simulations and the random sampling in gambling games like roulette. The method is now used as scientific tool for mathematical problems that are analytically intractable and for which experimentation is too time-consuming or costly.[131]

To perform a Monte Carlo analysis, a subjective probability distribution must be assigned to the given, uncertain parameters. The distribution can be uniform (where data is limited but uncertainty is low), triangular (when a midpoint exists), log-uniform or log-normal (when uncertainty exceeds a factor of 10).[132] Other distributions such as normal or empirical are also common as they often reflect the distribution of real-world data. Various others like Poisson, Weibull, or discrete distributions can be used as well.

Table 8.11 Investment cost projections relative to 2020.

	2015	2020	2025	2030	2035	2040	Comment
Pumped hydro	100%	100%	100%	101%	102%	103%	Experience rate projection
Compressed air	100%	100%	100%	100%	100%	100%	Assumption
Flywheel	165%	100%	70%	53%	44%	40%	Assumption
Lithium ion	197%	100%	52%	33%	25%	22%	Experience rate projection
Sodium sulphur	165%	100%	70%	53%	44%	40%	Assumption
Lead acid	165%	100%	70%	53%	44%	40%	Assumption
Vanadium redox flow	226%	100%	59%	42%	34%	31%	Experience rate projection
Hydrogen	122%	100%	55%	38%	29%	25%	Experience rate projection
Supercapacitor	165%	100%	70%	53%	44%	40%	Assumption

In this book, simple random sampling is used as Monte Carlo method. That means for each iteration, a random sample is taken from within the distribution that is specified for an uncertain parameter. Other methods like Latin Hypercube Sampling can be more efficient, in terms of required iterations for a meaningful result, but are less straightforward to understand.[132]

For the lifetime cost calculation, a Monte Carlo simulation is conducted to account for the uncertainty of technology input parameters. A normal distribution is attributed to a technology parameter, x, based on its central estimate, μ, and standard deviation, σ. A normal distribution is assumed to best reflect the variation of input parameters.

The parameters subject to variation are:

- Specific investment cost ($C_{p,inv}$, $C_{e,inv}$)
- Specific O&M cost ($C_{p,om}$, $C_{e,om}$)
- Specific replacement cost ($C_{p,rep}$, $C_{e,rep}$)
- Round-trip efficiency (η_{RT})
- Cycle life ($Life_{cyc}$)

For simplicity, the same standard variation is assumed for all technologies and parameters, increasing over time. Table 8.12 shows this variation as coefficient of variation, which is the standard deviation, σ, divided by the mean, μ. This metric allows expressing the value for all technologies and parameters.

The Monte Carlo analysis simulates 1000 lifetime cost calculations per technology applying the variation to the respective parameters. Extreme values are excluded by taking the 90th percentile of the results and excluding the 10th percentile.

Table 8.12 Coefficient of variation assumed for each technology and parameter subject to variation.

	2015	2020	2025	2030	2035	2040
Coefficient of variation	20%	25%	30%	35%	40%	45%

8.5.5 Probability analysis

The probability (*P*) of a technology exhibiting lowest lifetime cost in each application reflects the frequency with which each technology exhibits minimum cost when accounting for uncertainty in the Monte Carlo simulation.

If lifetime cost for technology *A*, *B*, and *C* are $\{a_1; a_2; \ldots; a_{1000}\}$, $\{b_1; b_2; \ldots; b_{1000}\}$, and $\{c_1; c_2; \ldots; c_{1000}\}$ respectively, up to *N* technologies, then

$$P(a_i = \min LCOS) = P(a_i < b_k, k \in [1;1000]) \cdot P(a_i < c_k, k \in [1;1000]) \cdot \ldots \tag{29}$$

$$P(A = \min LCOS) = \frac{1}{1000^N} \cdot \sum_{i=1}^{1000} \left| a_i < b_k, k \in [1;1000] \right| \cdot \left| a_i < c_k, k \in [1;1000] \right| \cdot \ldots \tag{30}$$

with $|X|$ the cardinality of set *X*.

So, if the maximum of the lifetime cost distribution for technology *A* is below the minimum of all other technologies, technology *A* is set as the cheapest option with 100% probability (see Figure 8.3a). However, if the intersection between the LCOS distributions is not an empty set, the probability is lower (see Figure 8.3b). The approach then counts the occurrences when technology *A* exhibits lower lifetime cost than all other technologies and divides by all occurrences (1000^N) to arrive at the probability of the technology exhibiting the lowest lifetime cost.

8.5.6 Lifetime cost landscape

To explore the sensitivity of lifetime cost to discharge duration and annual cycles, the LCOS or ACC of each technology is determined for each year between 2015 and 2040 using the

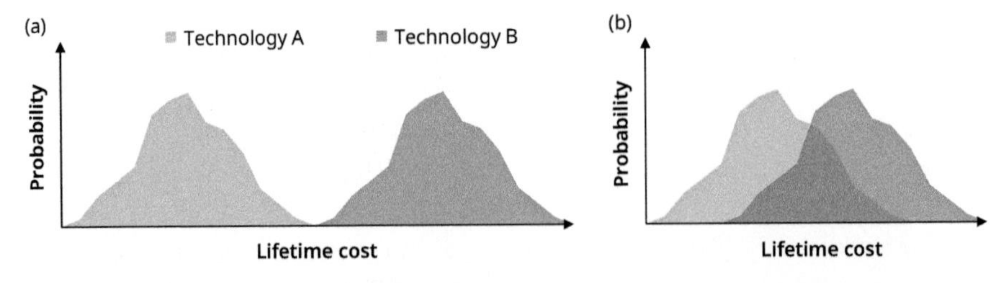

Figure 8.3 Schematic for probability assessment. (a) Technology A has 100% probability of lower lifetime cost than technology B. (b) Technology A has 83% probability of lower lifetime cost than technology B.

central estimate for technology inputs, fixed electricity price (50 USD/MWh), and discount rate (8%), while varying discharge duration and annual full equivalent discharge cycle requirements. Discharge duration and cycle requirements are varied in 1,000 steps on a logarithmic scale between 0.25 to 1,024 hours and 1 to 10,000 cycles respectively.

Technologies with the lowest lifetime cost at a specific discharge duration—annual cycle combination is represented by the respective colour. Lighter areas are contested between at least two technologies, while darker areas indicate a strong cost advantage of the prevalent technology. White spaces mean LCOS of at least two technologies differ by less than 5%.

8.6 Market value

In Chapter 6, the revenue that storage may earn from providing services is assessed across four major world markets. Revenue is harmonized according to the service requirements, in terms of cycling frequency and discharge duration to create a 'landscape chart' for revenue, which is then compared to the levelized cost and annuitized capacity cost from Chapter 5 to analyse the potential profitability of storage across the full spectrum of application requirements. Arbitrage is then considered in more depth, introducing an algorithm for finding the profit-maximizing storage dispatch and applying this to historical electricity prices from 36 markets.

This section outlines the methodology to assess the economic market value for electricity storage in various power system applications and the capacity required to integrate variable or relatively inflexible low-carbon generation.

8.6.1 Market value for any application

The economic market value of electricity storage in providing various applications is assessed using review data compiled by Balducci for the US and Housden for Great Britain, Germany, and Australia.[133,134] These values were matched to the discharge duration and annual cycle requirement ranges that were used in the review to differentiate between applications.[133,135]

Due to the lack of data for long-term storage applications, the value of seasonal storage was modelled using the storage dispatch algorithm developed by Ward et al. based on wholesale price data from each market (see section 8.6.2).[136] The algorithm finds the maximum profit that could be achieved by buying and selling power into the wholesale day-ahead market, assuming energy-only pricing (i.e. no capacity market payments). It was run for an 80% and 30% efficient technology with discharge durations of 512, 768, and 1,024 hours, returning discharge frequencies between 1.75 and 4.65 full equivalent cycles per year and a value range of 1–179 USD/kW-year.

The ranges of economic values, duration and frequency requirements for each application were used to determine the economic value of any potential application with a discharge duration between 0.25 to 1,024 hours and 1 to 10,000 annual discharge cycles. Duration

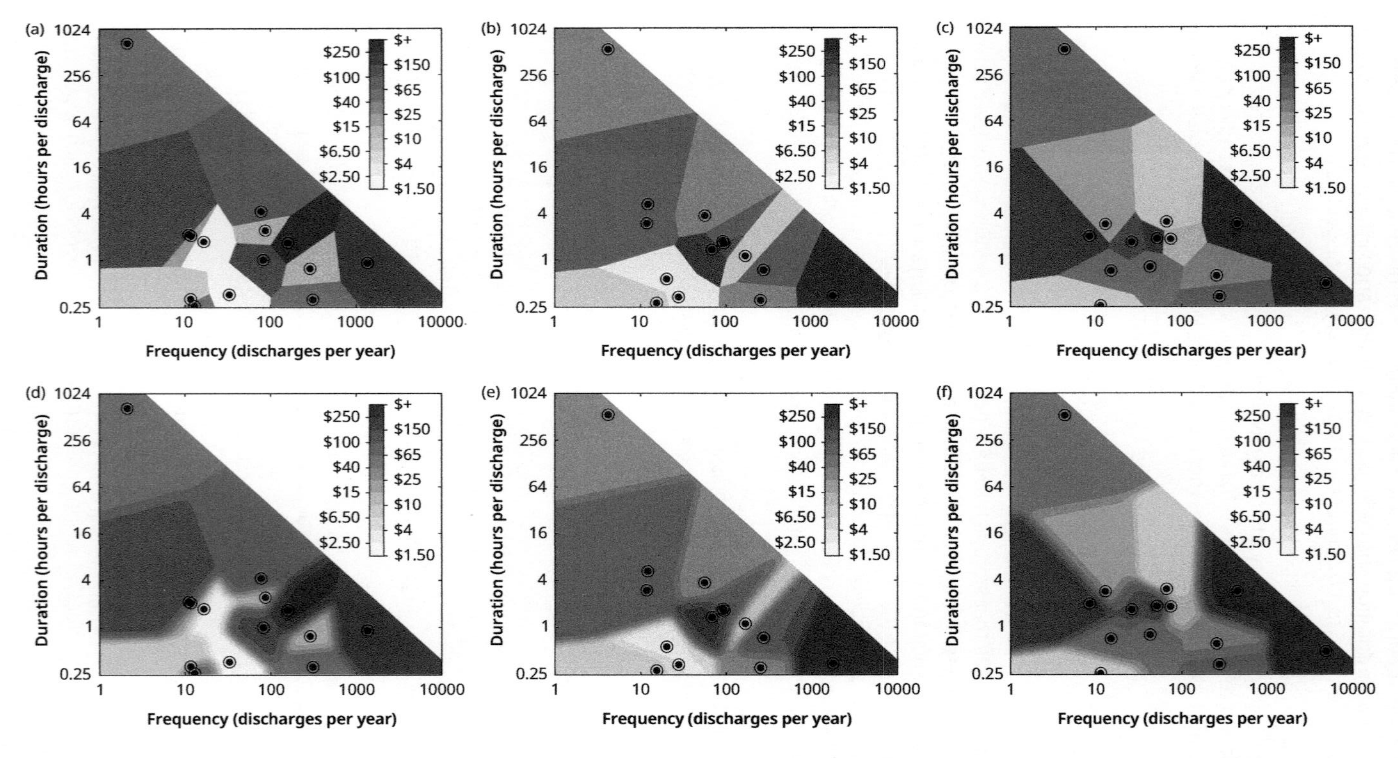

Figure 8.4 Allocation of market values to the entire frequency-duration space based on the variation in value, duration and frequency requirements of specific applications and a nearest neighbours' algorithm. Top panels (a, b, c): Three sample Monte Carlo trial results where each of the 13 applications is assigned an economic value, discharge duration and frequency from within their given range and each point on the duration-frequency matrix is assigned the value of its nearest application. Bottom panels (d, e, f): Gaussian smoothing kernel applied to Monte Carlo trial result.

and frequency were varied in 1,000 steps on a logarithmic scale to obtain this spectrum. A Monte Carlo simulation with 1,000 trials sampled across three dimensions of uncertainty: (i) the identified economic values and within the (ii) discharge and (iii) frequency ranges for each application.[133,135] Thus, for each trial there were 13 applications with a unique economic value and requirements for discharge duration and frequency. Each point on the duration-frequency matrix was then assigned the value of its nearest application, creating a Voronoi diagram for each Monte Carlo trial, as shown in Figure 8.4(a–c).

The discrete nature of the results from individual Monte Carlo trials lead to sharp discontinuities between the values of adjacent cells, so the resulting data from each trial were smoothed using a Gaussian kernel across the frequency-duration space (i.e. so that the data in each cell become an average of the original data across all surrounding cells). This is transformation is shown in Figure 8.4(d–f).

Finally, mean, 25th and 75th percentiles were determined for the entire spectrum based on the 1,000 Monte Carlo trials.

This analysis was conducted for economic market values (MV) in power (USD/kW-year) as well as energy terms (USD/MWh). Conversion was performed with the product of discharge duration, DD, and annual cycles, Cyc_{pa}, of the respective application.

$$MV_{energy} = \frac{MV_{power} \times 10^3}{\left(DD \times Cyc_{pa}\right)} \tag{31}$$

In an alternative approach, economic market values were not sampled from the explicit literature values identified, but instead randomly chosen between 25th and 75th percentiles of those values. This was performed to test the robustness of the analysis.

8.6.2 Arbitrage value

The operation of storage systems is widely modelled using historical time series of power prices or modelled prices produced by electricity market models, where the aim is to maximize profit from arbitrage.[137–142] Storage operation is also incorporated directly into numerous power and energy system models, where the aim is instead to minimize total system cost.[143–146] The latter is relevant to transmission system operators and system planners, whereas the former is more relevant to investors and developers of storage.

Linear optimization techniques are commonly used to find the schedule of charging and discharging, also called the dispatch of storage systems.[137,138,142,147–149] Several other methods have been explored, including Monte Carlo (testing random dispatch schedules and iterating towards ones that maximize profit),[140] approaches from game theory (to consider behaviour of storage operators in markets with limited competition),[150–152] the use of calibrated moving averages (to model the role in smoothing renewables output or demand),[153] and functional algorithms (which hard-wire the 'rules' for finding maximum profit).[147,154]

The model we demonstrate here fits into the fourth category: it uses simple logic to identify the most profitable actions to take, within the constraints of the system's operation. For a given time series of prices, it finds the dispatch schedule that maximizes the profit from arbitrage. The algorithm was first proposed by Lund and was extended by Staffell and Ward.[136,154,155]

Its key advantages are speed and simplicity. The hourly dispatch schedule across a whole year can be found within seconds (as can be demonstrated on the 'Storage Dispatch' tab of <www.EnergyStorage.ninja>). It also does not require general-purpose optimization solvers (such as CPLEX or Gurobi), and can be run as a standalone algorithm or easily incorporated into other models. The disadvantage is that the algorithm is not guaranteed to find the global optimum solution when considering thousands of time periods (e.g. 1 year of hourly data), and may instead settle on one of many similar local optima. However, the local optima Ward finds are within 0.2% of the global optimum.[136]

The algorithm works by pairing periods of time together when charging and discharging should occur. It begins with the periods of maximum and minimum price, as these will yield the greatest arbitrage profit. Those prices are then removed from the time-series as the storage is now operating at its limits (charging or discharging at maximum power). It then seeks the next best periods to pair together, starting from the next-highest price for discharging and repeats the process. It may not be possible to connect this next-highest price with the absolute lowest price remaining in the time series because of constraints on the system's state of charge. The algorithm prevents any pairings that would either raise the state of charge over the maximum storage capacity (because charging occurs before discharging, and at some point between the two the system is already fully charged); or ones which would lower the state of charge below zero (because recharging occurs after discharging, and at some point the system is already depleted). The impact of these constraints is shown in the example of Figure 6.9. Arbitrage profit is then calculated as the total revenue from discharging (energy sold in each period multiplied by the price in that period) minus the total cost of charging (energy bought in each period multiplied by its price), minus any marginal costs of operation which can be specified by the user. A summary of the algorithm's operation is given in Figure 8.5, while full descriptions of the algorithm along with open-access code implementations are available from Staffell and Ward.[136,154]

The analysis of arbitrage value presented in Chapter 6.3 is based on historical electricity prices in 36 markets around the world, which were chosen based on data availability and quality (specifically having data with hourly resolution or better, covering at least two years, with less than 1% missing or corrupted values). This includes 25 national markets in Europe, five regional markets in the US, five in Australia, and national average prices Japan. A full list of coverage and data sources is given in Halttunen.[156] We use day-ahead wholesale prices for each market, following Zipp,[157] as these markets have the highest liquidity and greater similarity in their setup (hence more representative prices than real-time spot markets), along with greater data availability. All prices were converted to USD using market exchange rates with daily resolution.[158] Scandinavia, the US, and Italy use nodal or zonal pricing, meaning that prices within a market at any given time can differ strongly by location. Prices were aggregated to market level using load-weighted averages, which will tend to homogenize prices in these markets, and thus may underestimate the value of arbitrage. The analysis presented here could be repeated with prices from specific markets or nodes within a market using the tools at <www.EnergyStorage.ninja>.

We model the impact of storage on power prices, and the impact of price cannibalisation on storage operation, by coupling the algorithm described above with a simple electricity

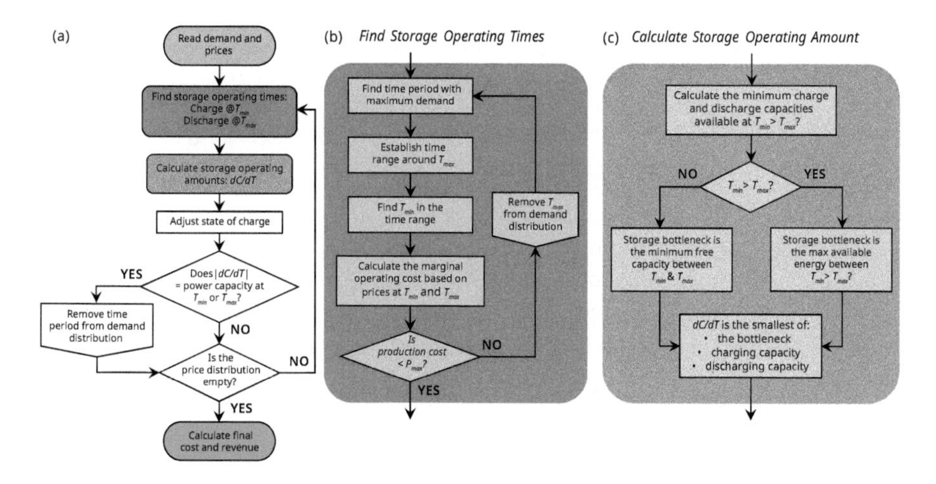

Figure 8.5 An algorithm for finding the profit-maximizing dispatch of energy storage for a given time-series of prices. The algorithm finds the change in the state of charge (dC/dT) at all time periods by iteratively identifying the time of minimum and maximum prices (T_{min} and T_{max}), assessing the selling price (P_{max}) relative to the cost of production (a function of P_{min} and the storage performance parameters), and operating constraints that restrict when charging and discharging can occur (to prevent state of charge exceeding its technical limits). See Ward for further details.[136]

market model. A structural market model, MOSSI,[159–161] is used to discover the least-cost dispatch of power stations and storage to meet half-hourly demand in the British electricity market. A supply curve is created from the short-run marginal cost of every power station, and the intersection of this with the demand curve in each half-hour period of time is used to determine the generation mix and resulting wholesale price. The model ignores inter-temporal constraints on power station operation (such as extended start-up and shut-down times), and ancillary service markets (which account for ~1% of the cost of generation).[160] In doing so, it gains the advantage of speed, thus allowing one year of half-hourly dispatch to be calculated within ~2 seconds, versus several minutes to hours from more complex market models. A particular focus on the model is to generate an accurate representation of market prices, both in terms of their hourly level and also their variability, which is of particular importance for valuing arbitrage services.[136,160]

Storage can be integrated into this (and any other type of model) by an iterative hard-linking approach.[136] The storage capacity is divided into a number of smaller units (e.g. of 10 MW each). The market model is first run without any storage to generate an initial series of power prices. The storage algorithm is then run with this set of prices, and used to determine the dispatch profile for the first unit of storage. The output of this storage is then subtracted from demand, and the market model rerun to generate a new series of prices. The second unit of storage is then dispatched with these prices, and so on. Revenues, costs, and profits for the entire storage fleet are calculated from the final prices which emerge after all storage units have been dispatched. Even though the first units of storage are dispatched using the wrong prices (as they were calculated without the influence of subsequent storage units), this is corrected for by later iterations, and with a sufficiently small step size, errors in the final dispatch (and thus profit) are negligible.[136] Alternative methods are possible, such as providing the storage algorithm with knowledge of the relationship between storage output and price (i.e. the market-wide supply curve), so that it implicitly chooses the correct dispatch for the entire storage fleet first time. This is shown by Ward to yield the same end results with lower computational time, at the cost of more complexity for the user in integrating the storage algorithm into the electricity market model.[136]

8.6.3 Finances of energy storage projects

The finances of energy storage projects are described by standard profitability metrics: net present value (NPV), internal rate of return (IRR), and payback period.

NPV determines the present value of the sum of all cash flows over the life of a project. All revenue and cost cash flows are discounted and summed. The sum of all discounted cost is then subtracted from the sum of all discounted revenues (equation (32)). If the NPV is greater than 0, the project will be profitable.

$$NPV = \sum_{n}^{N} \frac{revenue(n)}{(1+r)^n} - \sum_{n}^{N} \frac{cost(n)}{(1+r)^n} \quad (32)$$

Using the variables already defined in the previous sections as well as the price for each unit of energy discharged P_e (e.g. USD/MWh) and the price for each unit of power provided per year P_p (e.g. USD/kW-year), the NPV formula for energy storage projects can be tabulated as equation (33).

$$NPV = \left(\sum_{n}^{N} \frac{P_e(n) \cdot E_{out}(n)}{(1+r)^n} + \sum_{n}^{N} \frac{P_p(n) \cdot C_{p,nom}(n)}{(1+r)^n} \right) - \left(\sum_{n}^{N} \frac{Investment}{(1+r)^n} + \sum_{n}^{N} \frac{Replacement}{(1+r)^n} + \sum_{n}^{N} \frac{O\&M}{(1+r)^n} + \sum_{n}^{N} \frac{Charging}{(1+r)^n} + \frac{End\ of\ life}{(1+r)^{N+1}} \right) \quad (33)$$

The internal rate of return represents the discount rate r that returns an NPV of 0. It must be calculated iteratively through trial and error or by using software programmed to calculate IRR.

$$NPV = 0 = \sum_{n}^{N} \frac{revenue(n)}{(1+IRR)^n} - \sum_{n}^{N} \frac{cost(n)}{(1+IRR)^n} \quad (34)$$

The payback period describes the time it takes to recover the cost of an investment. It is the year (and month, depending on modelling granularity) at which the cumulative cash flows (revenue and cost) turn positive. If there is an initial investment and fixed cash flows thereafter, the payback period can be determined as follows.

$$Payback\ period\ (in\ years) = \frac{Initial\ investment}{Annual\ cash\ flow} \quad (35)$$

However, for energy storage projects as well as most other investments, cash flows are not constant. Therefore, models that determine the cash flows in each time period are required to determine the payback period.

Payback period can be determined as simple or discounted payback period. Simple payback period is determined without discounting future cash flows. Discounted payback period is determined by discounting future cash flows.

Further details on these standard methodologies to determine the finances of an investment can be obtained from textbooks on corporate and project finance or respective online webpages.[162,163]

8.7 System value

Chapter 7 looks into the system value that energy storage provides in integrating variable and inflexible low-carbon power sources into power systems.

8.7.1 Great Britain case study

The value of electricity storage in enabling low-carbon power systems is investigated in a meta-analysis by reviewing multiple academic, industry and government studies conducted that model the future evolvement of Great Britain's (GB) power system.

The value that electricity storage offers to power systems is a function of three study dimensions: the power system set-up (e.g. renewable, nuclear, flexibility capacity), model type (e.g. temporal and spatial resolution, technology detail), and input assumptions (e.g. technology cost, carbon and fuel prices). An additional study would likely be limited to one viewpoint on all of these aspects and could not present a consensus view. Instead, considering openly available studies in a meta-study approach allows heterogeneity across all study dimensions and enables identification of trends and a consensus view (if one exists).

The GB system is suitable for assessing the system value of electricity storage due to:

- High data availability from multiple studies by various institutions
- Ambitious targets for decarbonization of the power system[164]
- Increasing penetration of low-carbon electricity[1]
- Limited interconnection to neighbouring countries (4 GW in 2020)[1]

The chosen studies employ power system models and optimize for lowest cost under carbon emission and technology penetration constraints. The study by Edmunds et al. (see Table 7.1) is not considered because installed electricity storage capacities do not reflect requirements for system adequacy but the specific potential of two newly proposed pumped hydro storage sites only.[165]

All scenarios in the reviewed studies are assessed for installed capacity and generation of all electricity technologies (i.e. coal, gas, wind, solar, hydro, nuclear, biomass, geothermal, waste, wave, electricity storage, demand-side response (DSR), interconnection, open cycle gas turbine (OCGT), oil, diesel), and peak demand.

Electricity technologies are grouped into three categories:

1. Bulk generation capacity: Coal, gas combined cycle gas turbine (CCGT), wind, solar, nuclear, biomass, geothermal, waste, wave
2. Flexibility capacity: Electricity storage, DSR, interconnection, OCGT, oil, diesel, hydro
3. Firm reliable capacity: All, except wind and solar (and wave)

Missing information were inquired directly from the authors of the studies. If CCGTs are modelled with less than 100 full load hours per year, they are categorized as flexibility and not bulk generation capacity. Respective values for the existing GB power system in 2020 for comparison are taken from the Digest of UK energy statistics report.[1]

The resulting data are assessed for modelled electricity storage, flexibility, and total capacity relative to the share of wind, solar, and nuclear power capacity and energy generation. These three low-carbon technologies are chosen as dependencies because of their variable (i.e. solar, wind) or relatively inflexible (i.e. nuclear) generation pattern, creating the need for flexibility capacity such as electricity storage (see Chapter 1).

To account for varying peak demand values across power systems, the requirements for electricity storage, flexibility, and total capacity are normalized for this factor. The modelled dispatchable capacity relative to peak demand is also assessed.

In addition, the impact of the following variables on electricity storage and flexibility requirements in general were tested: nuclear penetration, electricity storage discharge duration, the ratio of wind to solar.

8.7.2 Global power system

The analysis of global flexibility capacity requirements in the so-called thought experiment is based on projections for the power sector of the Integrated Assessment Modelling Consortium (IAMC) to 2100 used for the IPCC 1.5 °C report in scenarios with a 50% likelihood to limit average global surface temperature increase to below 2 °C.[166–168]

Global non-coincident peak demand in each year was determined by dividing total global annual electricity demand by 8,760 hours to compute average demand and multiplying that by 1.8. This factor of 1.8 is the ratio between average and peak demand, identified by comparing average demand in 2005 to 80% of total generation capacity. The year 2005 was chosen because wind and solar capacity only formed 1% of total generation capacity. It is also implied that only 80% of the generation capacity is supposed to meet peak demand, assuming a system margin of 20%. The ratio between average (440 GW) and noncoincident peak demand (770 GW) in the US is also 1.8,[169] which confirms the ratio assumed in this analysis.

An implicit assumption in the 'thought experiment' is that hourly demand profiles remain unchanged compared to 2005. However, selected studies suggest this profile will change in future, projecting an increase of 20–25% in peak demand relative to average demand by 2050 in selected countries,[170] due to increasing electrification of heat and transport.

Projections for hydro- and oil-based generation capacity are considered as flexibility capacity, in addition to 2020 installation levels of electricity storage (175 GW), interconnection

(180 GW) and demand-side response (40 GW) since these are not projected into the future.[9,11,171]

The results are compared to the flexibility capacity modelled in the study by Jacobson et al. for a 100% wind-, water-, and sunlight-based energy system in 2050.[172]

8.7.3 Energy perspective

The analysis on the role of electricity storage in integrating wind and solar generation is based on the approach laid out by Weitemeyer et al. in 2015.[173]

It assumes representative data on power generation from wind $W(t_i)$ and solar $S(t_i)$ resources and load data $L(t_i)$ at discrete times t_i. The accumulated energy generated or consumed is based on the generation/load data and the interval between t_i and t_{i+1}. In the present analyses this interval is one hour.

In order to express wind and solar generation relative to electricity demand, it is normalized using

$$\langle X(t)\rangle_t := \frac{\sum_{i=1}^{N} X(t_i)}{N} \tag{36}$$

to yield

$$w_t := \frac{W_t}{\langle W(t)\rangle_t}\langle L(t)\rangle_t \tag{37}$$

$$s_t := \frac{S_t}{\langle S(t)\rangle_t}\langle L(t)\rangle_t \tag{38}$$

If γ sets the total electricity produced from wind and solar and α or $\alpha-1$ set the respective shares of wind and solar generation, then the mismatch in generation from wind and solar and energy demand at time t can be defined as

$$\Delta_{\alpha,\gamma}(t) := \gamma(\alpha w(t) + (1-\alpha)s(t)) - L(t) \tag{39}$$

Without any energy storage, wind and solar generation need to be curtailed in times of overproduction $(\Delta_{\alpha,\gamma}(t) > 0)$, whereas negative mismatches $(\Delta_{\alpha,\gamma}(t) < 0)$ need to be balanced by flexible generation. The total amount of curtailed energy relative to electricity demand can be determined by the overproduction function *OP*

$$OP(\alpha,\gamma) = \frac{\langle max[0, \Delta_{\alpha,\gamma}(t)]\rangle_t}{\langle L(t)\rangle_t} \tag{40}$$

The share of wind and solar energy generation that can be consumed or integrated in electricity demand accounting for curtailment can then be calculated as renewable integration function *RE*.

$$RE(\alpha,\gamma) = \frac{\gamma(\alpha w(t) + (1-\alpha)s(t))_t - OP(\alpha,\gamma)L(t)_t}{L(t)_t} = \gamma - OP(\alpha,\gamma) \tag{41}$$

Adding energy storage capacities of limited size H^{max} (relative to $\langle L(t)\rangle_t$) and round-trip efficiency η allows to determine the storage state-of-charge time series $H^{\eta}_{\alpha,\gamma}(t)$ as

$$H^{\eta}_{\alpha,\gamma}(t)=\begin{cases} if\Delta_{\alpha,\gamma}(t)\geq 0 \\ min\left[H^{max},H^{\eta}_{\alpha,\gamma}(t-1)+\eta\Delta_{\alpha,\gamma}(t)\right] \\ if\Delta_{\alpha,\gamma}(t)<0 \\ max\left[0,H^{\eta}_{\alpha,\gamma}(t-1)+\Delta_{\alpha,\gamma}(t)\right] \end{cases} \tag{42}$$

The initial charging level of the storage $H^{\eta}_{\alpha,\gamma}(t=0)$ has to be specified when the approach is applied to actual data.

The total amount of unusable energy due to storage size limitations and efficiency losses can be expressed as the adapted overproduction function OP^{η}_{H} (relative to $\langle L(t)\rangle_t$)

$$OP^{\eta}_{H}(\alpha,\gamma)=\frac{\langle max\left[0,\Delta_{\alpha,\gamma}(t)-(H^{\eta}_{\alpha,\gamma}(t)-H^{\eta}_{\alpha,\gamma}(t-1)\right]\rangle_t}{\langle L(t)\rangle_t} \tag{43}$$

The respective share of wind and solar electricity generation that can be integrated with electricity demand, that is consumed, can then be determined as adapted renewable integration function RE^{η}_{H}

$$RE^{\eta}_{H}(\alpha,\gamma)=\gamma-max\left[y-1,OP^{\eta}_{H}(\alpha,\gamma)\right] \tag{44}$$

Table 8.13 indicates the data sources for wind and solar generation and electricity load in the respective power markets. All data are at hourly resolution, and aggregated to national level, not considering limitations of the electricity network, exports/imports, or other limitations of the existing power station portfolio. Historically metered data were used for load, as this is reasonably static, with only 1–2% change from year to year. In contrast, wind and solar generation data were taken from the Renewables.ninja model,[174,175] simulating the output of the present-day fleet of wind and solar farms operating with the weather from

Table 8.13 Data sources and time frame for solar and wind reanalysis and electricity load data of the power markets studied.

Market	Solar and wind data	Load data	Time frame
Great Britain	Renewables.ninja[174,175]	National Grid[176,177]	1991–2019
Germany	Renewables.ninja	ENTSO-E[178]	2006–2014
France	Renewables.ninja	ENTSO-E[178]	2006–2014
Australia (Queensland)	Renewables.ninja	AEMO[179]	2013–2015
Japan	Renewables.ninja	Various[156]	2012–2015
US (PJM)	Renewables.ninja	PJM[180]	2013–2015

previous years. This has two advantages over using historical metered data, which represents a continually evolving fleet of farms (and thus does not provide consistency between years), and is typically only available for more recent years (and thus does not give a long time-series to explore year-to-year variability).

All time-series are normalized to total electricity load. Hence, the factor $\gamma = 1$ corresponds to scenarios where the total electricity produced by wind and solar resources in the respective time frame is identical to the overall electricity load in that time frame.

For the analysis of natural gas reserves in the EU and the US, data on the respective fill levels provided by the Aggregated Gas Storage Inventory for the EU on a daily and by the Energy Information Administration for the US on a weekly basis is used.[181,182] The total storage volume is determined by using the average working gas volume in 2021 for the EU and the peak demonstrated for the US in 2021.[181,183] Maximum charge/discharge capacity is determined by dividing the daily fill level changes by 24 hours (EU) or the weekly fill level changes by 168 hours (US). These are averages and may not represent the actual maximum charge/discharge rates. However, these are still reasonable approximations given that these energy stores are used for very long duration interseasonal storage. Energy-to-power ratio or discharge duration of the respective natural gas reserves is determined by dividing total storage volume over maximum charge/discharge rate.

8.8 References

1. BEIS. *Digest of UK Energy Statistics* (London: Department for Business, Energy & Industrial Strategy 2021).
2. Staffell I. *The Energy and Fuel Data Sheet* (2011). Available at: <https://www.academia.edu/1073990/The_Energy_and_Fuel_Data_Sheet> accessed 5 November 2022.
3. Nierop S, Humperdinck S. *International Comparison of Fossil Power Efficiency and CO2 Intensity* (Utrecht: Ecofys, 2018).
4. Next Green Car. Electric Vehicle Market Statistics (2022). Available at: <https://nextgreencar.com/electric-cars/statistics/> accessed 5 November 2022.
5. Argonne National Laboratory. BatPac Version 3.0. Excel Spreadsheet (2015). Available at: <https://www.anl.gov/cse/batpac-model-software> accessed 5 November 2022.
6. Ortiz L. *Grid-Scale Energy Storage Balance of Systems 2015–2020: Architectures, Costs and Players* (Austin: GTM Research, 2015).
7. Augustine C and Blair N. *Storage Futures Study: Storage Technology Modeling Input Data Report* (Washington, DC: U.S. Department of Energy, 2021).
8. Frith J. *Lithium-Ion Batteries: The Incumbent Technology Platform for Coal Regions in Transition* (London: BloombergNEF, 2019).
9. International Hydropower Association (2022). Available at: <https://www.hydropower.org/> accessed 5 November 2022.
10. Wood Mackenzie. *Global Energy Storage Outlook H1 2021* (Edinburgh: Wood Mackenzie, 2021).
11. International Energy Agency. *Energy Storage* (2022).
12. Sandia National Laboratories. DOE Global Energy Storage Database (2022). Available at: https://sandia.gov/ess-ssl/gesdb/public/index.html accessed 5 November 2022.

13. NGK Insulator Ltd. Energy Storage System Products (2022). Available at: <https://www.ngk-insulators.com/en/product/search-business/battery/> accessed 5 November 2022.
14. Consortium for Battery Innovation. Lead Battery Market Data (2021). Available at: <https://batteryinnovation.org/resources/lead-battery-market-data/> accessed 5 November 2022.
15. Akhil A, Huff G, Currier AB, Kaun BC, *et al. DOE/EPRI 2013 Electricity Storage Handbook in Collaboration with NRECA. Sandia Report* (Albuquerque: Sandia National Laboratories, 2013).
16. Rastler DM. *Electric Energy Storage Technology Options: A White Paper Primer on Applications, Costs and Benefits* (Palo Alto: Electric Power Research Institute, 2010).
17. Lazard. *Lazard's Levelized Cost of Storage Analysis—version 2.0* (New York: Lazard Ltd, 2016).
18. Wright TP. 'Factors Affecting the Cost of Airplanes' (1936) 3 *Journal of the Aeronautical Sciences* 122–8.
19. Farmer JD and Lafond F. 'How Predictable is Technological Progress?' (2015) 45 *Research Policy* 647–65.
20. The Boston Consulting Group. *Perspectives on Experience* (Boston: Boston Consulting Group, 1968).
21. Junginger M, van Sark W, and Faaij A. *Technological Learning in the Energy Sector: Lessons for Policy, Industry and Science* (Cheltenham: Edward Elgar Publishing, 2010).
22. Nagy B, Farmer JD, Bui QM and Trancik JE. 'Statistical Basis for Predicting Technological Progress' (2013) 8 *PLoS One* e52669.
23. Arrow KJ. 'The Economic Learning Implications of By Doing' (1962) 29 *The Review of Economic Studies* 155–73.
24. International Energy Agency and OECD. *Experience Curves for Energy Technology Policy*. (Paris: OECD / IEA, 2000).
25. McDonald A and Schrattenholzer L. 'Learning Rates for Energy Technologies' (2001) 29 *Energy Policy* 255–61.
26. Jamasb T. 'Technical Change Theory and Learning Curves: Patterns of Progress in Electricity Generation Technologies' (2007) 28 *The Energy Journal* 51–71.
27. Kahouli-Brahmi S. 'Technological Learning in Energy-Environment-Economy Modelling: A Survey' (2008) 36 *Energy Policy* 138–62.
28. Rubin ES, Azevedo IML, Jaramillo P, and Yeh S. 'A Review of Learning Rates for Electricity Supply Technologies' (2015) 86 *Energy Policy* 198–218.
29. Gross R, Heptonstall P, Greenacre P, Candelise C, *et al. Presenting the Future—An Assessment of Future Costs Estimation Methodologies in the Electricity Generation Sector* (London: UK Energy Research Center, 2013).
30. Abernathy WJ and Kenneth W. 'Limits of the Learning Curve' (1974) Sept-Oct *Harvard Business Review* 109–19.
31. Junginger M, Lako P, Lensink S, and Weiss M. Technological learning in the energy sector. *Netherlands Programme on Scientific Assessment and Policy Analysis Climate Change (WAB)* 1–190 (Bilthoven: Netherlands Environmental Assessment Agency, 2008). Available at: <https://www.pbl.nl/sites/default/files/downloads/500102017_0.pdf>accessed 5 November 2022.
32. Cohen W and Levinthal D. 'Innovation and Learning: The Two Faces of R&D' (1989) 99 *The Economic Journal* 569–96.
33. Hall G and Howell S. 'The Experience Curve from the Economist's Perspective' (1985) 6 *Strategic Management Journal* 197–212.

34. Nemet GF. 'Beyond the Learning Curve: Factors Influencing Cost Reductions in Photovoltaics' (2006) 34 *Energy Policy* 3218–32.
35. Ferioli F, Schoots K, and van der Zwaan BCC. 'Use and Limitations of Learning Curves for Energy Technology Policy: A Component-Learning Hypothesis' (2009) 37 *Energy Policy* 2525–35.
36. Wiesenthal T, Dowling P, Morbee J, Thiel C, *et al. Technology Learning Curves for Energy Policy Support. JRC Scientific and Policy Reports* (Geneva: European Commission, 2012).
37. Zheng C and Kammen DM. 'An Innovation-focused Roadmap for a Sustainable Global Photovoltaic Industry' (2014) 67 *Energy Policy* 159–69.
38. Kittner N, Lill F, and Kammen DM. 'Energy Storage Deployment and Innovation for The Clean Energy Transition' (2017) 2 *Nature Energy* 17125.
39. Watanabe C, Moriyama K, and Shin JH. 'Functionality Development Dynamism in a Diffusion Trajectory: A Case of Japan's Mobile Phones Development' (2009) 76 *Technological Forecasting and Social Change* 737–53.
40. Staffell I and Green R. 'The Cost of Domestic Fuel Cell Micro-CHP Systems' (2013) 38 *International Journal of Hydrogen Energy* 1088–102.
41. Söderholm P and Sundqvist T. 'Empirical Challenges in the Use of Learning Curves for Assessing the Economic Prospects of Renewable Energy Technologies' (2007) 32 *Renewable Energy* 2559–78.
42. Kouvaritakis N, Soria A, and Isoard S. 'Modelling Energy Technology Dynamics: Methodology for Adaptive Expectations Models with Learning By Doing and Learning By Searching' (2000) 14 *International Journal of Global Energy Issues* 104–15.
43. Seebregts AJ, Kram T, Scaeffer GJ, and Stoffer A. *Endogenous Technological Learning: Experiments with MARKAL* (Petten: Netherlands Energy Research Foundation, 1998).
44. Epple D, Argote L, and Devadas R. 'Organizational Learning Curves: A Method for Investigating Intra-Planet Transfer of Knowledge Acquired Through Learning by Doing' (1991) 2 *Organization Science* 58–70.
45. OECD. Consumer Price Indices (CPIs) (2021). Available at: <https://stats.oecd.org/Index.aspx?DataSetCode=PRICES_CPI> accessed 5 November 2022.
46. OECD. National Accounts Statistics: PPPs and Exchange Rates (2021). Available at: <https://doi.org/10.1787/data-00004-en> accessed 5 November 2022.
47. OECD. Purchasing Power Parity Exchange Rates (2017). Available at: <https://data.oecd.org/conversion/purchasing-power-parities-ppp.htm> accessed 5 November 2022.
48. Chen H, Ngoc T, Yang W, Tan C, *et al.* 'Progress in Electrical Energy Storage System: A Critical Review' (2009) 19 *Progress in Natural Science* 291–312.
49. International Energy Agency. *Technology Roadmap—Energy Storage* (Paris: IEA, 2014).
50. International Energy Agency. *Projected Costs of Generating Electricity* (Paris: IEA, 2005).
51. International Energy Agency. *Projected Costs of Generating Electricity* (Paris: IEA, 2010).
52. International Energy Agency. *Projected Costs of Generating Electricity* (Paris: IEA, 2015).
53. International Energy Agency. *Projected Costs of Generating Electricity* (Paris: IEA, 2020).
54. Sandia National Laboratories. DOE Global Energy Storage Database 2016 (Albuquerque: Sandia National Laboratories, 2016).
55. Matteson S and Williams E. 'Residual Learning Rates in Lead-Acid Batteries: Effects on Emerging Technologies' (2015) 85 *Energy Policy* 71–9.
56. Thornton I, Rautiu R, and Brush S. *Lead: The Facts* (IC Consultants Ltd, 2001).

57. Exide Technologies. Exide AGM—Automobil. Available at: <https://www.exidegroup.com/de/de/product/exide-agm-lv> accessed 5 November 2022.
58. Bundesverband Solarwirtschaft e.V. (BSW-Solar). *Solarstromspeicher-Preismonitor Deutschland—2. Halbjahr 2021 [German]* (Berlin: Bundesverband Solarwirtschaft e.V., 2021).
59. Bräutigam A. *Batteries for Stationary Energy Storage in Germany: Market Status & Outlook* (Berlin: Germany Trade & Invest, 2016).
60. Wedepohl D. *Über 50.000 Solarbatterien in Betrieb* (Berlin: Bundesverband Solarwirtschaft e.V., 2017).
61. Kairies K-P, Haberschusz D, Magnor D, Leuthold M, *et al. Wissenschaftliches Mess- und Evaluierungsprogramm Solarstromspeicher—Jahresbericht 2015* (in German) (Achen: Institut für Stromrichtertechnik und Elektrische Antreibe—RWTH Aachen, 2015).
62. PV Magazine. 'Storage & Smart Grids: Market Overview Storage Systems' (2014) 8 *PV Magazine* 74–7.
63. Pitt A, Buckland R, Antonio PD, Lorenzen H, *et al. Citi GPS: Global Perspectives & Solutions—Investment Themes in 2015* (New York City: Citigroup, 2015).
64. Pillot C. *The Rechargeable Battery Market and Main Trends 2016–2025* (Paris: Avicenne, 2017).
65. Matteson S and Williams E. 'Learning Dependent Subsidies for Lithium-Ion Electric Vehicle Batteries' (2015) 92 *Technological Forecasting and Social Change* 322–31.
66. Pillot C. *The Rechargeable Battery Market and Main Trends 2013–2025* (Paris: Avicenne, 2014).
67. BloombergNEF. Battery Pack Prices Fall to an Average of $132/kWh, but Rising Commodity Prices Start to Bite (2021). Available at: <https://about.bnef.com/blog/battery-pack-prices-fall-to-an-average-of-132-kwh-but-rising-commodity-prices-start-to-bite/> accessed 5 November 2022.
68. International Energy Agency. *Global EV Outlook* (Paris: IEA, 2021).
69. Figgener J, Haberschusz D, Kairies K-P, Wessels O, *et al. Wissenschaftliches Mess-und Evaluierungsprogramm Solarstromspeicher 2.0* (Aachen: Rheinisch-Westfälische Technische Hochschule (RWTH), 2017).
70. Tepper M. *Solarstromspeicher-Preismonitor Deutschland 2017* (in German) (Berlin: Bundesverband Solarwirtschaft e.V., 2017).
71. Hocking M, Kan J, Young P, Terry C, *et al. F.I.T.T. for Investors: Welcome to the Lithium-ion Age* (Frankfurt am Main: Deutsche Bank, 2016).
72. CleanTechnica. 140 MWh German Energy Storage Project Coming From LG Chem (2015). Available at: <https://cleantechnica.com/2015/11/25/140-mwh-german-energy-storage-project-coming-lg-chem/> accessed 5 November 2022.
73. Frith J and Goldie-Scot L. *Lithium-Ion Batteries: A Matter of Life and Death* (London: BloombergNEF, 2018).
74. Sandia National Laboratories. DOE Global Energy Storage Database 2018 (Albuquerque, Sandia National Laboratories, 2018).
75. Kalhammer FR, Kopf BM, Swan DH, Roan VP, *et al. Status and Prospects for Zero Emissions Vehicle Technology—Report of the ARB Independent Expert Panel 2007* (Sacramento: State of California Air Resources Board, 2007).
76. Toyota Motor Corporation. 'With 8 Million Units Sold, Toyota Proves Hybrids Have Staying Power' (Toyota: Toyota motor corporation, 21 April 2015). Available at: <https://global.toyota/en/detail/9163695>. accessed 5 November 2022.

77. Best Hybrid Batteries. Prius Battery Specifications (2013). Available at: <https://www.besthybridbatteries.com/categories/hybrid-batteries-by-make/toyota/prius> accessed 5 November 2022.
78. EVsRoll. Hybrid Car Statistics (2012). Available at: <http://www.evsroll.com/Hybrid_Car_Statistics.html> accessed 5 November 2022.
79. Gerssen-Gondelach SJ and Faaij APC. Performance of Batteries for Electric Vehicles on Short and Longer Term' (2012) 212 *Journal of Power Sources* 111–29.
80. SAE International. 'Don't Count NiMH Power Out, Says Battery Guru' (2010) 118(3) *Automotive Engineering International* 14–15.
81. Tortora AC. 'Experiences and Initial Results from Terna's Energy Storage Projects'. *7th Grid+Storage Workshop* (Rome: Terna S.p.A., 2015).
82. Rial P. Sodium-Sulfur Battery Powers NGK's Unique Wind Energy (2008). Available at: <https://www.bloomberg.com/news/articles/2008-04-10/sodium-sulfur-batteries-power-ngk-insulators-unique-wind-energy> accessed 5 November 2022.
83. Abe H. *NAS Battery Energy Storage System* (Nagoya: NGK Insulators Ltd, 2013).
84. NGK Insulators Ltd. NAS Battery System Cost Reduction Roadmap (Plan) (Nagoya: NGK Insulators Ltd, 11 August 2016).
85. Indicative Manufacturer Quote. Obtained 17 March 2016 at Energy Storage Trade Show in Dusseldorf, Germany.
86. Mustang RB. *Energy Storage Pricing Survey* (Albuquerque: Sandia National Laboratories, 2019).
87. Spoerke ED, Gross MM, Small LJ, and Percival SJ. 'Chapter 4: Sodium-based battery technologies' in *U.S. DOE Energy Storage Handbook* (Albuquerque: Sandia National Laboratories, 2021).
88. Mizutani T. *Battery Energy Storage for Increasing Renewable Energy* (Nagoya: NGK Insulators Ltd., 2015).
89. NGK Insulator Ltd. *Installation Record* (2016). Available at: <https://www.ngk-insulators.com/en/product/nas-solutions.html> accessed 5 November 2022.
90. Tamakoshi T. 'Sodium-Sulfur (NAS) Battery' in DOE Long Duration Energy Storage Workshop (Albuquerque: Sandia National Laboratories, 2021.
91. Zhang H. 'Rechargeables: Vanadium Batteries will be Cost-effective' (2014) 508 *Nature* 319.
92. Lazard. *Lazard's Levelized Cost of Storage analysis—Version 3.0* (New York: Lazard Ltd, 2017).
93. Peacock B. 'Another Vanadium Redox Flow Battery to be Installed in Australia' (2021). Available at: <https://www.pv-magazine.com/2020/12/23/another-vanadium-redox-flow-battery-to-be-installed-in-australia/> accessed 5 November 2022.
94. Sánchez-Díez E, Ventosa E, Guarnieri M, Trovò A, *et al.* 'Redox Flow Batteries: Status and Perspective Towards Sustainable Stationary Energy Storage' (2021) 481 *Journal of Power Sources* 228804.
95. Schoots K, Ferioli F, Kramer GJ, and van der Zwaan BCC. 'Learning Curves for Hydrogen Production Technology: An Assessment of Observed Cost Reductions' (2008) 33 *International Journal of Hydrogen Energy* 2630–45.
96. Smolinka T, Günther M, and Garche J. *Stand und Entwicklungspotenzial der Wasserelektrolyse zur Herstellung von Wasserstoff aus regenerativen Energien* (in German). (Freiburg and Ulm: Fraunhofer ISE/FCBAT, 2011).
97. Steinmüller H, Reiter G, Tichler R, Friedl C, *et al. Power to Gas—eine Systemanalyse. Markt- und Technologiescouting und -analyse* (in German) (Johannes Kepler Universität Linz/ Montanuniversität Loeben/ TU Wien, 2014).

98. Glenk G and Reichelstein S. 'Economics of converting renewable power to hydrogen' (2019) 4 *Nature Energy* 216–22.
99. Tengler M. *Green Hydrogen: Time to Scale Up European Hydrogen Forum* (London: BloombergNEF, 2020).
100. Staffell I, Scamman D, Velazquez Abad A, Balcombe P, *et al.* 'The Role of Hydrogen and Fuel Cells in the Global Energy System' (2019) 12 *Energy & Environmental Science* 463–91.
101. Agency for Natural Resources and Energy. 'Future Resource and Fuel Policy—Issues and Direction of Response' 今後の資源・燃料政策の課題と 対応の方向性 (案) (Tokyo: Ministry of Economy, Trade and Industry, 2021).
102. Hydrogenics. Investor Presentation (Mississauga: Hydrogenics, 2014).
103. VDMA. *International Technology Roadmap for Photooltaic (ITRPV)* (Frankfurt am Main: Verband Deutscher Maschinen- und Anlagenbau e. V., 2021).
104. Mayer JN, Philipps S, Hussein NS, Schlegl T, *et al. Current and Future Cost of Photovoltaics* (Freiburg and Berlin: Fraunhofer ISE/Agora Energiewende, 2015).
105. Ashby MF and Polyblank J. *Materials for Energy Storage Systems—A White Paper* (Cambridge: Granta Design, 2012).
106. Sullivan JL and Gaines L. 'Status of Life Cycle Inventories for Batteries' (2012) 58 *Energy Conversion and Management* 134–48.
107. Koj JC, Schreiber A, Zapp P, and Marcuello P. 'Life Cycle Assessment of Improved High Pressure Alkaline Electrolysis' (2015) 75 *Energy Procedia* 2871–77.
108. Notter DA, Kouravelou K, Karachalios T, Daletou MK, *et al.* 'Life Cycle Assessment of PEM FC Applications: Electric Mobility and μ-CHP' (2015) 8 *Energy and Environmental Science* 1969–85.
109. Galloway RC and Dustmann C-H. 'ZEBRA Battery—Material Cost Availability and Recycling' in *Proceeding of the International Electric Vehicle Symposium (EVS-20)* 19–28 (Washington DC: EDTA, 2003).
110. Bloomberg Professional [Online]. Subscription Service. Available at: <https://www.bloomberg.com/professional/solution/bloomberg-terminal/> accessed 5 November 2022.
111. Pillot C. *The Rechargeable Battery Market and Main Trends 2018–2030* (Paris: Avicenne, 2019).
112. Sandalow D, McCormick C, Rowlands-Rees T, Izadi-Najafabadi A, *et al. Distributed Solar and Storage—ICEF Roadmap 1.0* (Tokyo: Innovation for Cool Earth Forum, 2015).
113. International Energy Agency. *Electric and Plug-in Hybrid Roadmap* (Paris, IEA, 2010).
114. BloombergNEF. *Electric Vehicle Outlook* (London: BloombergNEF, 2020).
115. BloombergNEF. *New Energy Outlook* (London: BloombergNEF, 2021).
116. Wu G, Inderbitzin A, and Bening C. 'Total Cost of Ownership of Electric Vehicles Compared to Conventional Vehicles: A Probabilistic Analysis and Projection Across Market Segments' (2015) 80 *Energy Policy* 196–214.
117. Needell ZA, McNerney J, Chang MT, and Trancik JE. 'Potential for Widespread Electrification of Personal Vehicle Travel in the United States' (2016) 1 *Nature Energy* 16112.
118. U.S. Department of Energy. Spot Prices for Crude Oil and Petroleum Products (2022). Available at: <https://www.eia.gov/dnav/pet/pet_pri_spt_s1_m.htm> accessed 5 November 2022.
119. Hill N, Varma A, Harries J, Norris J, *et al. A Review of the Efficiency and Cost Assumptions for Road Transport Vehicles to 2050* (Carlsbad (US): AEA Technology, 2012).
120. U.S. Energy Information Administration (EIA). *Electric Power Monthly* (Washington DC: U.S. Energy Information Administration, 2017).

121. Morgan MG. 'Use (and Abuse) of Expert Elicitation in Support of Decision Making for Public Policy' (2014) 111 *PNAS* 7176–84.

122. Kynn M. 'The "Heuristics and Biases" Bias in Expert Elicitation' (2008) 171 *Journal of the Royal Statistical Society. Series A: Statistics in Society* 239–64.

123. Bistline JE. 'Energy Technology Expert Elicitations: An Application to Natural Gas Turbine Efficiencies' (2014) 86 *Technological Forecasting and Social Change* 177–87.

124. Tversky A and Kahneman D. 'Judgment under Uncertainty: Heuristics and Biases' (1974) 185 *Science* 1124–31.

125. Anadon LD, Baker E, Bosetti V, and Aleluia Reis L. 'Expert Views—and Disagreements—about the Potential of Energy Technology R&D' (2016) 136 *Climatic Change* 677–91.

126. Schmidt O, Gambhir A, Staffell I, Hawkes A, *et al.* 'Future Cost and Performance of Water Electrolysis: An Expert Elicitation Study' (2017) 42 *International Journal of Hydrogen Energy* 30470–92.

127. *Expert Elicitation Task Force—White Paper* (Washington DC: U.S. Environmental Protection Agency, 2011).

128. Meyer MA and Booker JM. *Eliciting and Analyzing Expert Judgment, A Practical Guide* (Washington DC: American Statistical Association and the Society for Industrial and Applied Mathematics, 2001).

129. Frith J. *2018 Lithium-Ion Battery Price Survey* (London: Bloomberg New Energy Finance, 2018).

130. Mackay DJC. *Introduction to Monte Carlo methods* (Amsterdam: Kluwer Academic Press, 1998).

131. Harrison RL. 'Introduction To Monte Carlo Simulation' (2010) 1204 *AIP Conference Proceedings* 17–21.

132. Hoffman FO and Hammonds JS. *An Introductory Guide to Uncertainty Analysis in Environmental and Health Risk Assessment* (Washington DC: U.S. Department of Energy, 1992).

133. Balducci PJ, Alam MJE, Hardy TD, and Wu D. 'Assigning Value to Energy Storage Systems at Multiple Points in an Electrical Grid' (2018) 11 *Energy & Environmental Science* 1926–44.

134. Housden J. *An Evaluation of Energy Storage Profitability in the United Kingdom, Germany, Australia and the United States* (London: Imperial College London, 2019).

135. Akhil A, Huff G, Currier AB, Kaun BC, *et al. DOE/EPRI 2015 Electricity Storage Handbook in Collaboration with NRECA* (Albuquerque: Sandia National Laboratories, 2015).

136. Ward KR and Staffell I. 'Simulating Price-aware Electricity Storage Without Linear Optimisation' (2018) 20 *Journal of Energy Storage* 78–91.

137. Sioshansi R, Denholm P, Jenkin T, and Weiss J. 'Estimating the Value of Electricity Storage in PJM: Arbitrage and Some Welfare Effects' (2009) 31 *Energy Economics* 269–77.

138. McConnell D, Forcey T, and Sandiford M. 'Estimating the Value of Electricity Storage in an Energy-Only Wholesale Market' (2015) 159 *Applied Energy* 422–32.

139. Denholm P and Sioshansi R. 'The Value of Compressed Air Energy Storage with Wind in Transmission-Constrained Electric Power Systems' (2009) 37 *Energy Policy* 3149–58.

140. Barbour E, Wilson IAG, Bryden IG, McGregor PG, *et al.* 'Towards an Objective Method to Compare Energy Storage Technologies: Development and Validation of a Model to Determine the Upper Boundary of Revenue Available from Electrical Price Arbitrage' (2012) 5 *Energy & Environmental Science* 5425–36.

141. Sousa JAM, Teixeira F, and Faias S. 'Impact of a Price-Maker Pumped Storage Hydro Unit on the Integration of Wind Energy in Power Systems' (2014) 69 *Energy* 3–11.

142. López Prol J, and Schill WP. 'The Economics of Variable Renewable Energy and Electricity Storage' (2021) 13 *Annual Review of Resource Economics* 443–67.

143. Chang M, Thellufsen JZ, Zakeri B, Pickering B, *et al.* 'Trends in Tools and Approaches for Modelling the Energy Transition' (2021) 290 *Applied Energy* 116731.

144. Kittner N, Castellanos S, Hidalgo-Gonzalez P, Kammen DM, *et al.* 'Cross-sector Storage and Modeling Needed for Deep Decarbonization' (2021) 5 *Joule* 2529–34.

145. Cebulla F, Naegler T, and Pohl M. 'Electrical Energy Storage in Highly Renewable European Energy Systems: Capacity Requirements, Spatial Distribution, and Storage Dispatch' (2017) 14 *Journal of Energy Storage* 211–23.

146. Gils HC, Scholz Y, Pregger T, Luca de Tena D, *et al.* 'Integrated Modelling of Variable Renewable Energy-Based Power Supply in Europe' (2017) 123 *Energy* 173–88.

147. Connolly D, Lund H, Finn P, Mathiesen BV. *et al.* 'Practical Operation Strategies for Pumped Hydroelectric Energy Storage (PHES) Utilising Electricity Price Arbitrage' (2011) 29 *Energy Policy* 4189–96.

148. Walawalkar R, Apt J, and Mancini R. 'Economics of Electric Energy Storage for Energy Arbitrage and Regulation in New York' (2007) 35 *Energy Policy* 2558–68.

149. Bradbury K, Pratson L, and Patiño-Echeverri D. 'Economic Viability of Energy Storage Systems Based on Price Arbitrage Potential in Real-Time U.S. Electricity Markets' (2014) 114 *Applied Energy* 512–19.

150. Sioshansi R. 'Welfare Impacts of Electricity Storage and the Implications of Ownership Structure' (2010) 31 *The Energy Journal* 173–98.

151. Schill WP and Kemfert C. 'Modeling Strategic Electricity Storage: The Case of Pumped Hydro Storage in Germany' (2011) 32 *The Energy Journal* 59–87.

152. Crampes C and Moreaux M. 'Pumped Storage and Cost Saving' (2010) 32 *Energy Economics* 325–33.

153. Fattori F, Anglani N, Staffell I, and Pfenninger S. 'High Solar Photovoltaic Penetration in the Absence of Substantial Wind Capacity: Storage Requirements and Effects on Capacity Adequacy' (2017) 137 *Energy* 193–208.

154. Staffell I and Rustomji M. 'Maximising the Value of Electricity Storage' (2016) 8 *Journal of Energy Storage* 212–25.

155. Lund H, Salgi G, Elmegaard B, and Andersen AN. 'Optimal Operation Strategies of Compressed Air Energy Storage (CAES) on Electricity Spot Markets with Fluctuating Prices' (2009) 29 *Applied Thermal Engineering* 799–806.

156. Halttunen K, Staffell I, Slade R, Green R, *et al.* 'Global Assessment of the Merit-Order Effect and Revenue Cannibalisation for Variable Renewable Energy' (2020) *SSRN Electronic Journal*. Available at: <https://dx.doi.org/10.2139/ssrn.3741232>.

157. Zipp A. 'The Marketability of Variable Renewable Energy in Liberalized Electricity Markets—An Empirical Analysis' (2017) 113 *Renewable Energy* 1111–21.

158. OFX. Historical Exchange Rates (2022). Available at: <https://www.ofx.com/en-gb/forex-news/historical-exchange-rates/> accessed 5 November 2022.

159. Staffell I and Green R. 'Is There Still Merit in the Merit Order Stack? The Impact of Dynamic Constraints on Optimal Plant Mix' (2016) 31 *IEEE Transactions on Power Systems* 43–53.

160. Ward KR, Green R, and Staffell I. 'Getting Prices Right in Structural Electricity Market Models' (2019) 129 *Energy Policy* 1190–206.

161. Green R and Staffell I. 'The Contribution of Taxes, Subsidies, and Regulations to British Electricity Decarbonization' (2021) 5 *Joule* 2625–45.

162. Marney J-P and Tarbert H. *Corporate Finance for Business* (Oxford: Oxford University Press, 2011).

163. Investopedia. Financial Terms Dictionary (2022). Available at: <https://www.investopedia.com/financial-term-dictionary-4769738> accessed 5 November 2022.

164. UK Government. *Climate Change Act 2008* (2008). Available at: <https://www.legislation.gov.uk/ukpga/2008/27/contents>.

165. Edmunds RK, Cockerill TT, Foxon TJ, Ingham DB, *et al.* 'Technical Benefits of Energy Storage and Electricity Interconnections in Future British Power Systems' (2014) 70 *Energy* 577–87.

166. Huppmann D, Kriegler E, Krey V, Riahi K, *et al.* IAMC 1.5°C Scenario Explorer and Data hosted by IIASA (2019). Available at: <https://iiasa.ac.at/models-and-data/iamc-15degc-scenario-explorer> accessed 5 November 2022.

167. Masson-Delmotte V, Zhai P, Pörtner H-O, Roberts D, *et al. IPCC Special Report: Global Warming of 1.5 C* (Paris: Intergovernmental Panel on Climate Change, 2018).

168. Huppmann D, Rogelj J, Kriegler E, Krey V, *et al.* 'A New Scenario Resource For Integrated 1.5 °C Research' (2018) 8 *Nature Climate Change* 1027–30.

169. U.S. Energy Information Administration (EIA). *Electric Power Monthly* (2019).

170. Boßmann T and Staffell I. 'The Shape of Future Electricity Demand: Exploring Load Curves in 2050s Germany and Britain' (2015) 90 *Energy* 1317–33.

171. International Energy Agency. *World Energy Outlook* (Paris: IEA, 2018).

172. Jacobson MZ, Delucchi MA, Bauer ZAF, Goodman SC, *et al.* '100% Clean and Renewable Wind, Water, and Sunlight All-Sector Energy Roadmaps for 139 Countries of the World' (2017) 1 *Joule* 108–21.

173. Weitemeyer S, Kleinhans D, Vogt T, andAgert C. 'Integration of Renewable Energy Sources in Future Power Systems: The Role of Storage' (2015) 75 *Renewable Energy* 14–20.

174. Pfenninger S and Staffell I. 'Long-term Patterns of European PV Output Using 30 Years of Validated Hourly Reanalysis and Satellite Data' (2016) 114 *Energy* 1251–65.

175. Staffell I and Pfenninger S. 'Using Bias-Corrected Reanalysis to Simulate Current and Future Wind Power Output' (2016) 114 *Energy* 1224–39.

176. Staffell I. 'Measuring the Progress and Impacts of Decarbonising British Electricity' (2017) 102 *Energy Policy* 463–75.

177. National Grid ESO. Data Finder and Explorer (2022). Available at: <https://www.nationalgrideso.com/industry-information/industry-data-and-reports/data-finder-and-explorer> accessed 5 November 2022.

178. ENTSO-E. Transparency Platform (2022). Available at: <https://transparency.entsoe.eu/> accessed 5 November 2022.

179. AEMO. Aggregated Price and Demand Data (2022). Available at: <https://aemo.com.au/energy-systems/electricity/national-electricity-market-nem/data-nem/aggregated-data> accessed 5 November 2022.

180. PJM. Data Miner 2—Hourly Load: Metered (2022). Available at: <http://dataminer2.pjm.com/feed/hrl_load_metered> accessed 5 November 2022.

181. AGSI. Gas Infrastructure Europe (2022). Available at: <https://agsi.gie.eu/>. accessed 5 November 2022.

182. U.S. Energy Information Administration (EIA). Weekly Natural Gas Storage Report (2022). Available at: <https://ir.eia.gov/ngs/ngs.html> accessed 5 November 2022.

183. U.S. Energy Information Administration (EIA). U.S. Natural Gas Storage Capacity has Remained Flat over the Past Eight Years (2021). Available at: <https://www.eia.gov/todayinenergy/detail.php?id=48216> accessed 5 November 2022.

9 Conclusion
Wrapping it all up

The ambition of this book is to provide you with a comprehensive toolkit to assess how the benefits of energy storage stack up against its costs. This conclusion reinforces how the individual chapters contribute to this ambition, reflects on the wider implications of the derived insights for energy storage, and draws out how you can use the insights and methods to assess whether energy storage makes money.

The first part of the book introduced you to energy storage: the need for it, the means of assessing it, and the many technologies and applications in the sector.

Chapter 1 provided you with the wider context of why electricity storage is needed and the potential scale that is required. It showed that electricity storage provides the required flexibility when decarbonizing the electricity and the transport sectors with renewable generators and electric vehicles. This underlines the multiple roles that electricity storage will play in the twenty-first-century energy system and gives you confidence that, regardless of specific market forecasts, there is a fundamental need for terawatt-hours of storage capacity in getting to net-zero emissions, which is orders of magnitude more than is available today. This can help you to clearly communicate the expected need for energy storage in future energy and transport systems.

Chapter 2 provided the fundamental grounding in both technology metrics and financial performance metrics that are used to assess energy storage technologies. It showed that a comprehensive set of metrics is needed to reflect the unique characteristic of energy storage technologies to consume, store, and deliver energy. It introduced application-specific lifetime cost as the appropriate approach for assessing energy storage technologies, considering all relevant cost and performance metrics. This chapter therefore equips you with the 'language' to interact with other stakeholders in the energy storage industry and enables you to verify the validity and comprehensiveness of storage cost assessments.

Chapter 3 structured the range of technologies, industry activity, and the many applications that exist for energy storage. It showed the existing dominance of selected technologies in stationary and transport applications, and the evolution of application requirements in future, which may create room for alternative technologies. It also provided insights on the future market size for energy storage technologies and concentration of power within specific companies and countries. As such, this chapter provides you with the holistic understanding of the sector required to engage with other stakeholders.

The second part of the book dived into the methods for assessing the cost and value of energy storage, and the insights that flow from them.

Monetizing Energy Storage. Oliver Schmidt & Iain Staffell, Oxford University Press. © Oliver Schmidt & Iain Staffell (2023).
DOI: 10.1093/oso/9780192888174.003.0009

Chapter 4 introduced experience curves as an objective method for measuring historical cost reductions and projecting future investment costs for energy storage technologies. Applying this method provides strong evidence that technology costs are falling rapidly across several energy storage technologies as the industry scales up. As a result, the one technology that brings most capacity to market is likely to be the most cost-effective one. Limitations to this approach are that cost reductions flatten once raw material cost dominate, radical step-change innovations may not be captured, and that the method is ill-suited to provide short-term price forecasts. In the short term, prices fluctuate around the long-term trend due to supply and demand dynamics in raw materials, components, and manufacturing capacities. For you, this method is a tool to derive your own objective long-term price forecasts for distinct technologies based on ever-changing market growth projections. The insights derived in this chapter can help you classify and interrogate statements on current and projected investment cost for different technologies.

Chapter 5 used lifetime cost, the method of choice for assessing the cost of energy storage. It applied this method to the nine most widely deployed technologies in 13 archetypical applications using future investment cost estimates derived in Chapter 4. The strongest drivers of lifetime cost were shown to be investment cost, application requirements, and financing conditions. Lifetime cost can be minimized if capital efficiency is optimized, i.e. total cost is minimized and distributed over as many charge–discharge cycles and as many lifetime years as possible. By 2030, lithium ion is projected to provide the lowest cost of any technology for applications with short to medium discharge duration and most power-focused applications. Alternative technologies may struggle to gain market share and achieve cost reductions through scale-up in these applications due to the existing dominance of lithium ion and its continued wide-scale deployment. By understanding and applying the lifetime cost method yourself, you can conduct economic assessments of energy storage technologies using bespoke technology cost and performance assumptions. The insights from this chapter on dominant technologies in different application categories can act as a first guide in focusing technology selection when planning storage projects.

Chapter 6 shifted focus from cost to value assessment. It reviewed the revenue potential of energy storage applications in different markets around the world, and introduced the method to compare these revenue potentials with lifetime costs to assess profitability. Energy arbitrage promises to be one of the largest markets in future. Analysis of its revenue potential revealed the impact of technology performance characteristics on profitability and the role of electricity price variability as predictor for profitability. The concept and many approaches of 'revenue stacking' to increase profitability were introduced, along with the method to conduct financial appraisal of energy storage projects. The analyses highlighted that there is no outstanding best market or application, rather 'pockets of opportunity' where profitability is likely. There is no universal business model to profit from energy storage, bespoke plans are required, and multiple revenue streams should be considered to optimize economic performance and hedge against potential revenue cannibalization through new assets. These insights allow you to fast-track your identification of profitable energy storage business cases, and the tools the chapter provides enable you to assess their profitability properly.

Chapter 7 focused on the system value of energy storage: what benefits it provides to society by reducing the total cost of providing secure, low-carbon electricity. A comprehensive review of previous studies showed that across academic and industry studies there is a consensus that electricity storage reduces the total system cost of achieving high penetrations of variable renewable energy compared to conventional flexibility technologies. The chapter then introduced methods to assess the amount of energy storage power capacity and energy capacity needed to fully decarbonize power systems. Applying these methods to various power systems confirmed the substantial scale-up of energy storage needed to support deep decarbonization. It also showed that long-duration storage with discharge durations beyond 16 hours will be required in systems where more than 90% of electricity comes from variable renewable generators. These insights enable you to assess the relevance of distinct energy storage technologies at different stages of power system decarbonization, and equip you with an understanding of the amounts of storage that may be needed, and by when. You can also use the derived methods to determine the amount of energy storage needed in the power systems of your interest.

The third part of the book, Chapter 8, provided additional depth on the methods applied and the input data used. It serves to support you in replicating the analyses performed in the previous chapters for the technologies, applications, and markets of your choice using your own assumptions and hypotheses.

Finally, it is unavoidable that the numbers presented in this book will become outdated as the storage industry continues to grow and innovate. They also do not represent every novel technology that is currently at the early stages of development, or perhaps has not even been invented yet. That is why we have built the <www.EnergyStorage.ninja> website to complement all the analyses found within this book. The website features up to date technology cost, performance, and deployment data. Its tools are flexible and interactive, allowing you to model results with your own data and consider new technologies, applications, and markets, taking the analyses we explain in this book in new directions.

There is no shortage of profound statements about the 'next big thing', be it artificial intelligence, automation, 3D printing, or plant-based foods. Energy storage may seem mundane in comparison, with limited ability to transform and improve the human condition or generate a vast and profitable new industry. However, there is an inevitable need for energy storage driven by the economic and environmental shift towards clean electricity and vehicles.

Energy storage, like many other technologies and services, can provide flexibility to energy systems; however, it is unique among them in that mass manufacture, economies of scale, and the power of learning by doing have already achieved substantial cost reductions, and will continue doing so in the coming decades. The energy storage industry is in its exponential growth phase across multiple sectors (residential, utility, transport) with deployment doubling year on year. New business opportunities become viable with ongoing cost reductions, but the breadth, heterogeneity, and complexity of the technologies, the applications they serve, and the markets they operate within mean it is no easy task to exploit these business opportunities successfully.

We hope this book has helped to increase transparency on how to monetize energy storage, both from the developer/investor perspective of making a profitable business case, and from the societal/governmental perspective of minimizing the cost of transitioning to clean energy. Energy storage will undoubtedly play a major role in this transition! We believe the toolkit and the insights this book provides, alongside the <www.EnergyStorage.ninja> website, will empower you to actively shape how energy storage takes this role, and benefit from it.

Glossary

Term	Definition
Ancillary services	Services that support the generation and transmission/distribution of electricity.
Annuitized capacity cost	Cost per unit of electric power provided by an electrical device for a specified time period.
Anode	Electrode in an electrochemical cell where electrons are released (i.e. electrons flow out).
Balance of plant	Supporting facilities and components in a storage system.
Balancing services	Services designed to match supply and demand.
Baseline	Operating point in terms of power capacity of an energy storage system.
Baseload	Electrical energy delivered or consumed at a constant rate.
Behind-the-meter	Position of an energy system behind the end user's electric meter with electricity not having to pass through it before consumption.
Black start	Application where energy storage restores power plant operations after a network outage without external power supply
Bulk generation capacity	Power generators that operate either whenever the resource is available (i.e. wind, solar, wave), or near-continuous, (i.e. nuclear, coal, combined-cycle gas, biomass, geothermal, waste).
Calendar life	Number of years before end of usable life at no operation.
Capital cost	See investment cost.
Cathode	The electrode in an electrochemical cell where electrons are consumed (i.e. electrons flow in).
Cell	Single, smallest energy- or charge-storing unit in a battery system.
Charging cost	Cost to charge a storage system with energy.
Congestion management	Application where energy storage mitigates risk of overloading and optimizes operation of existing infrastructure by smoothing renewable output and demand patterns.
Construction time	Duration to construct a storage system from groundbreaking to completion.
Cost floor	Lower limit below which cost is unlikely to fall.
Cumulative installed capacity	All capacity ever installed up to a certain point in time.
Cycle	Represents a single charge and discharge of an energy storage device.
Cycle life	Number of full charge-discharge cycles before end of life threshold is reached.
Degradation	Rate of loss in energy capacity incurred by cycles and/or time lapse.

Demand reduction	Application where energy storage reduces demand supplied by the network during periods of highest network charges.
Demand side response	Proactive changes in power consumption by consumers to accommodate variability in power supply.
Depth-of-discharge	Amount of energy capacity that can be charged/discharged without severely degrading nominal energy capacity.
Discount rate	Rate at which future costs and revenues are devalued in present-day terms.
Electricity storage	Energy storage that consumes electric energy and delivers electric energy.
Electrolyte	Medium containing ions that is electrically conducting through the movement of those ions, but not conducting electrons.
End-of-life cost	Cost to dispose technology at its end-of-life.
End-of-life threshold	State-of-health at which storage system is taken out of operation.
Energy applications	Applications where electricity storage systems are reimbursed per MWh of energy delivered.
Energy arbitrage	Application where energy storage purchases (renewable) power in low-price periods and sells in high-price periods
Energy density - gravimetric	Nominal energy capacity divided by system mass.
Energy density - volumetric	Nominal energy capacity divided by system volume.
Energy-to-power ratio	Usable energy capacity divided by nominal power capacity.
Experience curve	Trendline through historic product price data against cumulative installed capacity of respective product.
Experience rate	Percentage change in product price with each doubling in cumulative installed capacity.
Firm reliable capacity	Power generators that can actively increase or decrease generation capacity (i.e. all except solar, wind and wave).
Flexibility options	Services to balance and maintain the quality of electricity on the grid.
Flexible generation	Electrical energy generation technologies that are able to rapidly increase or reduce their output.
Flow battery	Batteries that use two liquid electrolytes as energy carriers.
Frequency regulation	Application where energy storage automatically corrects the continuous changes in supply or demand within the shortest market interval.
Frequency response	Application where energy storage automatically stabilizes network frequency after instantaneous changes in supply or demand.
Front-of-the-meter	Position of an energy system in front of the end user's electric meter with electricity having to pass through the meter before consumption.
Half-cell	Structure that contains one electrode and a surrounding conductive electrolyte.
Investment cost (specific)	Cost to add energy capacity or power capacity respectively.
Investment cost (total)	Cost of all energy capacity and power capacity components.
Levelized cost of energy	Cost per unit of electric energy generated for a specified time period.

Levelized cost of storage	Cost per unit of electric energy discharged from an energy storage device for a specified time period.
Lifetime cost	Cost metric accounting for all cost incurred throughout the lifetime of a technology as well as all relevant performance parameters.
Max. C-rate	Maximum rate to discharge storage system relative to its usable energy capacity.
Min. discharge duration	Duration to discharge usable energy capacity at nominal power. *Inverse of max. C-rate*.
Net zero	No net greenhouse gas emissions, i.e. human-made emissions to the atmosphere are balanced by removals from it.
Nominal energy capacity	Rated amount of energy that can be discharged from an energy storage device.
Nominal power capacity	Rated amount of power that can be charged and discharged by an energy storage device.
O&M cost	Cost to operate, insure, and periodically service technology components.
Pack	Structure consisting of multiple electrochemical cells and balance-of-plant components.
Peak capacity	Application where energy storage ensures availability of sufficient generation capacity at all times.
Pockets of opportunity	Time-limited opportunities for highly profitable energy storage use cases.
Power applications	Applications where electricity storage systems are reimbursed per kW-year of power provided.
Power density - gravimetric	Nominal power capacity divided by system mass.
Power density - volumetric	Nominal power capacity divided by system volume.
Power reliability	Application where energy storage fills sustained gaps between supply and demand.
Power system flexibility	The extent to which a power system can modify electricity production or consumption in response to variability, expected or otherwise.
Power-to-energy ratio	Nominal power capacity divided by usable energy capacity.
Power-to-Power	Energy storage approach that consumes and discharges electrical energy.
Power-to-X	Energy storage approach that consumes electrical energy and discharges another form of energy (e.g. thermal, chemcial).
Product price	Price payable for a product. See investment cost (total).
Redox	Chemical reaction in which reduction and oxidation occur at the same time.
Renewables integration	Application where energy storage stores large amounts of excess renewable electricity to be used at a later time.
Replacement cost	Cost to replace major technology components not accounted for in O&M cost.
Replacement interval	Time interval for replacement of major technology components.
Reserves	Amount of a geological commodity that has been discovered, has a known size, and can be extracted at a profit.

Resources	Amount of a geologic commodity that exists in both discovered and undiscovered deposits.
Response time	Time between idle state and maximum power.
Revenue cannibalization	Worsening of the profitability of all existing storage due the smoothing of power prices.
Round-trip efficiency	Proportion of energy discharged over energy required to charge for a full charge-discharge cycle.
Sealed battery	Batteries that use two electrodes as energy carriers.
Seasonal storage	Application where energy storage compensates longer-term supply disruptions or seasonal variability in supply and demand.
Self-consumption	Application where energy storage increases self-consumption of energy produced by non-dispatchable distributed (renewable) generators.
Self-discharge	Unavoidable loss of state-of-charge while a storage system is idle.
Shared Socioeconomic Pathways	Scenarios of projected socioeconomic global changes up to 2100 that are used to derive greenhouse gas emissions scenarios with different climate policies.
State of charge	Fraction of energy stored at a point in time.
State of health	Actual energy capacity relative to nominal energy capacity.
Stationary storage	Energy storage system providing services from a fixed location.
System value	The value of a technology to the energy system as a whole.
T&D deferral	Application where energy storage defers, reduces or avoids transmission and/or distribution network upgrades when peak power flows exceed existing capacity.
Usable energy capacity	Energy capacity that can be discharged accounting for depth-of-discharge.
Utilisation	Actual operation relative to maximum possible operation.
Variability	Difference between supply (or demand) peak and trough.

Index

Note: figures, tables and boxes are indicated by *f*, *t* and *b*.

M

N

O

P

R

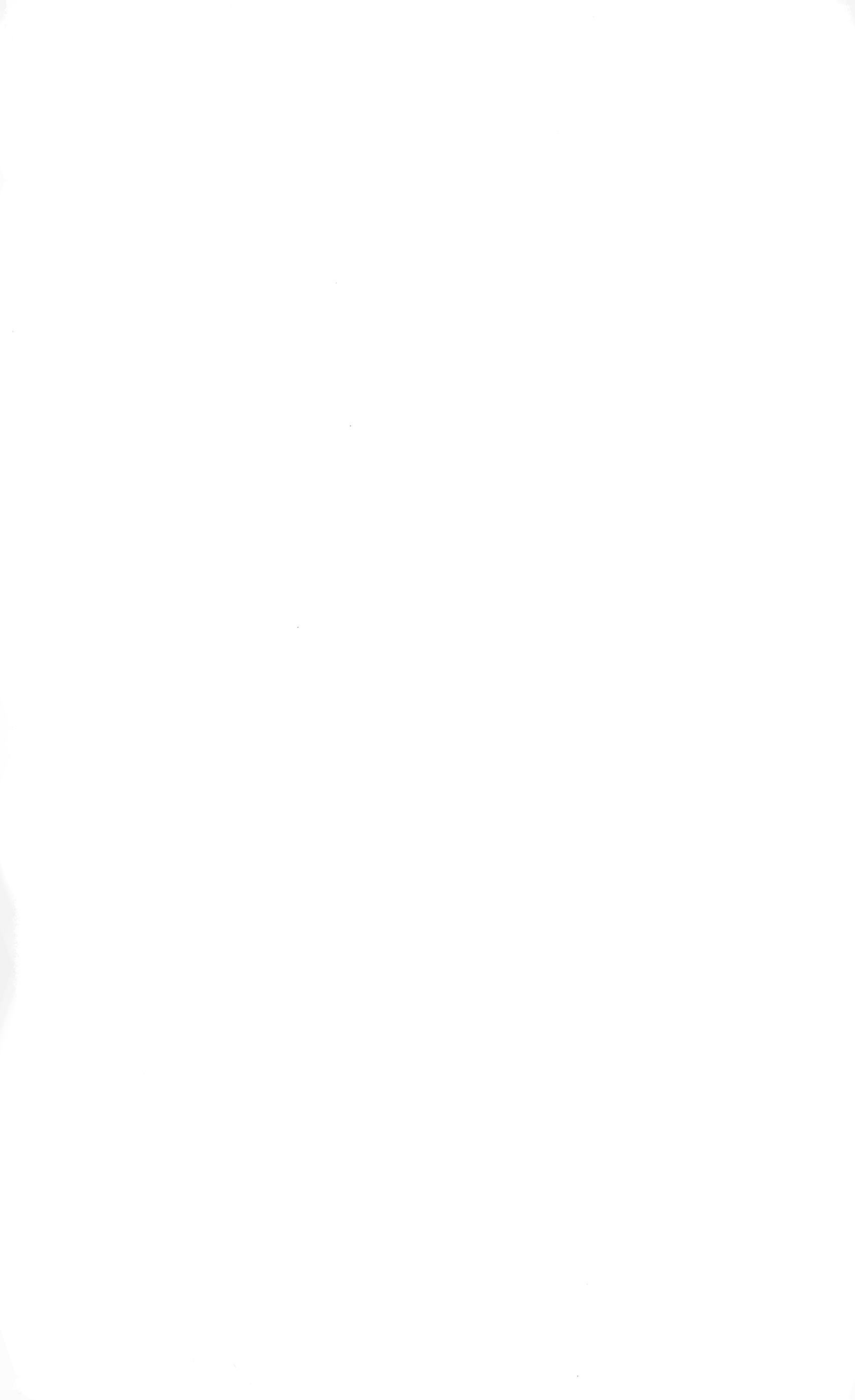